Teachin

Teaching Statistics

A Bag of Tricks

Andrew Gelman
Columbia University

Deborah Nolan
University of California, Berkeley

OXFORD
UNIVERSITY PRESS

OXFORD
UNIVERSITY PRESS

Great Clarendon Street, Oxford OX2 6DP

Oxford University Press is a department of the University of Oxford.
It furthers the University's objective of excellence in research, scholarship,
and education by publishing worldwide in

Oxford New York

Auckland Cape Town Dar es Salaam Hong Kong Karachi
Kuala Lumpur Madrid Melbourne Mexico City Nairobi
New Delhi Shanghai Taipei Toronto

With offices in

Argentina Austria Brazil Chile Czech Republic France Greece
Guatemala Hungary Italy Japan South Korea Poland Portugal
Singapore Switzerland Thailand Turkey Ukraine Vietnam

Oxford is a registered trade mark of Oxford University Press
in the UK and in certain other countries

Published in the United States
by Oxford University Press Inc., New York

First published 2002
Reprinted 2003, 2004 (with corrections), 2005, 2006, 2007, 2008

British Library Cataloguing in Publication Data
Data available

Library of Congress Cataloging in Publication Data
Data available

ISBN 978-0-19-857224-4 (Pbk)

10

Typeset by Newgen Imaging Systems (P) Ltd., Chennai, India
Printed in Great Britain
on acid-free paper by the
MPG Books Group, Bodmin and King's Lynn

To Caroline and Dave

Preface

How to use this book

This book is intended for teachers of college and high school statistics courses. The different chapters cover demonstrations, examples, and projects for a succession of topics in statistics and probability. (We distinguish between *demonstrations*, which involve active student participation, and *examples*, which are conducted by the instructor.)

This book contains more material than could possibly be used in a single course; we suggest you read through it all and then try out some of the ideas. Pick and choose what works for you.

Part I of the book presents a large selection of activities for introductory statistics courses. Part II gives some tips on what works and what doesn't, how to set up effective demonstrations and examples, and how to inspire students to participate in class and work effectively in homeworks, exams, and group projects. A sample course plan is presented in Chapter 12 to illustrate how to integrate these materials into a semester-long statistics course. Part III presents demonstrations, examples, and projects for some more advanced courses on topics including decision theory, Bayesian statistics, sampling, and mathematical probability and statistics.

General acknowledgments and comment on originality

This book is collected as much as it is authored, and we have cited references wherever we are aware of related writings on the topics. However, we suspect that even the demonstrations and examples that we have made up ourselves have been independently invented many times. Thus, we do not claim this book has value because of "originality"; in fact, even if all the material here had appeared in the statistics teaching literature, we would find it useful to collect it all in one place.

We present each demonstration, example, and project in self-contained format. If you are interested in further reading on any of these topics, or details about previous presentations of these ideas, see the Notes at the end of the book.

This is intended to be a collection of the "good stuff": we do not attempt to have a demonstration or example for every topic in statistics or to provide a complete set of teaching materials for students. Rather, we've thrown in the best examples we know about, including some we've developed ourselves. In many cases, we are particularly happy with the form of the demonstrations we have constructed; even if the basic idea has appeared several times in print, we have tried in our particular version to focus on the statistical essentials while at the same time creating scenarios that allow for personal student involvement.

Our teaching methods are (we hope) continually subject to improvement. If you have any comments, suggestions, or ideas, please email them to us at gelman@stat.columbia.edu or nolan@stat.berkeley.edu.

Specific acknowledgments

We would like to thank, in addition to the authors of the articles and books cited in our references, all the people who have told us about teaching demonstrations and helped us collect and understand this material. Mark Glickman coauthored an article that included several of the demonstrations in Part I of this book. Anna Men, Steve Warmerdam, and Michelle Bautista contributed to the statistical literacy project described in Chapter 6, including preparing much of the material in the course packets, and Jason Chan, Inchul Hong, and Keay Davidson provided helpful comments. The material in Chapter 15 on teaching mathematical probability was adapted from an article cowritten by Ani Adhikari, and these teaching aids were developed as part of the Mills Summer Mathematics Institute, an intensive six-week program for undergraduate women, and at a similar program for minority students held at the University of California. The laboratory assignments for a theoretical statistics course in Chapter 16 were developed in collaboration with Terry Speed, who coauthored the article and book on which much of this chapter was based. All computations were performed using the statistical packages R, S-Plus, and Stata.

We also thank Tom Belin, Eric Bradlow, Valerie Chan, Herman Chernoff, Rajeev Dehejia, Phil Everson, Marjorie Hahn, David Krantz, Jim Landwehr, Jim Liebman, Tom Little, Xiao-Li Meng, Roger Purves, Jerry Reiter, Seth Roberts, Caroline Rosenthal, Andrea Siegel, Hal Stern, Stephen Stigler, Ron Thisted, Howard Wainer, and Alan Zaslavsky for helpful comments and ideas. We especially thank Phillip Price for many useful suggestions.

Many thanks go to the students who have helped us put together this book. Natalie Jomini and Christina Wells collected data, drew diagrams, and helped construct the index. Kimmy Szeto drew diagrams, and Meg Lamm and Chandler Sims found references and obtained permissions.

We would also like to thank the following sources of financial support: the National Science Foundation for grants DMS-9404305, DMS-9720490, DUE-9950161, and Young Investigator Award DMS-9496129/9796129, CDC Investment Management Corporation, the Undergraduate Research Apprenticeship Program at the University of California, Berkeley, and also the general educational resources of Columbia University and the University of California. We thank our instructors and teaching assistants for sharing many good ideas with us. Most of all, we thank all the students who have taught us so much about teaching during the past fifteen years.

All royalties from the sales of this book will be donated to nonprofit educational organizations.

New York
Berkeley
April 2002

A. G.
D. N.

Contents

1 Introduction

1.1 The challenge of teaching introductory statistics

We have taught introductory statistics to college students for several years. There are a variety of good textbooks at this level, all of which cover the material pretty well. However, we found it a challenge to keep students motivated in class. Statistics is problem-solving. Watching the instructor solve a problem on the blackboard is not as effective or satisfying for students as actively involving themselves in problems. To improve participation, we and our colleagues have been collecting and developing tools, tricks, and examples for statistics teaching.

Many teachers have their own special teaching methods but they are not always disseminated widely; as a result, some of the best ideas (for example, the survey of family sizes presented here on page 56 and the coin-flipping demonstration on page 105) have been rediscovered several times. We put together this book to collect all the good ideas that we have heard about or developed in our own courses.

By collecting many demonstrations and examples in one place and focusing on the techniques used to involve students as active participants, we intend this book to be a convenient resource for instructors of introductory probability and statistics at the high school and college level. Where possible, we give references to earlier descriptions of these demonstrations, but we recognize that many of them have been used by teachers long before they appeared in any of these cited publications.

This book contains a range of teaching tools. What we find most important about a teaching tool is the basic idea, along with details about how to implement it in class. This can often be explained in a few paragraphs. We try to do this where possible but without skimping on the little details that keep the students involved.

1.2 Fitting demonstrations, examples, and projects into a course

The demonstrations, examples, and project ideas presented here are most effective for relatively small classes (fewer than 60 students); with a large lecture course, some of the demonstrations can be done in the discussion sections. These materials are not intended to stand alone. We use them along with a traditional statistics text, lectures, homeworks, quizzes, and exams.

The chapters in Part I of the book cover the topics roughly in order of when they occur during the semester, with enough so that there can be a class-participation activity of some sort during every lecture. Many of the activities are relevant to multiple statistical concepts, and so for convenience we have listed them on pages 6–8 by subject. Part III presents some demonstrations and activities for more advanced courses in statistics. In between, we discuss issues of implementing the class-participation activities, along with a detailed schedule of how they fit in to a course. Finally, background information on many of the examples is given in the Notes at the end of the book.

We have found student-participation demonstrations to be effective in dramatizing concepts that students often find difficult (for example, numeracy, conditional probability, the difference between an experiment and a survey, statistical and practical significance, the sampling distribution of confidence intervals). Students are made aware that they and others are subject to cognitive illusions (see, for example, the United Nations demonstration on page 66, the coin-flipping demonstration on page 105, and the lie detection example on page 113). In addition, the experiments that involve data-gathering illustrate general concerns of bias and variance (for example, in the age-guessing example on page 11 and the candy weighing on page 120) and also involve important practical issues such as time trends (shooting baskets on page 134), displaying data (guessing exam scores on page 25), experimental protocol (weighted coins on page 115 and helicopter design on page 259), and the relation between models and data (average family size on page 56). One reason that we believe these demonstrations are important is that the active settings emphasize that statistics is, in reality, a participatory process with many actors (typically, different people design a study, collect data, are experimental subjects, analyze data, interpret results, and so forth).

Perhaps most important, the demonstrations get all the students involved and help to create an environment where students feel free to participate and ask questions in class.

The goal of class-participation tools is to improve learning within class and also to encourage more learning outside of class. Our demonstrations are intended to involve students in traditional lecture material and are not intended as a substitute for more detailed student-involved investigations or group projects. In Section 5.2 we include step-by-step instructions on how to run a class survey project, and in Section 5.3.3 we provide directions for running a simple taste-testing experiment. Chapter 11 gives advice on coordinating projects in which groups of students work on projects of their own design.

At the center of any course on introductory statistics are worked examples illustrating the important concepts and methods in real-data situations that are interesting to students. Good statistics texts are full of such examples, and a good lecture course will introduce new examples rather than simply working out material already in the textbook. Interesting examples can often be found directly from newspapers, magazines, and scientific articles; Chapter 6 presents some examples that have worked well in our classes.

There are many other excellent examples in statistics textbooks and other sources. We discuss various places to find additional material in Section 11.5.

1.3 What makes a good example?

We enjoy discussing in class the issue of what makes a good example, opening up some of our teaching strategies to the students.

When a topic is introduced, we like the first example to be simple so that the mechanics of the method are transparent. It is also good to prepare a handout showing the steps of the procedure so that the students can follow along without having to scribble everything down from the blackboard. You can put some fill-in-the-blanks on the handout so the students have to pay attention during the exposition.

It can be good to use fake data for a first example and to discuss how you set up the fake data and why you did it the way you did. For example, we first introduce the log-log transformation with the simple example of the relation between the area of a square and its circumference (page 33). Once students are familiar with the log-log transformation, we use it to model real data and to introduce some realistic complexity (for example, the data on metabolic rates on page 35).

1.4 Why is statistics important?

Finally, it is useful before studying these demonstrations and examples to remind ourselves why we think it is important for students to learn statistics. We use statistics when we make decisions, both at an individual and a public level, and we use statistics to understand the world. Many of the major decisions affecting the lives of everything on this planet have some statistical justification or basis, and the methods we teach are relevant to understanding these decisions. These points are well illustrated by the kinds of studies reported in the press that students chose to read about for a course project (see page 79). For example, a pre-med student who ate fish frequently was interested in the observational study described in "Study finds fish-heavy diet offers no heart protection," *New York Times*, April 13, 1995. Another student who commuted to school wanted to learn more about a commuter survey reported on in the article entitled, "For MUNI riders, familiarity breeds contempt, study says," *San Francisco Examiner*, August 27, 1998, and yet another chose to follow up on the article, "Audit finds stockbrokers treat women differently," *San Francisco Chronicle*, March 22, 1995, because of her political convictions. Of course, studies of college students, such as "More students drinking to get drunk, study finds," *San Francisco Examiner*, September 11, 1998, generate interest among most students.

Studies reported in the newspaper typically use statistics to analyze data collected by survey, experiment, or observationally (examples of each of these appear in Chapter 5). But in some cases, the simple act of collecting and reporting numerical data has a positive effect. For example, around 1990, Texas began requiring every school and school district in the state to report average test scores for all grades in each of several ethnic groups. This provides an incentive for schools to improve performance across these sub-populations.

Another example of the value of public statistics is the Toxics Release Inventory, the result of laws in 1984 and 1990 that require industrial plants to release information on pollutants released and recycled each year. Both the Texas schools and the environmental releases improved after these innovations. Although one cannot be sure of causal links in these observational settings (see page 72), our point is that gathering and disseminating information can stimulate understanding and action.

1.5 The best of the best

This book collects our favorite demonstrations, activities, and examples. But among these we have some particular favorites that get the students thinking hard about problems from unexpected angles. If you only want to use a few of our activities, we suggest you try these first.

- Where are the cancers? (page 13)
- World record times for the mile run (page 20)
- Handedness of students (page 21)
- Guessing exam scores (page 25)
- Who opposed the Vietnam War? (page 27)
- Metabolic rates of animals (page 35)
- Tall people have higher incomes (pages 39 and 137)
- Sampling from the telephone book (page 48)
- How large is your family? (page 56)
- An experiment that looks like a survey (page 66)
- Randomizing the order of exam questions (page 68)
- Real vs. fake coin flips (page 105)
- What color is the card? (page 111)
- Weighing a "random" sample (page 120)
- Where are the missing girls? (page 121)
- Coverage and noncoverage of confidence intervals (pages 126–128)
- Shooting baskets (page 134)
- Examples of lying with statistics (page 147)
- How many quarters are in the jar? (page 204)
- What is the value of a life? (page 210)
- Subjective probability intervals (page 216)
- Helicopter design project (page 259)

1.6 Our motivation for writing this book

This book grew out of our shared interest in teaching. We describe three events in our teaching career that influenced its creation.

The first time that one of us taught a statistics course, we took ten minutes halfway through the course and asked the students for anonymous written comments. As we feared, most of comments were negative; the course was in fact a disaster, largely because we did not follow the textbook closely, and the homeworks were a mix of too easy and too hard. Our lectures were focused too much on concepts and not enough on skills, with the result that the students learned little of either. This experience inspired us, in the immediate time frame, to organize our courses better (and, indeed, our courses were much more successful after that). We were also inspired to begin gathering the demonstrations, examples, and teaching tools that ultimately were collected in this book, as a means of getting the students more actively involved in their own learning.

Another time, one of us was asked to teach a special seminar in probability for a small number of advanced undergraduate students who showed promise for graduate study. We were asked to break from the traditional lecture style and have students read original research papers even though they had no probability background before starting the seminar. This was a tall order, but the students did not disappoint us. The experience opened our eyes to the variety and level of work that students could tackle with the right kind of guidance, and as a result, we started to experiment with projects and advanced work in our regular courses.

Then in 1994, we jointly led an undergraduate research project in statistical literacy. The students' enthusiasm for the project surprised and inspired us, and led to the design of our course packets in statistical literacy (Chapter 6). This project further sparked discussions on ideas for making classroom lectures more fun for students and on ways to design meaningful course projects, and after years of sharing ideas and techniques, we decided to write this book to provide a resource for other instructors of statistics. We hope you find it useful.

Table 1.1 Demonstrations, examples, and projects that can be used in the first part of an introductory statistics course. See also Tables 1.2–1.3 and Chapter 12. Some activities are listed more than once because we return to them throughout the course.

Concept		Activities
Introducing data collection	2.1:	Guessing ages
	2.5:	Collecting handedness data
Numeracy	2.3:	Estimating a big number
Relevance of statistics	1.4:	Why is statistics important?
	2.2:	Where are the cancers?
	2.4:	What's in the news?
Time plots, interpolation and extrapolation	3.2.1:	World record times for the mile run
	3.8.2:	World population
Histograms	3.3.2:	Handedness of students
	3.3.3:	Soft drink consumption
Means and medians	3.4.1:	Average soft drink consumption
	3.4.2:	The average student
Scatterplots	3.5.1:	Guessing exam scores
	9.2:	Psychometric analysis of exam scores
Two-way tables	3.5.2:	Who opposed the Vietnam War?
Normal distribution	3.6.1:	Heights of men and women
	3.6.2:	Heights of conscripts
	3.6.3:	Scores on two exams
Linear transformations	3.7.1:	College admissions
	3.7.2:	Social and economic indexes
	3.7.3:	Age adjustment
Logarithmic transformations	3.8.1:	Amoebas, squares, and cubes
	3.8.2:	World population
	3.8.3:	Metabolic rates
	5.1.2:	First digits and Benford's law
Linear regression with one predictor	4.1.2:	Tall people have higher incomes
	4.1.3:	World population
Correlation	4.2.1:	Correlations of body measurements
	4.2.2:	Exam scores and number of pages written
	9.2:	Psychometric analysis of exam scores
Regression to the mean	4.3.1:	Memory quizzes
	4.3.2:	Scores on two exams
	4.3.2:	Heights of mothers and daughters

Table 1.2 Demonstrations, examples, and projects that can be used in the middle part of an introductory statistics course. See also Tables 1.1–1.3 and Chapter 12. Some activities are listed more than once because we return to them throughout the course.

Concept		Activities
Introduction to sampling	5.1.1:	Sampling from the telephone book
	5.1.2:	First digits and Benford's law
	5.1.5:	Simple examples of sampling bias
	5.1.6:	How large is your family?
	8.1:	Weighing a "random" sample
Applied survey sampling	5.1.3:	Wacky surveys
	5.1.4:	An election exit poll
	5.2:	Class projects in survey sampling
	6.6.2:	1 in 4 youths abused, survey finds
	14:	Activities in survey sampling
Experiments	5.3.1:	An experiment that looks like a survey
	5.3.2:	Randomizing the order of exam questions
	5.3.3:	Soda and coffee tasting
	6.6.1:	IV fluids for trauma victims
	10.5:	Ethics and statistics
	16.4:	Designing a paper "helicopter"
Observational studies	4.1.2:	Tall people have higher incomes
	5.4.1:	The Surgeon General's report on smoking
	5.4.2:	Large population studies
	5.4.3:	Coaching for the SAT
	6.6.3:	Monster in the crib
	9.1:	Regression of income on height and sex
Statistical literacy	6:	Statistical literacy and the news media
	10:	Lying with statistics
Probability and randomness	7.2:	Random numbers via dice or handouts
	7.3.1:	Probabilities of boy and girl births
	7.3.2:	Real vs. fake coin flips
	7.3.3:	Lotteries
Conditional probability	7.3.1:	Was Elvis an identical twin?
	7.5.1:	What color is the other side of the card?
	7.5.2:	Lie detectors and false positives
Applied probability modeling	7.4.1:	Lengths of baseball World Series
	7.4.2:	Voting and coalitions
	7.4.3:	Probability of space shuttle failure
	7.6:	Crooked dice and biased coins
	9.3:	Success rates of golf putts
	13.1.1:	How many quarters?
	13.1.2:	Utility of money
	13.1.4:	What is the value of a life?
	13.1.5:	Probabilistic answers to exam questions
	15.4:	Does the Poisson distribution fit real data?

Table 1.3 Demonstrations, examples, and projects that can be used in the last part of an introductory statistics course. See also Tables 1.1–1.2 and Chapter 12. Some activities are listed more than once because we return to them throughout the course.

Concept		Activities
Distribution of the sample mean	8.2.1:	Where are the missing girls?
	8.2.2:	Real-time gambler's ruin
	8.3.4:	Poll differentials
Bias and variance of an estimate	2.1:	Guessing ages
	8.1:	Weighing a "random" sample
	8.3.1:	Biases in age guessing
Confidence intervals	8.3.2:	An experiment that looks like a survey
	8.3.3:	Land or water?
	8.3.4:	Poll differentials
	8.4.1:	Coverage of confidence intervals
	8.4.2:	Noncoverage of confidence intervals
	8.6.1:	How good is your memory?
	8.6.2:	How common is your name?
	13.2.2:	Subjective probability intervals
	13.2.5:	Hierarchical modeling and shrinkage
Hypothesis testing	5.3.2:	Randomizing the order of exam questions
	7.3.2:	Real vs. fake coin flips
	8.3.5:	Golf: can you putt like the pros?
	8.5.1:	Land or water?
	8.5.1:	Evidence for the anchoring effect
	8.5.2:	Detecting a flawed sampling method
	8.5.3:	Taste testing projects
	8.5.4:	Testing Benford's law
	8.5.5:	Lengths of baseball World Series
Statistical power	8.2.1:	Where are the missing girls?
	8.7.1:	Shooting baskets
Multiple comparisons	8.7.2:	Do-it-yourself data dredging
	8.7.3:	Praying for your health
Multiple regression	9.1:	Regression of income on height and sex
	9.2:	Exam scores
	16.3:	Quality control
Nonlinear regression	9.3:	Success rate of golf putts
	9.4:	Pythagoras goes linear
	16.4:	Designing a paper "helicopter"
Lying with statistics and statistical communication	10.1:	Many examples from the news media
	10.2:	Selection bias
	10.4:	1 in 2 marriages end in divorce?
	10.5:	Ethics and statistics

Part I

Introductory probability and statistics

2

First week of class

On the first day of class, in addition to introducing the concepts of probability and statistics, and giving an overview of what the students will learn during the term, we include class-participation demonstrations and examples. Then we start with descriptive statistics, as discussed in the next chapter. We often begin the first class with an activity on variation, such as the age-guessing demonstration or the cancer example described here. In the first week, we also discuss numeracy, address the question of why statistics is important, and collect data on students to use later in the course. We like to get students involved right away and set the expectation for plenty of class participation.

2.1 Guessing ages

This demonstration illustrates concepts of data collection and variation. Ahead of time, we obtain 10 photographs of persons whose ages we know but will not be known by the students (for example, friends, relatives, or non-celebrities from newspapers or magazines) and tape each photo onto an index card. We number the cards from 1 to 10 and record the ages on a separate piece of paper.

As the students are arriving for the first class, we divide them into 10 groups, labeled as A through J, arranged in a circuit around the room. We pass out to each group:

- A large sheet of cardboard on which we have written the group's identifying letter,
- One of the ten photo cards, and
- A copy of the form shown in Fig. 2.1.

We then ask the students in each group to estimate the age of the person in their photograph and to write that guess on the form. Each group must come up with a single estimate, which forces the students to discuss the estimation problem and also to get to know each other. We then explain that each group will be estimating the ages of all 10 photographed persons, and that the groups are competing to get the lowest error. Each group passes its card to the next group (A passes to B, B passes to C, . . ., J passes back to A) and then estimates the age on the new photo, and this is continued until each group has seen all the cards, which takes about 20 minutes. We walk around the room while this is happening to keep it running smoothly.

Guessing ages. For each card your group is given, estimate the age of the person on the card and write your guess in the table below in the row corresponding to that numbered card. Later, you will be told the true ages and you can compute your errors. The error is defined as estimated minus actual age.

Card	Estimated age	Actual age	Error
1			
2			
3			
4			
5			
6			
7			
8			
9			
10			

Fig. 2.1 On the first day of class, divide the students into 10 groups and give one copy of this form to each group. They fill it out during the age-guessing demonstration (Section 2.1).

We then go to the blackboard and set up a two-way table with rows indicating groups and columns indicating cards. We have a brief discussion of the expected accuracy of guesses (students typically think they can guess to within about 5 years) and then, starting with card 1, we ask each group to give its guess, then we reveal the true age and write it at the bottom margin of the first column of the table. We do the same for card 2. At this point, there is often some surprise because the ages of some photos are particularly hard to guess (for example, we have a photo of a 29-year-old man who is typically guessed to be in his late thirties). We introduce the concept of error—guessed age minus actual age—and go back to the blackboard and write the errors in place of the guessed ages. For each of the remaining cards, we simply begin by writing the true age at the bottom, then for each entry in that column of the table we write the error of the guess for each group.

Figure 2.2 shows an example of a completed blackboard display with the results for all 10 cards. At this point we ask the students in each group to compute the average absolute error of their guesses (typically some students make a mistake and forget to take absolute values, which becomes clear because they get an absurdly low average absolute error, such as 0.9). Typical average absolute errors, for groups of college students guessing ages ranging from 8 to 80, have been under 5.0. We have found that—possibly because of the personal nature of age guessing, and because everybody has some experience in this area—the students enjoy this demonstration and take it seriously enough so that they get some idea of uncertainty, empirical analysis, and data display.

	Card										Avg. abs.	Students in group	
Group	1	2	3	4	5	6	7	8	9	10	error	#	Sex
A	+14	−6	+5	+19	0	−5	−9	−1	−7	−7	7.3	3	F
B	+8	0	+5	0	+5	−1	−8	−1	−8	0	3.7	4	F
C	+6	−5	+6	+3	+1	−8	−18	+1	−9	−6	6.1	4	M
D	+10	−7	+3	+3	+2	−2	−13	+6	−7	−7	6.0	2	M/F
E	+11	−3	+4	+2	−1	0	−17	0	−14	+3	5.5	2	F
F	+13	−3	+3	+5	−2	−8	−9	−1	−7	0	5.1	3	F
G	+9	−4	+3	0	+4	−13	−15	+6	−7	+5	6.6	4	M
H	+11	0	+2	+8	+3	+3	−15	+1	−7	0	5.0	4	M
I	+6	−2	+2	+8	+3	−8	−7	−1	+1	−2	4.0	4	F
J	+11	+2	+3	+11	+1	−8	−14	−2	−1	0	5.3	4	F
Truth	29	30	14	42	37	57	73	24	82	45			

Fig. 2.2 Errors in age guesses from a class of eighth graders, as displayed on the blackboard. For the first two cards, we wrote the guessed and true ages, then we switched to writing the true age and the estimation error for each group. At the end, the groups were compared using the average of the absolute values of the errors. The sexes and group sizes are shown because one of the students in the class commented that the girls' guesses were, on average, more accurate.

The age-guessing demonstration is extremely rich in statistical ideas, and we return to it repeatedly during the term. The concepts of bias and variance of estimation are well illustrated by age guessing: we have found the variance of guesses to be similar for all the photos, but the biases vary quite a bit, since some people "look their age" and some do not (see Fig. 2.2). Later in the course, important points of experimental design can be discussed referring to issues such as the choice of photographs, the order in which each group gets the cards, randomization, and the practical constraints involved in running the experiment. This is also an interesting example because, even if the experiment is randomized, it does not yield unbiased estimates of ages. In addition, the data from the study can be used as examples in linear regression, the analysis of two-way tables, statistical significance, and so forth.

2.2 Where are the cancers?

We often conclude the first day of class by passing out copies of Fig. 2.3; this is a map of the United States, shading the counties with the highest rates of kidney cancer from 1980–1989. We ask the students what they notice about the map; one of them points out the most obvious pattern, which is that many of the counties in the Great Plains but relatively few near the coasts are shaded. Why is this? A student asks whether these are the counties with more old people. That could be the answer, but it is not—in fact, these rates are age-adjusted. Any other ideas? A student notes that most of the shaded counties are in rural areas; perhaps the health care there is worse than in major cities. Or perhaps people in rural areas have less healthy diets, or are exposed to more harmful chemicals. All these are possibilities, but here is a confusing fact: a map of the

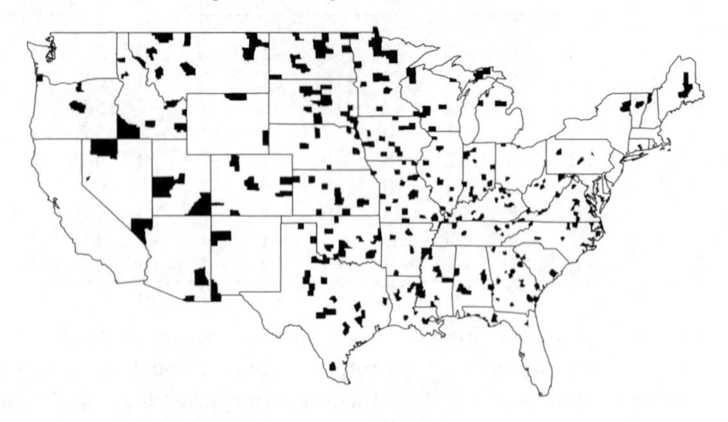

Fig. 2.3 The counties of the United States with the highest 10% age-standardized death rates for cancer of kidney/ureter for U.S. white males, 1980–1989. Hand this map out to the students and ask why most of the shaded counties are in the center-west of the country. See Section 2.2 for discussion.

lowest age-adjusted kidney cancer death rates also mostly highlights rural areas. We show the students Fig. 2.4.

At this point, the students are usually stumped. To help them, we consider a county with 100 people: if it has even one kidney cancer death in the 1980s, its rate is 1 per thousand per year, which is among the highest in the nation. Of course, if it has no kidney cancer deaths, its rate will be lowest in the nation (tied with all the other counties with zero deaths). The observed rates for smaller counties are much more variable, and hence they are much more likely to be shaded, even if the true probability of cancer in these counties is nothing special. If a small county has an observed rate of 1 per thousand per year, this is probably random fluctuation, but if a large county such as Los Angeles has a very high rate, it is probably a real phenomenon.

This example can be continued for use in a course on statistical modeling, decision analysis, or Bayesian statistics, as we discuss in Section 13.2.1. In addition, Section 3.7.3 describes a class demonstration involving age adjustment.

2.3 Estimating a big number

For this activity, we ask the class to guess how many school buses operate in the United States. Some students will speak up and guess; we then ask the students to pair up and for each pair to write a guess on a sheet of paper. (We have students guess to warm them up for the class discussion that follows, and students typically work more systematically and give more reasonable answers when they work in pairs or small groups.) As a motivation, we will give a prize (for example, a school-insignia T-shirt) to the pair of students whose guess is

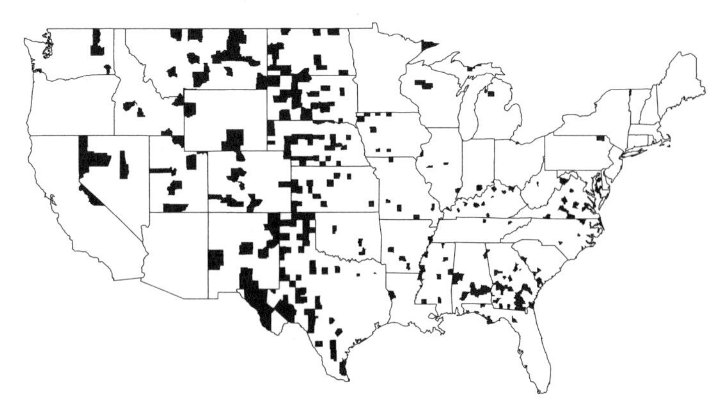

Fig. 2.4 The counties of the United States with the *lowest* 10% age-standardized death rates for cancer of kidney/ureter for U.S. white males, 1980–1989. Surprisingly, the pattern is somewhat similar to the map of the highest rates, shown in Fig. 2.3. Hand this map out to the students only after they have discussed the previous map.

closest to the true value. How can we check whether these guesses are reasonable and improve upon them? We lead the class through a discussion: how many people are there in the United States, how many children of school age, how many of them ride the bus, how many students per bus, and so forth. The resulting estimate should have a great deal of uncertainty, but it will probably be closer to the truth than most of the original guesses, as well as focusing our attention on what parts of the problem we understand better than others. The students are learning "numeracy" and the propagation of uncertainty, which are important in practical uses of statistics. Further discussion here can lead to additional statistical issues, including the reliability of data sources and the design of data collection (how could we more accurately estimate the different factors in our estimate). We repeat the class-participation exercise a few times during the term using other uncertain quantities (for example, how many Smiths are listed in the telephone book of Oakland, California, a city of 400 000 people) when we have five minutes available at the end of class. (The numbers of school buses and Smiths are given on page 265.)

2.4 What's in the news?

News reports with shocking or humorous titles can make for good ice-breakers in the first week of class. We choose news stories that report on statistical results from recent studies, and hand out, or put on an overhead projector, excerpts from these news reports. We read the title and short excerpt aloud and get a discussion going by asking students to describe the main goal of the study and to identify the subjects of the study, including how they are enrolled in the study and the

A8 San Francisco Chronicle ★★★★★

Infant Deaths

Low weights not solely to blame

New York Times

A new study of more than 7.5 million births has challenged the assumption that low birth weights per se are the cause of the high infant mortality rate in the United States. Rather, the new findings indicate, prematurity is the principal culprit.

Being born too soon, rather than too small, is the main underlying cause of stillbirth and infant deaths within four weeks of birth.

Each year in the United States about 31,000 fetuses die before delivery and 22,000 newborns die during the first 27 days of life.

The United States has a higher infant mortality rate than those in 19 other countries, and this poor standing has long been attributed mainly to the large number of babies in this country who are born too small, including a large proportion who are born "small for date," or weighing less than they should for the length of time they were in the womb.

The researchers found that American-born babies, on average, weigh less than babies born in Norway, even when the length of pregnancy is the same. But for a given length of pregnancy, the lighter American babies are no more likely to die than are the slightly heavier Norwegian babies.

The researchers, directed by Dr. Allen Wilcox of the National

1 in 4 youths abused, survey finds

CHICAGO One in four adolescents said they were physically or sexually abused within the past year, according to a new survey.

The telephone survey of 2,000 children ages 10 to 16, suggests, "We're not doing a very good job of counting and tracking the problem," said David Finkelhor, a sociologist at the University of New Hampshire and co-author of the study in the October issue of Pediatrics.

The assault rate reported by the participants was 15.6 percent, three times higher than the 5.2 percent reported in the National Crime Survey in 1991, the study said. The survey's rate for rape was 0.5 percent, five times higher than the federal estimate of 0.1 percent.

SAN FRANCISCO EXAMINER

Surgeons may opera

Calming tunes are the key, study says

By Brenda C. Coleman
ASSOCIATED PRESS

CHICAGO — Toscanini for a tonsillectomy. Bach for brain surgery.

Surgeons are likely to do a better job at the operating table with a little background music, a study suggests — but not if Aerosmith is blasting away, and your surgeon prefers Beethoven.

According to researchers in Wednesday's Journal of the American Medical Association, surgeons had lower blood pressure and pulse rates and performed better on nonsurgical mental exercises while listening to music.

"In 1889 Nietzsche wrote, 'Without music, life would be a mistake.' Over a century later our data prompt us to wonder if, without music, surgery would be a mistake," the report said.

Two Chicago-area surgeons

Fig. 2.5 Example of a display of fragments of newspaper articles compiled for the What's in the News class-participation exercise described in Section 2.4.

response measured on them. Figure 2.5 displays fragments of three stories that have generated plenty of student remarks: "1 in 4 youths abused, survey finds" (*San Francisco Examiner*, October 4, 1994), "Infant deaths tied to premature births" (*New York Times*, March 1, 1995), and "Surgeons may operate better with music: Toscanini for a tonsillectomy, Bach for brain surgery" (*San Francisco*

Examiner, September 21, 1994). In comparing the data collection techniques of these three studies, we point out that the children in the abuse study are sampled at random, the babies are part of a large observational study, and the surgeons participated in an experiment. The provocative titles catch students' attention, and our discussion of the reported studies is a springboard for introducing the types of studies they will learn about in the course.

2.5 Collecting data from students

Please indicate which hand you use for each of the following activities by putting a + in the appropriate column, or ++ if you would never use the other hand for that activity. If in any case you are really indifferent, put + in both columns. Some of the activities require both hands. In these cases the part of the task, or object, for which hand preference is wanted is indicated in parentheses.

Task	Left	Right
Writing		
Drawing		
Throwing		
Scissors		
Toothbrush		
Knife (without fork)		
Spoon		
Broom (upper hand)		
Striking match (hand that holds the match)		
Opening box (hand that holds the lid)		
Total		

Right − Left: Right + Left: $\frac{\text{Right} - \text{Left}}{\text{Right} + \text{Left}}$:

Create a Left and a Right score by counting the total number of + signs in each column. Your handedness score is (Right − Left)/(Right + Left): thus, a pure right-hander will have a score of score $(20-0)/(20+0) = 1$, and a pure left-hander will score $(0-20)/(0+20) = -1$.

Fig. 2.6 Handedness inventory. Have each student fill out this form and report his or her total score. Students can later divide into pairs and sketch their guesses of the histogram of these scores for the students in the class, as described in Section 3.3.2 and shown in Fig. 3.2.

It is traditional at the beginning of a statistics course to collect data on students—each student is given an index card to write the answers to a series of questions and then these cards are collected and used to illustrate methods throughout the semester. For example, when we teach sampling, we use the cards as a prop. They represent the class population, and we mix them up on the desk and have students pick cards at random for our sample (Section 14.1.2).

Also, when we teach regression and correlation, we bring to class a scatterplot of student data to study. For example, we compare a scatterplot of the height and hand span for the students in our class to Pearson's data on university students collected over one hundred years ago (Section 4.2.1).

Possible information that can be asked include questions on student activities outside the classroom, such as the amount of soda drunk yesterday (see Section 3.3.3) or the time spent watching television last night. Physical measurements, such as the span from the thumb to the little finger (see Section 4.2.1), can be measured on the spot by handing out a sheet of paper with a photocopy of a ruler on it. Students enjoy completing the handedness form in Fig. 2.6, which is the basis for a demonstration described in Section 3.3.2. Data can be displayed on a blackboard using a projected grid (see Section 3.1).

3 Descriptive statistics

Descriptive statistics is the typical starting point for a statistics course, and it can be tricky to teach because the material is more difficult than it first appears. We end up spending more class time on the topics of data displays and transformations, in comparison to topics such as the mean, median, and standard deviation that are covered more easily in the textbook and homework assignments. Data displays and transformations must be covered carefully, since students are not always clear on what exactly is to be learned. In presenting the activities in this chapter, we focus on specific tasks and topics to be learned. Also, as always, the activities are designed to be thought-provoking and relate in some way to the world outside the textbook.

3.1 Displaying graphs on the blackboard

A useful trick we have learned for displaying graphs is to photocopy a sheet of graph paper onto a transparency sheet and then project it onto the blackboard. This is helpful with histograms, scatterplots, least-squares lines, and other forms of data display and calculations because we can then display information on both the transparency and the blackboard. For example, we can plot data on the transparency and then have students draw a regression line on the blackboard. Or we can display one set of axis labels on the transparency and another on the blackboard.

A related trick, when gathering data from students, is to pass around the class an indelible marker and a transparency sheet with axes and grid already labeled, with instructions for each student to mark his or her data point on the graph (for example, plotting mother's height vs. his or her own height). Different colors of marker can be passed out with instructions, for example, for males and females to use different colors.

3.2 Time series

Time series are probably the simplest datasets for students to visualize and display. We present here a time-series extrapolation problem with a surprise twist. Section 3.8.2 describes how we involve the class in another time series example—world population over the past 2000 years—when covering logarithms.

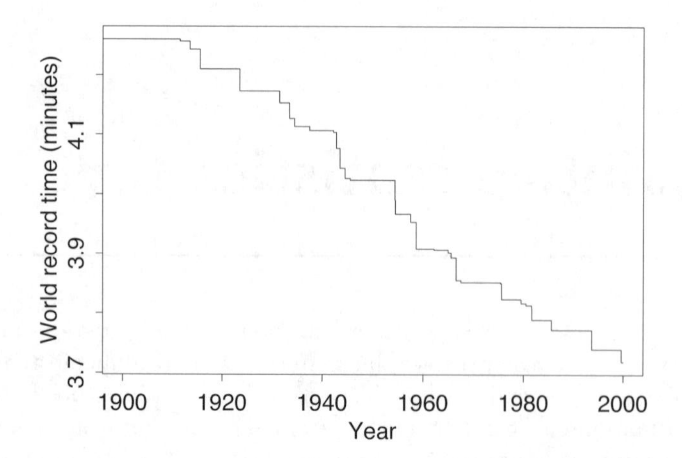

Fig. 3.1 World record times in the mile run since 1900. Blow up this graph onto a transparency and display it on the blackboard with the portion after 1950 covered up. Then ask the students to extrapolate to the year 2000. It is surprising that the approximately linear trend continues unabated.

3.2.1 World record times for the mile run

We blow up and photocopy Fig. 3.1 on a transparency and project it on the blackboard, but with the right half of the graph covered, so that the students see the world record times for the mile run from 1900 to about 1950. We then ask how well a straight line fits the data (reasonably well) and get a student to come to the blackboard and draw a straight-line fit with chalk on the blackboard, extending to the year 2000 and beyond. We discuss interpolation and extrapolation: is the straight-line prediction for the year 2100 reasonable? What about the year 2000? The students agree that these extrapolations will be too optimistic since there is some minimum time that will never be attained, and they agree on an extrapolated curve that flattens out a bit between 1950 and 2000. We draw this extrapolation on the blackboard, and then prepare to reveal the covered-up half of the curve, so that the students can compare their extrapolation to what actually happened.

We then reveal the right half of the curve: amazingly, the linear extrapolation fits pretty well all the way to the year 2000—in fact, the time trend is actually slightly steeper for the second half of the century than for the first half. We were pretty shocked when we first saw this. Of course, we wouldn't expect this to work out to the year 2050 ...

3.3 Numerical variables, distributions, and histograms

3.3.1 Categorical and continuous variables

When introducing numerical variables, we begin with some simple variables such as sex (categorical), age (continuous), and number of siblings (numerical and discrete) that are defined on individuals. We do not bother to gather data from students here—the idea is simply to define the concepts based on familiar examples. We then introduce some more complicated examples such as income (continuous but with a discrete value at zero) and height of spouse (continuous but undefined if you have no spouse).

Other discussion-provoking examples are not hard to find. For example, consider the following table of variables labeled as "categorical" in an introductory statistics textbook:

Variable	Possible categories
Dominant hand	Left-handed, right-handed
Regular church attendance	Yes, no
Opinion about marijuana legislation	Yes, no, not sure
Eye color	Brown, blue, green, hazel

Of all of these, we would consider only eye color to be reasonably modeled as categorical. Handedness is more properly characterized as continuous: as Figs. 2.6 and 3.2 indicate, many people fall between the two extremes of pure left- and pure right-handedness. Church attendance could be measured by a numerical frequency (for example, number of times per year), which would be more informative than simply yes/no. The three options for opinion about marijuana legislation could be coded as 1, 0, and 0.5, and one could imagine further identifying intermediate preferences with detailed survey questions.

Histograms are a key topic at the beginning of a statistics course—they are an important way of displaying data and also introduce the fundamental concept of a statistical distribution. Unfortunately, they can be hard to teach to students, partly because they seem so simple and can be confused with other sorts of bar graphs. (For a histogram or distribution, it is necessary to display counts or proportions, which means that some population of items needs to be defined.) We have developed an activity for this topic that we describe below in the context of some examples that have worked well in our classes.

3.3.2 Handedness

In the first week of class, we collect data from students, including their handedness scores, which range from -1 to 1 (see Fig. 2.6). Then, at the point when we are introducing histograms and distributions, we divide the class into pairs, give each pair a sheet of graph paper, and give them two minutes to draw their guess of the histogram of the handedness scores of the students in their class. We then present one group's histogram on the blackboard (it typically looks like Fig. 3.2a), and invite comments from the class. Since they have all just worked on the problem, many students are eager to participate. We correct the drawn histogram following the suggestions of the class, then display the actual histogram, which we have prepared ahead of time. An example from a recent class appears in Fig. 3.2b. We also discuss with the students the perils of bins that are too narrow (losing the shape of the distribution amid the small sample size) or too wide (thus missing some of the detail about the distribution).

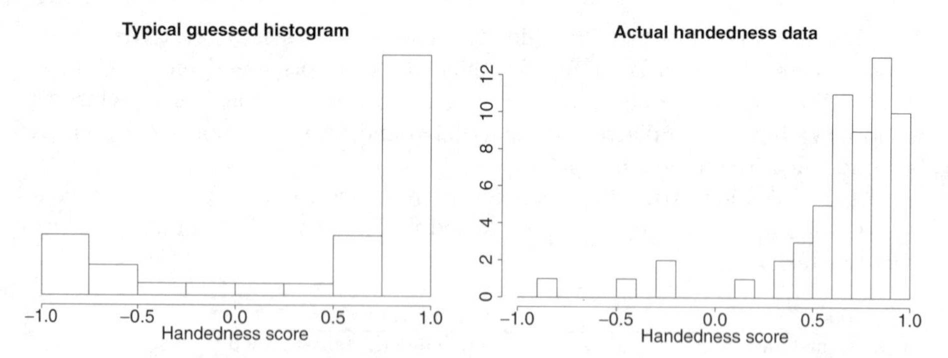

Fig. 3.2 (a) A guess from a group of students of the histogram of handedness scores in their class (see Fig. 2.6); (b) actual data. After completing their handedness forms but *before* seeing this sort of graph of real data, the students should try in pairs to guess the shape of the distribution of responses.

3.3.3 Soft drink consumption

How much soda do people drink?, we ask the students in our class. After some discussion, it is clear that the answer is expressed better as a distribution than a single number. Once again, we divide students into pairs and ask them to sketch on graph paper their guesses of the distribution of daily soft drink consumption (in ounces) by persons in the United States (for concreteness, we just consider regular (nondiet) sodas), then we display one group's guess on the blackboard. Drawing this distribution is tricky because there will be a spike at zero (corresponding to the people who drink no soda) which cannot be put on the same scale as the continuous histogram. We suggest solving this problem by drawing a spike at zero and then labeling the graph (for example, 70% of the mass of the distribution is at 0 and 30% is in the rest of the distribution).

We then compare the students' guesses to published estimates of soda consumption. For example, Fig. 3.3 shows the responses of almost 50 000 adults in a random-sampled nutrition survey in 1995, giving their daily consumption of regular (nondiet) soft drinks. If we have collected data from the students on this topic we compare it to the students' guesses too. For example, Fig. 3.4 shows a histogram of the regular soda consumed by students in a statistics class the day before this material was covered in class.

3.4 Numerical summaries

We introduce the mean and median through the histograms of distributions we have already examined.

3.4.1 Average soft drink consumption

The mean regular soft drink consumption can be computed from the distribution as follows. First, we visually estimate the mean of the nonzero part of the distribution. We can do this by identifying the median and then recognizing that,

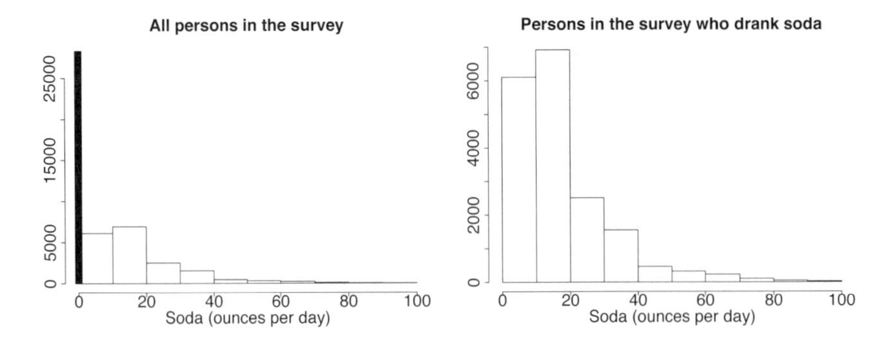

Fig. 3.3 Histograms of number of ounces of regular (nondiet) soft drinks consumed per day from a national survey of 46 709 persons in the United States, for (a) all respondents, and (b) respondents who reported positive consumption. The spike in the first histogram indicates the 61% of the people who reported consuming no regular soft drinks. The graphs exclude the fewer than 1% of respondents who reported consuming more than 100 ounces per day.

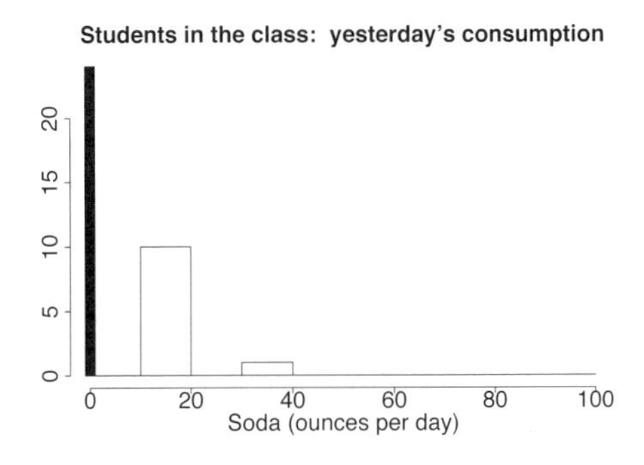

Fig. 3.4 Histogram of number of ounces of regular (nondiet) soda consumed by students the day before this topic was covered in class. The spike in the histogram indicates the 69% who reported consuming no regular soft drinks that day. Compare to the survey data in Fig. 3.3a.

since the distribution is skewed to the right, the mean will be larger than the median. We then compute the mean of the entire distribution, accounting for the spike at zero. For example, if the mean of the nonzero part of the distribution (that is, the average daily soda consumption for adults who drink soda) is 20 ounces, and 70% of the mass of the distribution is at zero, then the average daily soda consumption for all persons is,

$$70\% \cdot 0 + 30\% \cdot (20 \text{ ounces}) = 6 \text{ ounces}.$$

We can compare the students' guesses to published estimates of soda consumption. For example, in the 1995 nutrition survey (Fig. 3.3), 61% of the people reported drinking no regular (nondiet) soft drinks, and of the remaining 39%, the mean consumption was 18.3 ounces per day, which yields an overall mean of 7.2 ounces per day.

But other sources give different estimates. For example, the *Statistical Abstract of the United States* reports average soft drink consumption as 43.2 gallons per year in 1991. If we multiply by 128 and divide by 365.24, this yields 15.1 ounces per day of all soft drinks, and multiplying by 75% (the proportion of soft drink consumption that is nondiet, according to the survey) gives an estimate of 11.3 ounces per day of regular soft drinks. As another example, in a class survey, we find that 24 of the 35 students drank no regular soda the day before the survey was taken, and the remaining students drank an average of 16.4 ounces, which gives an overall average of 5.1 ounces of regular soda consumed that day. The students can discuss why the different sources give such different estimates.

In any case, the published estimates can be used as a check on the students' guesses (Section 3.3.3). In particular, if a student's histogram yields an average of 24 ounces per day, it is probably too high by at least a factor of 2. This is a rich example because we can discuss both the shape and the numerical values of the distribution.

3.4.2 The average student

Today, none of us wants to be average, but in 1835 the average man was a symbol of an egalitarian society. According to Quetelet, a nineteenth century statistician, "If an individual at any given epoch in society possessed all the qualities of the average man, he would represent all that is great, good, or beautiful."

We like to summarize the results from our class survey and describe the average student. For one of our classes, the individual who, as Quetelet puts it, possesses all the qualities of the average student, is 67.3 inches tall, watches 58 minutes of television per day, drinks 5.1 ounces of regular soda and 1.8 ounces of diet soda per day, comes from a family of 2.7 children, and is 0.49 male. From this amusing account of an average student, we discuss averages and medians for continuous and discrete numerical variables.

3.5 Data in more than one dimension

When we introduce scatterplots to students, we also explain that graphs can display data in many more than two dimensions. On a scatterplot itself, different symbols can be used for different points (for example, to illustrate males and females in Fig. 3.5). Another categorical variable can be coded by color, and the size of each point can code a continuous variable. Even more information can be shown using arrows (as in a plot of wind speeds and directions), and one can go further by placing multiple scatterplots in a single display, in which the positions of the scatterplots themselves convey information. We do not discuss such complicated graphs further in the course, but we want students to think about and be aware of such possibilities.

To allow students to understand some of the possibilities and challenges of multidimensional data displays, we set up prepared examples for class discussion.

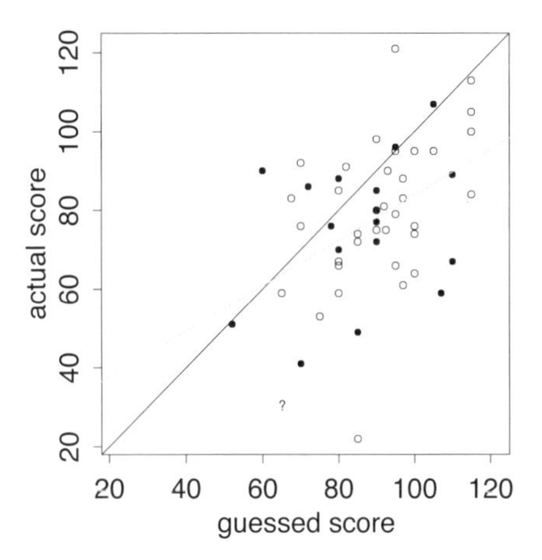

Fig. 3.5 Actual vs. guessed midterm exam scores for a class of 53 students. Each symbol represents a student; empty circles are men, solid circles are women, and ? has unknown sex. The 45° line represents perfect guessing, and the dotted line is the linear regression of actual score on guessed score. Both men and women tended to perform worse than their guesses.

Here we give one example each of continuous and discrete data.

3.5.1 Guessing exam scores

We sometimes include a question at the end of an exam asking the student to guess his or her total score on the other questions of the exam. As an incentive, the student receives five points extra credit if the guess is within ten points of the actual score. When the students complete their exams, we keep track of the order in which they are handed in, so that we can later check to see if students who finish the exam early are more or less accurate in their self-assessments than the students who take the full hour. When grading the exams, we do not look at the guessed score until all the other questions are graded. We then record the guessed grade, actual grade, and order of finish for each student. We have three reasons for including the self-evaluation question: it forces the students to check their work before turning in the exam; it teaches them that subjective predictions can have systematic bias (in our experience, students have tended to be overly optimistic about their scores); and the students' guesses provide us with data for a class discussion, as described below.

Figure 3.5 displays the actual and guessed scores (out of a possible score of 125) for each student in a class of 53, with students indicated by solid circles (women), empty circles (men), and ? for a student with unknown sex. (This student had an indeterminate name, was not known by the teaching assistants, and dropped the course after the exam.) The points are mostly below the 45° line, indicating that most students guessed too high. Perhaps surprisingly, men did

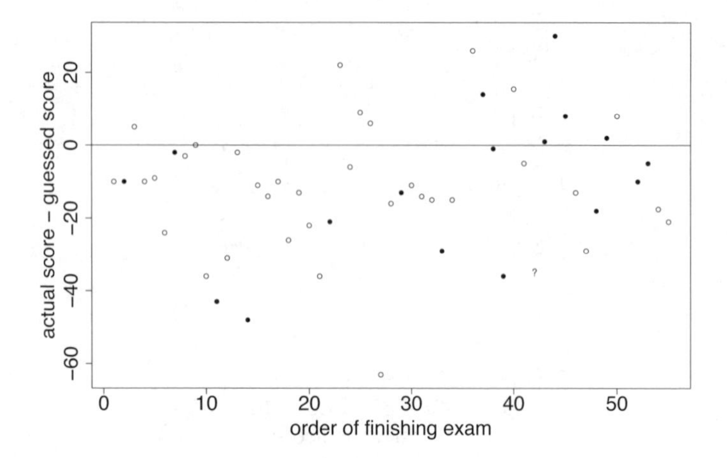

Fig. 3.6 Difference between actual and guessed midterm exam scores, plotted against the order of finishing the exam. The exact order is only relevant for the first 20 or 25 students, who finished early; the others all finished within five minutes of each other at the end of the class period. Each symbol represents a student; empty circles are men, solid circles are women, and ? has unknown sex. The horizontal line represents perfect guessing. The students who finished early were highly optimistic, whereas the other students were less biased in their predictions.

not differ appreciably from women. The dotted line shows the linear regression of actual score on guessed score and displays the typical "regression to the mean" behavior (see Section 4.3).

A class discussion should bring out the natural reasons for this effect. Figure 3.6 shows the difference between actual and guessed scores, plotted against the order of finish. Many of the first 20 or 25 students, who finished early, were highly overconfident; whereas the remaining students, who took basically the full hour to complete the exam, were close to unbiased in their predictions. Perhaps this suggests that students who finish early should take more time to check their results. (The students who finished early did, however, have higher than average scores on the exam.) Other studies have found that students' test performances are overestimated by their teachers as well.

The data also have a detective-story aspect that can be fun to discuss. For example, why do the guesses scores max out at 115? Since students got extra credit for a guess within 10 points of their exam score, and the exam was only worth 125 points, it would not make sense to guess higher than 115. (In fact, however, all four students who guessed 115 were overconfident about their grades.) What about the student with uncertain sex who guessed 65 and scored 30? How could someone guess so poorly? There could be a logical motivation, based on the following reasoning: if he or she scored below 55 on the exam, he or she would drop the course anyway. The extra credit points would then only be useful with

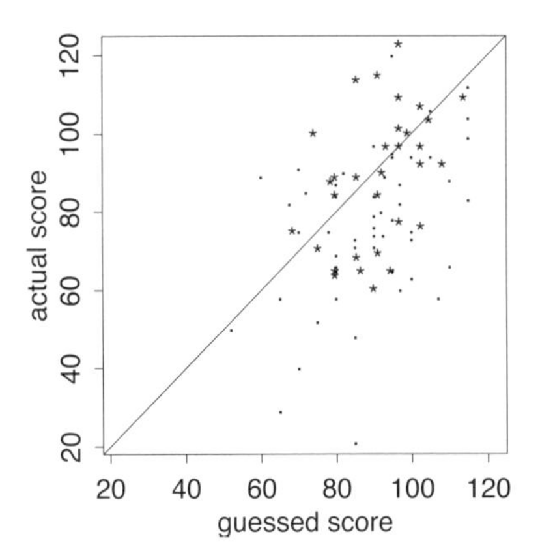

Fig. 3.7 Actual vs. guessed midterm exam scores for students in two terms of introductory statistics classes. The dots represent students in the first term; the asterisks represent students in the second term, who were shown the data from the first term (Figs. 3.5 and 3.6) a week before the exam. The students in the second term gave predictions that were less biased. A square scatterplot is used because the horizontal and vertical axes are on the same scale.

a score above 55, hence the guess of 65. In fact, the student did drop the course after the low exam score.

When teaching this course again, we varied the procedure by handing out Figs. 3.5 and 3.6 a week before the midterm exam, discussing the overconfidence phenomenon, and warning them that the same question would appear on their exam. We were encouraged to find that, thus prepared, the students' guesses were less biased than those of the earlier class. Figure 3.7 displays the results for the unprepared class (indicated by dots, the same data as displayed in Fig. 3.5) and the prepared class (indicated by asterisks).

3.5.2 Who opposed the Vietnam War?

To give our students practice in working with discrete data, we hand out copies of the partially filled table in Fig. 3.8 and ask them to guess the numbers to fill in the blanks. The table summarizes public opinion about the Vietnam War in 1971 among adults in the United States as classified by their education level. We give the students two minutes to fill out the table (in pairs, of course) and then discuss the results. We ask one student to present his or her pair's guessed table on the blackboard, and we begin by checking it for arithmetic: the numbers in each column must add up to 100% and the numbers in the top and bottom row must average, approximately, to 73% and 27%, respectively. (More precisely, if the numbers in the table are

p_1	p_2	p_3
$1 - p_1$	$1 - p_2$	$1 - p_3$

In January 1971 the Gallup poll asked: "A proposal has been made in Congress to require the U.S. government to bring home all U.S. troops before the end of this year. Would you like to have your congressman vote for or against this proposal?"
Guess the results, for respondents in each education category, and fill out this table (the two numbers in each column should add up to 100%):

	Adults with:			
	Grade school education	High school education	College education	Total adults
% for withdrawal of U.S. troops (doves)				73%
% against withdrawal of U.S. troops (hawks)				27%
Total	100%	100%	100%	100%

Fig. 3.8 In-class assignment for students to get experience in discrete-data displays. Students should work on this in pairs filling out the table. We discuss their guesses and then reveal the true values (see page 266).

and the proportion of adults with grade school, high school, and college education are $\lambda_1, \lambda_2, \lambda_3$ (with $\lambda_1 + \lambda_2 + \lambda_3 = 1$), then $\lambda_1 p_1 + \lambda_2 p_2 + \lambda_3 p_3 = 0.73$. Since $\lambda_1, \lambda_2, \lambda_3$ are each approximately 1/3, the average $(p_1 + p_2 + p_3)/3$ must be approximately 0.73.)

We then ask the student at the blackboard to explain his or her numbers and ask the other students in the class for comments. Finally, we present the true numbers, which typically differ dramatically from students' guesses (see page 266). This is an interesting example because it connects the mathematics of tabular displays to an interesting historical and political question. The key link is that, if some group supports withdrawal by more than 73% (the national average), then some other group must support it by *less* than 73%. Thus, the table of percentages (technically, the conditional distribution of the survey response conditional on education) focuses on differences among the educational groups, which is of political interest.

3.6 The normal distribution in one and two dimensions

We illustrate the one-dimensional normal distribution with examples that are twists on the typical example of men's heights. We also examine students' exam scores from previous classes; these scores provide information that catches the attention of our current students, and they are alert to ways in which exam data depart from normality (for example, in being bounded, discrete, and possibly skewed or with outliers). We do not spend too much time on examining such distributions, however, since exploratory analysis is much more fun in two dimensions, as we illustrated with predicted and actual exam scores in Section 3.5.1 and also with repeated exams in Section 3.6.3 below.

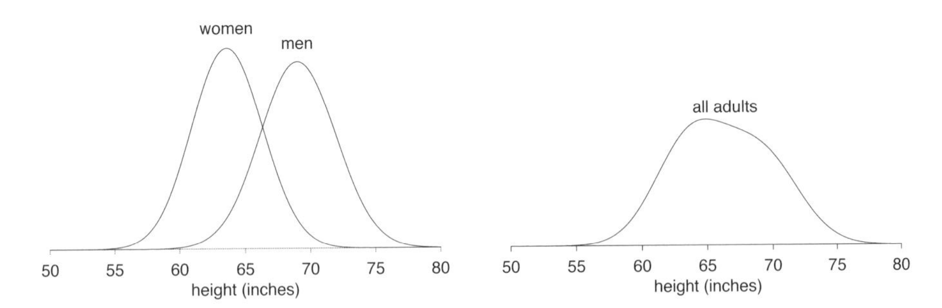

Fig. 3.9 (a) Distributions of heights of adult men and women in the United States. (b) Distribution for all adults (52% women, 48% men). The combined distribution has only one mode because the mean heights of men and women are so close together.

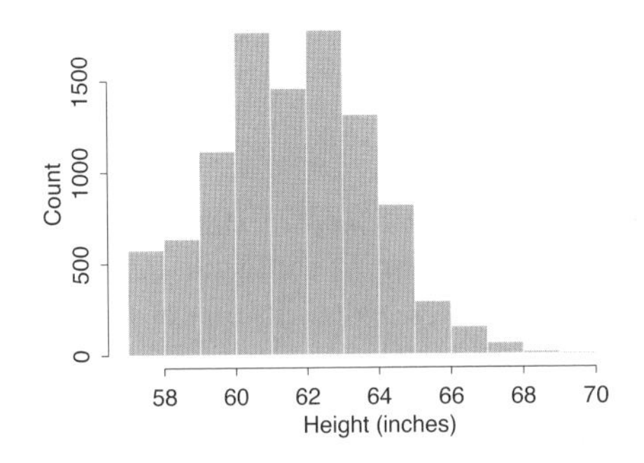

Fig. 3.10 Distribution of heights of conscripts from Doubs, France. The distribution appears to be bimodal, but it is an artifact of the conversion of the measurements from centimeters to inches (see Section 3.6.2).

3.6.1 Heights of men and women

A simple and standard example of the normal distribution is the height of adult men or women. In the United States, men's heights have mean 69.1 inches (175.5 centimeters) and standard deviation 2.9 inches (7.4 cm), women's have mean 63.7 inches and standard deviation 2.7 inches, and both distributions are well approximated by the normal. The normal distribution is understandable here if we think of the height of a man or woman as a sum of many small factors, so that the Central Limit Theorem applies.

Perhaps surprisingly, the combined distribution of the heights of all adults is *not* bimodal—the means of the two sexes are close enough that the modes overlap (see Fig. 3.9). However, data from any specific sample can be bimodal (see Fig. 4.5 on page 44).

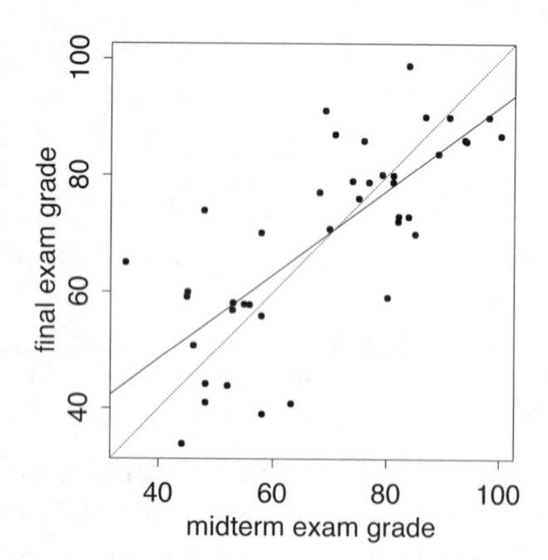

Fig. 3.11 Grades on midterm exams (each student is indicated by a dot on the graph), a familiar example with which to illustrate the regression effect. The solid line is the least-squares prediction, $(y - 70) = 0.72(x - 69)$, which has a slope that is visibly less steep than that of the $45°$ dotted line. The correlation of the data in this plot is 0.75.

3.6.2 Heights of conscripts

The distribution of the heights of military recruits from Doubs, France, in the 1850s shows two modes (Fig. 3.10), and at that time it was hypothesized that the two modes came from combining the distributions of two groups of men—Celts and Burgundians. After the students have seen that the combined distribution of the heights of men and women is not bimodal, they are suspicious of this claim. We ask them to come up with a reason for how this could happen.

In this case, rounding is the culprit. The original measurements were reported to the nearest centimeter, but these measurements were converted to inches and plotted to the nearest inch, creating an artificial dip at the center of the distribution. The interval from 61 to 62 inches contains counts from two-centimeter bins, whereas the intervals 59–60, 60–61, 62–63, and 63–64 all contain counts from three-centimeter bins.

3.6.3 Scores on two exams

Scatterplots of students' scores on two successive exams is a familiar example with which to illustrate the regression effect (see, for example, Fig. 3.11). The regression line of the second exam on the first typically has a slope less than 1, the students who score the highest on the first exam typically do worse on the second exam ("regression to the mean"), and so forth. Many students are more interested in this example than in the traditional bivariate normal example of parents' and children's heights. Students commonly see exam scores represented as univariate distributions (for example, mean, median, and standard deviation of scores, stem-and-leaf plots) but the bivariate display stimulates new thoughts. We return to this example in Section 4.3.2.

3.7 Linear transformations and linear combinations

When we cover linear and logarithmic transformations in our college statistics course, we're really covering high school algebra. But this is often new to the students since it is an application of algebra, not simply solving mathematical problems. We attempt to motivate them to learn this difficult material by connecting it to real problems (beyond the usual transformation examples such as exchange rates or converting degrees Fahrenheit to Celsius).

3.7.1 College admissions

Students are always interested in what is needed to get into college. We use the idea of a college admissions formula to explain linear transformations: for example, total score = SAT + 1000 · GPA. We start by prompting the students to guess the mean and standard deviation of the Scholastic Assessment Test (SAT) scores of the applicants to the university—this is a good way to remind them of the rule of thumb that approximately two-thirds of the distribution falls within ± 1 standard deviation. A similar exercise works for high school grade point average (GPA). From there, they are asked to compute the mean and standard deviation of the total scores of the applicants. After some discussion, the students realize that the standard deviation of the total score depends on the correlation of SAT and GPA in the population.

3.7.2 Social and economic indexes

Various social and economic topics lend themselves well to discussion in a statistics class. Opinion polls are a popular source of examples; these are familiar enough that there is no need for us to discuss them further here. Another popular topic is census adjustment (see Section 14.6.3), but for an introductory course we prefer examples whose main interest lies outside of statistics, thus better demonstrating the pervasive influence of statistical ideas.

An example of a topic in economics is defining economic growth. There is some controversy about how to measure the size of the economy. The traditional measure, Gross Domestic Product (GDP), has some problems. For example, if you and your spouse divorce, have huge legal bills, and have to move to separate houses, that increases the GDP. But if you stay together and spend quiet afternoons at home, the GDP remains unchanged. Some economists have suggested alternative measures of the size of the economy to separately identify productive and unproductive activities. A discussion on this topic is relevant for a statistics class.

A related measurement problem arises when defining the rate of inflation. The Consumer Price Index is defined as the price of a standard "basket of goods," and 10% inflation, for example, means that the cost of these items has increased 10% in the past year. But not all the goods increase in price at the same rate. For example, the prices of high-end computer and electronic equipment typically decline over time—thus, including these products in the "basket of goods" causes the official rate of inflation to decrease. But it's relatively rich people who buy these items. Thus, the true rate of inflation might be lower for rich people than for poor people. More generally, the rates of inflation can be different in different sectors of the economy. So choices have to be made in defining the linear combination that is the official "inflation rate."

3.7.3 Age adjustment

Comparing the averages of two groups can be misleading if there are lurking variables. The idea of adjusting an average for these variables is easier to understand if the students simulate such data. To do this, the students act as our population, and we use random digits from dice (see Section 7.2) to simulate their ages and whether or not they have insomnia. The students are divided into two groups, and the setup is such that the incidence of insomnia is dramatically different in the two groups, but when we compare people of the same age, the incidence is the same for the two groups.

If, for example, the classroom has four rows of about 10 students each, we have each student sitting in the first row roll a die once and add the result (0–9) to 40 to get his or her age, then on the second roll of the die, if it lands 0 or 1 the student has insomnia. For the second row, the first toss is added to 50 to find the age of the person, and he or she is an insomniac if the second roll lands 0–2. The third roll consists of 60-year-olds who have a 30% chance of having the condition, and those in the fourth row are in their 70s and have a 40% chance.

We now divide the class into two groups cutting diagonally across the rows, so that the first group has one student from the first row, three from the second row, six from the third row, and nine from the last row, and the other group contains the remaining students. Then roughly one-third of the first group and one-fifth of the other group have insomnia. The students see the problem with this comparison right away, and we ask them to figure out how to fix it while keeping the same people in each group. We wind up computing the rate of insomnia by age in each group and then find a weighted average of these rates for each group. We discuss how the weights can come from the combined age distribution of the two groups, the age distribution of one of the groups, or the distribution from some other group.

We tell the students that a similar age adjustment was used in constructing the maps of cancer rates discussed in the first week of class (see Section 2.2).

3.8 Logarithmic transformations

Logarithms are a topic in mathematics, not statistics, but we teach them in our introductory course because they are so useful for understanding data, partly because they allow numbers that vary by several orders of magnitude to be viewed on a common scale, and more importantly because they allow exponential and power-law relations to be transformed into linearity. We illustrate in Section 3.8.1 with some simple examples and then follow up in Sections 3.8.2 and 3.8.3 with some applied examples with interesting twists.

3.8.1 Simple examples: amoebas, squares, and cubes

Logarithms and exponentials are unfamiliar to most students we have taught, so it makes sense to start with simple examples of exact transformations. We do all logarithms in the base-10 scale in our introductory classes.

For example, suppose you have an amoeba that takes one hour to divide, and then the two amoebas each divide in one more hour, and so forth. What is the equation of the number of amoebas, y, as a function of time, x (in hours)? It can be written as $y = 2^x$ or, on the logarithmic scale, $\log y = (\log(2))x = 0.30x$.

Suppose you have the same example, but the amoeba takes three hours to divide at each step. Then the number of amoebas y after time x has the equation, $y = 2^{x/3} = (2^{1/3})^x = 1.26^x$ or, on the logarithmic scale, $\log y = (\log(1.26))x = 0.10x$. The slope of 0.10 is one-third the earlier slope of 0.30 because the population is growing at one-third the rate.

In the example of exponential growth of amoebas, y is logged while x remains the same. For power-law relations, it makes sense to log both x and y, as we illustrate first with some simple geometrical examples. How does the area of a square relate to its circumference? If the side of the cube has length L, then the area is L^2 and the circumference is $4L$; thus

$$\text{area} = (\text{circumference}/4)^2.$$

Taking the logarithm of both sides yields,

$$\begin{aligned}\log(\text{area}) &= 2(\log(\text{circumference}) - \log(4))\\ \log(\text{area}) &= -1.20 + 2\log(\text{circumference}),\end{aligned}$$

a linear relation on the log-log scale.

After doing this, we ask students to work in pairs and graph the relation between the surface area and volume of a cube, which, in terms of the side length L, are $6L^2$ and L^3, respectively. On the original scale, this is

$$\text{surface area} = 6(\text{volume})^{2/3},$$

or, on the logarithmic scale,

$$\log(\text{surface area}) = \log(6) + \frac{2}{3}\log(\text{volume}).$$

The students are now prepared for more complicated examples using logarithms to model real data.

3.8.2 Log-linear transformation: world population

Other time series work well to illustrate other points. For example, logarithmic transformations can be used to explore world population from the year 0 to 2000 (see Fig. 3.12). We show this to students in stages: first, we present the data in tabular form (the first two columns of Fig. 3.10, which we put on the blackboard before the class begins), then we graph the raw data (Fig. 3.13a) and then make

Year	Population	$\log_{10}$ (population)	Residual	10^{residual}
1	170 million	8.230	.258	1.81
400	190	8.279	.037	1.09
800	220	8.342	−.171	.68
1200	360	8.556	−.227	.59
1600	545	8.736	−.318	.48
1800	900	8.954	−.236	.58
1850	1200	9.079	−.145	.72
1900	1625	9.200	−.047	.90
1950	2500	9.398	.107	1.28
1975	3900	9.591	.283	1.92
2000	6080	9.784	.459	2.88

Fig. 3.12 Time series of estimated world population and residuals from a logarithmic fit, which reveal that the population has been growing even faster than exponentially. After graphing the data above (see Fig. 3.13), we ask the students to guess the population in the year 1400; it was 350 million.

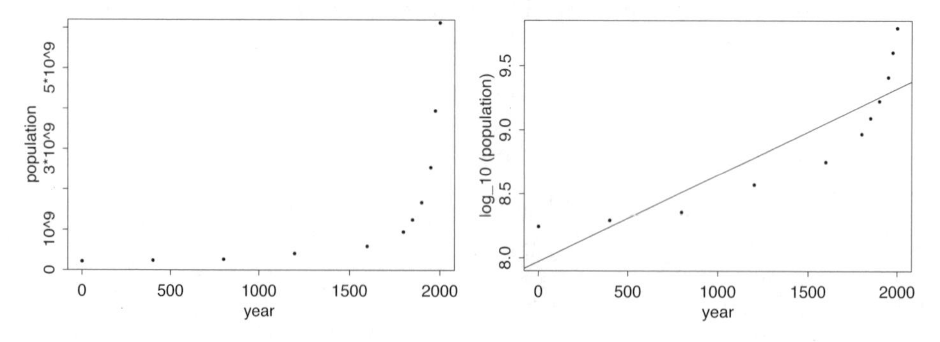

Fig. 3.13 World population over time, graphed on the original and logarithmic scales. This is a subtle example since the growth is faster than linear, even on the logarithmic scales. On the logarithmic graph, the least-squares regression line is drawn; the data in the original and logarithmic scales, and the residuals from the regression, are given in Fig. 3.12.

the graph on the logarithmic scale (Fig. 3.13b). On the raw scale, all you can see is that the population has increased very fast recently. On the log scale, convex curvature is apparent—that is, the rate of increase has itself increased.

Finally, we display the data for the year 1400, at which time the world population was 350 million (actually lower than the year 1200 population, because of plague and other factors), just to illustrate that even interpolation can sometimes go awry.

Another possible discussion topic is the source of the population numbers: for example, how would you estimate the population of the world in the year 1?

We return to this example in Section 4.1.3 to illustrate the use of regression lines on the logarithmic scale.

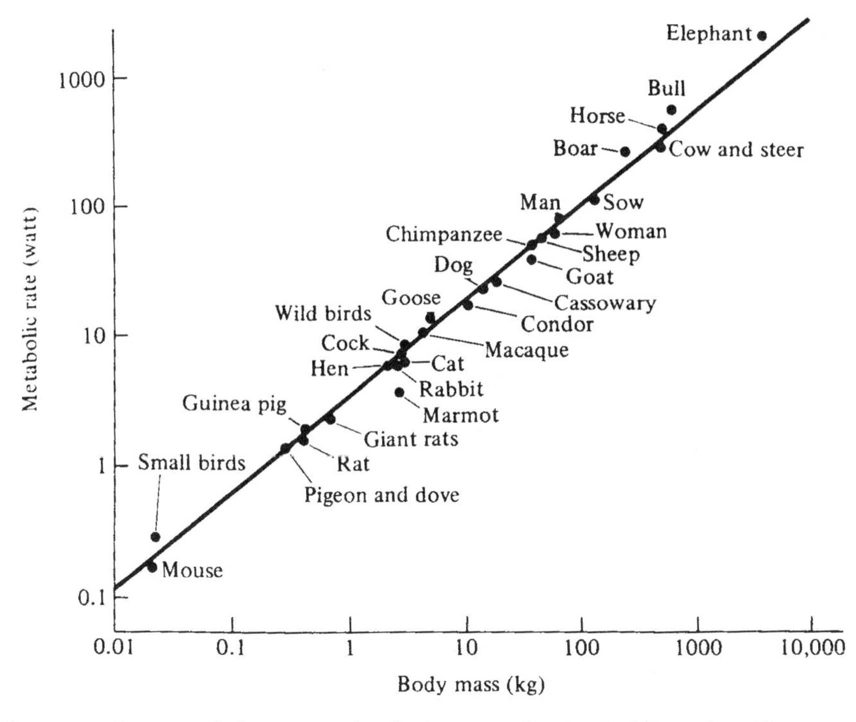

Fig. 3.14 Log metabolic rate vs. log body mass of animals. These data illustrate the logarithmic transformation, a difficult concept for students. The fitted line has a slope of about 0.75. See also Fig. 3.15.

3.8.3 Log-log transformation: metabolic rates

A rich source of examples when covering log-log transformations are biological scaling relations. For example, Fig. 3.14 displays data on log metabolic rate vs. body mass indicating an underlying linear relation. The scaling on the graph can be made vivid to the students by pointing to the 70-kilogram man who consumes about 100 watts; thus, a classroom with 100 men is the equivalent of a 10 000 watt space heater. By comparison, we ask the students to figure out the amount of heat given off by a single elephant (which weighs about 7000 kilograms according to the graph) or 10 000 rats (which together also weigh about 7000 kilograms). The answer is that the elephant gives off *less* heat than the equivalent weight of men, and the rats give off more. We then discuss how to understand this using logarithms.

What is the equation of the line in Fig. 3.14? The question is not quite as simple as it looks, since the graph is on the log-log scale, but the axes are labeled on the original scale. We start by relabeling the axes on the logarithmic (base 10) scale, as shown in Fig. 3.15a. We can then determine the equation of the line by identifying two points that it goes through: for example, when $x = -2$, $y = -0.9$, and when $x = 3$, $y = 2.8$. So, when x increases by 5 units, y increases by $2.8 - (-0.9) = 3.7$ units, and the slope of the line is $3.7/5 = 0.74$. Since the

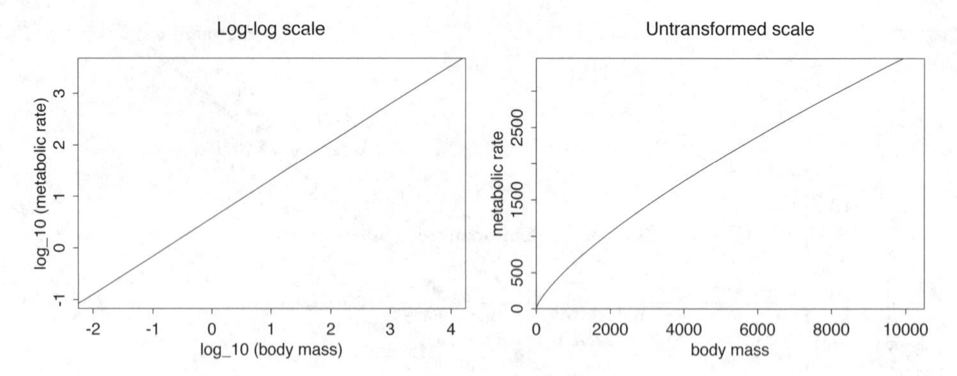

Fig. 3.15 Fitted curve (from data in Fig. 3.14) of metabolic rate vs. body mass of animals, on the log-log and untransformed scales. We give these to the students and ask them to put dots on each graph corresponding to Rat, Man, and Elephant (as displayed in Fig. 3.14).

line goes through the point (3, 2.8), its equation can be written as,

$$\begin{aligned} y - 2.8 &= 0.74(x - 3) \\ y &= 0.58 + 0.74x \\ \log\text{ (metabolic rate)} &= 0.58 + 0.74 \cdot \log\text{ (body mass)}. \end{aligned} \qquad (1)$$

We can exponentiate both sides of (1) to see the relation between metabolic rate and body mass on the untransformed scales:

$$\begin{aligned} 10^{\log(\text{metabolic rate})} &= 10^{0.58} 10^{0.74 \log(\text{body mass})} \\ (\text{metabolic rate}) &= 3.8\,(\text{body mass})^{0.74}. \end{aligned} \qquad (2)$$

This curve is plotted in Fig. 3.15b. So, for example, when body mass is 1 kilogram, metabolic rate is 3.8 watts. When body mass is 10 kilograms, metabolic rate is $3.8 \cdot 10^{0.74} = 21$ watts, when body mass is 100 kilograms, metabolic rate is $3.8 \cdot 100^{0.74} = 115$ watts, and so forth.

We want to focus on the slope in (1), which is the exponent in (2), so we write,

$$\begin{aligned} \log\text{ (metabolic rate)} &= a + 0.74 \log\text{ (body mass)} \\ (\text{metabolic rate}) &= A\,(\text{body mass})^{0.74}. \end{aligned}$$

For example, if you multiply body mass by 10, then you multiply metabolic rate by $10^{0.74} = 5.5$. If you multiply body mass by 100, then you multiply metabolic rate by $10^{0.74} = 5.5^2 = 30.2$, and so forth. These are all basic calculations, but it is useful to work through them in class.

Now we return to the rats and the elephant. The relation between metabolic rate and body mass is less than linear (that is, the exponent 0.75 is less than 1.0, and the line in Fig. 3.15b curves downward, not upward), which implies that the equivalent mass of rats gives off more heat, and the equivalent mass of elephant gives off less heat, than the men. This seems related to the general geometrical relation that surface area and volume are proportional to linear dimension to the second and third power, respectively, and thus surface area should be proportional to volume to the 2/3 power. Heat produced by an animal is emitted from its surface, and it would thus be reasonable to suspect metabolic rate to be proportional to the 2/3 power of body mass. Biologists have considered why the empirical slope is closer to 3/4 than to 2/3; the important thing here is to get students thinking about log transformations and power laws.

Our students are typically very uncomfortable with logarithms, and when covering this material we give them many simple exercises, such as dividing them in pairs to plot the lines $\log y = a + 0.74 \log x$ and $y = Ax^{0.74}$. We also focus their thinking by pointing out patterns in the animal data; for example, the males (Bull, Boar, Man, Cock) are all above the line and the females (Cow, Sow, Woman, Hen) are all below.

4 Linear regression and correlation

We follow descriptive statistics in our course with a descriptive treatment of linear regression with a single predictor: straight-line fitting, interpretation of the regression line and standard deviation, the confusing phenomenon of "regression to the mean," correlation, and conducting regressions on the computer. We illustrate various of these concepts with student discussions and activities.

This chapter includes examples of the sort that are commonly found in statistics textbooks, but our focus is on how to work them into student-participation activities rather than simply examples to be read or shown on the blackboard.

4.1 Fitting linear regressions

We have found that students have difficulty with the algebra of linear equations; for example, it is far from intuitive that the line going through the point $(\bar{x}, \bar{y})$ and with slope b has the equation, $y = \bar{y} + b(x - \bar{x})$. Thus, when introducing linear regression, we start with algebraic exercises with straight lines in various examples.

In this section we present some simple drills illustrating the mathematics of least-squares regression and then discuss two examples where least-squares fits are informative but not quite ideal for the data at hand. In presenting the examples, we focus first on the mechanics of the model—in particular, on the interpretation of the fitted line. We then introduce the concept of residual standard deviation, that not all points will fall exactly on the line. Finally, we look more carefully at the data and discuss with the students the limitations of the least-squares fits.

4.1.1 Simple examples of least squares

We use simple scatterplots to introduce straight-line fitting. We bring to class a handout of four scatterplots (Fig. 4.1), divide the students into pairs, and ask the students in each pair to draw the best-fitting line for predicting y from x from the data on each plot. We don't go into the exact definition of a best fit yet—we just tell them that we want to predict y from x and let them loose on the problem. Once they have drawn their lines, we have them compute their predictions for each y_i and the sum of squared errors. Then we pass out red pencils and ask each pair to mark a red $\times$ at the average y-value for each observed x (see Fig.

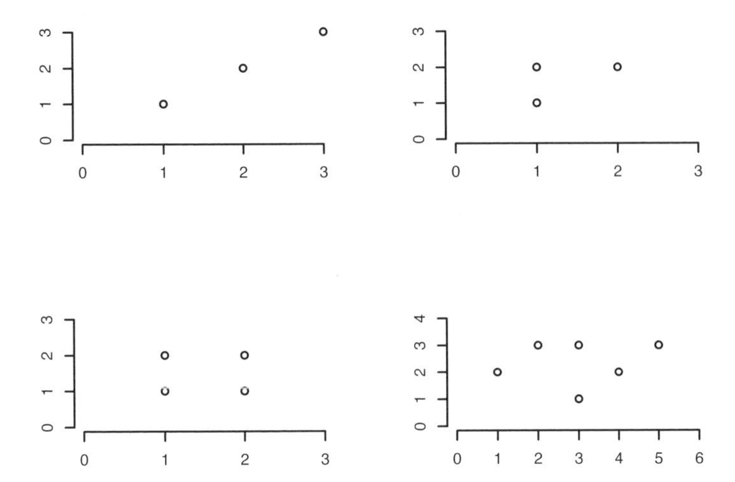

Fig. 4.1 We present these simple scatterplots to students to teach least-squares fitting of a line to points. Students sketch on each plot what they think is the best-fitting line for predicting y from x. The example is continued in Fig. 4.2.

4.2). Then they find the best-fitting line to the crosses and compute the sum of squared errors for predicting y_i from the red line.

For a couple of the scatterplots, the placement of the points suggests the wrong line to the naive line-fitter. The students are surprised to find the red line outperforms their line in these cases, and at this point we begin the discussion of minimizing square error in the y-direction.

4.1.2 Tall people have higher incomes

To demonstrate linear regression, many examples are possible. We use a regression of income on height, because it has a story with lurking variables to which we can return in the discussion of multiple regression at the end of the semester (see Section 9.1). Our focus here is on the interaction with the students more than the example itself.

Before class begins, we set up Fig. 4.3 on a transparency and project it onto the blackboard. We trace the lines and label the axes of the graph, then turn off the projector, but leave it in place.

We begin the discussion by asking students if they think that taller people have higher earnings (that is, income excluding unearned sources such as interest income). If so, by how much? We draw on the blackboard a pair of axes representing earnings and height, and a point at $(66.5, 20\,000)$: the average height of adults in the United States is about $5'6.5''$ and their average earnings (in 1990) were about \$20 000. We then draw a line through this central point with slope

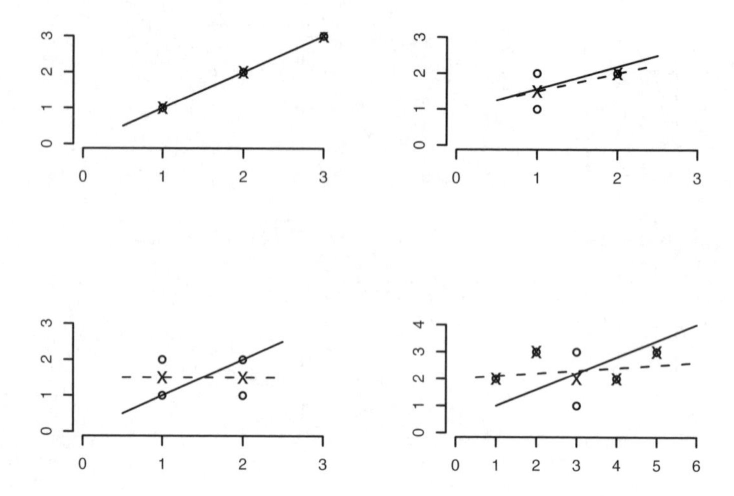

Fig. 4.2 The solid line in each plot (from Fig. 4.1) shows a student's guess of the best-fitting line to the set of points. The student then marks crosses at the average y-value for each x and draws the best-fitting line to the crosses. In some cases these lines are quite different, which provides a good lead-in to the method of least squares.

1560, carefully connecting the central point $(66.5, 20\,000)$ and two points on the regression line: $(56.5, 20\,000 - 10 \cdot 1560)$ and $(76.5, 2000 + 10 \cdot 1560)$). We pick points ± 10 inches from the central point just for convenience in calculation and display.

We explain that this line has equation $y - 20\,000 = 1560(x - 66.5)$, or $y = -84\,000 + 1560x$. We tell the students that this is the regression line predicting earnings from height. We now ask the students to work in pairs and sketch a scatterplot of data that are consistent with this regression line. They need one more piece of information: the standard deviation of the residuals, which is 19 000. We show this on the graph by two dotted lines, parallel to the regression line, with one line 19 000 above and the other line an equal distance below. Approximately 68% of the data should fall in this region, but plotting the points is tricky because of the constraint that earnings cannot be negative.

We then turn on the projector and display the graph of the actual survey data (Fig. 4.3) on the blackboard. (The survey may have its own problems of response and measurement error, but that is not our point here.) On the scale of the actual data, the regression slope ($1560 per inch of height) is small but undeniably positive. (We discuss statistical significance later in the course; see Section 9.1.1.)

We then ask the students how do we interpret the constant term in the regression: that is, the value $-84\,000$ in the equation, $y = -84\,000 + 1560x$ (see

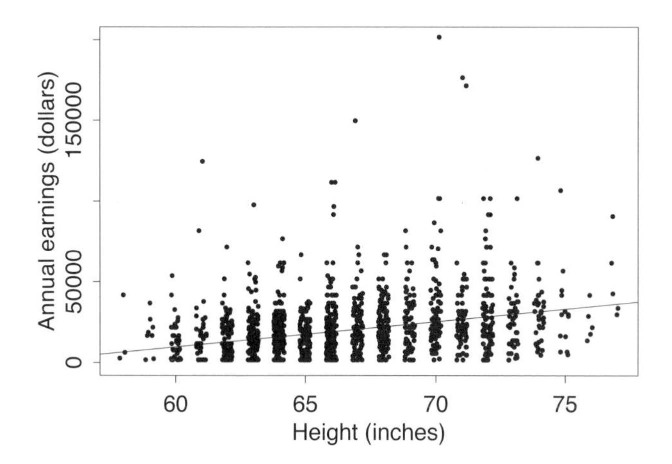

Fig. 4.3 Earnings vs. height for a random sample of adult Americans in 1990. The heights have been jittered slightly so that the points do not overlap.

the regression table at the bottom of Fig. 4.4)? The answer is, $-84\,000$ is the y-value of the regression line where $x = 0$—that is, the predicted value of income for an adult who is zero inches tall. In this example, such an extrapolation is meaningless. That is why we prefer to work with the form, $y = \bar{y} + b(x - \bar{x})$; in this case, $y = 20\,000 + 1560(x - 66.5)$.

The next question is how to interpret the result that taller people have higher earnings? The students realize that men are taller than women and tend to make more money; thus, sex is a lurking variable. We return to this example using multiple regression in Section 9.1. (In fact, it turns out that height is correlated with earnings even after controlling for sex.)

At the conclusion of the discussion, we hand out copies of Fig. 4.4, which shows what we had to do to clean the data file before running the regression. They can also use this as a template when doing their computer homework assignments. We explain to the students that linear regression is not the best model for this sort of data (economists might use a logarithmic model, or a tobit regression), but it is in some ways more useful to illustrate the concept in an example for which it is not completely appropriate.

4.1.3 Logarithm of world population

We return to the world population data (see Section 3.8.2) to illustrate linear modeling on a transformed scale. It is worth going through these calculations of linear regression predictions and errors on the logarithmic scale. We have found students to have great difficulty with this sort of problem on exams unless they have practiced it a lot.

A good way to understand the logarithmic model is with the least-squares line: for the world population data in Fig. 3.12, the years have mean $\bar{x} = 1406$,

```
. infix earn 203-208 height1 144 height2 145-146 using "wfw90.dat"
(2031 observations read)
. summ
Variable |      Obs        Mean    Std. Dev.         Min         Max
---------+-----------------------------------------------------------
    earn |     1380    20290.21    22247.52            0      400000
 height1 |     2031    5.138848    .4342736            4           9
 height2 |     2031    5.461841    6.692941            0          99
```

Now let's look at the height variables (in feet and inches).

```
. graph height1
. graph height2
```

Recode the missing data and create a total height variable in inches. Graph it.

```
. recode height1 9=.
(8 changes made)
. recode height2 99=.
(6 changes made)
. gen height = 12*height1 + height2
(8 missing values generated)
. graph height
```

The tallest person is 91 inches—that's 7 feet, 7 inches. We suspect this is a mistake. We'll drop this data point and then graph the data.

```
. drop if height==91
(1 observation deleted)
. graph earn height, xlabel ylabel
```

The graph showed an outlier at \$400 000. This is possible, but we are doubtful. We drop it and then run the regression and plot the data and regression line.

```
. drop if earn==400000
(1 observation deleted)
. regress earn height
  Source |       SS       df       MS                  Number of obs =     1379
---------+------------------------------               F(  1,  1377) =   137.21
   Model |  4.8773e+10     1  4.8773e+10               Prob > F      =   0.0000
Residual |  4.8948e+11  1377   355470204               R-squared     =   0.0906
---------+------------------------------               Adj R-squared =   0.0900
   Total |  5.3826e+11  1378   390606004               Root MSE      =    18854
------------------------------------------------------------------------------
    earn |      Coef.   Std. Err.       t     P>|t|     [95 Conf. Interval]
---------+--------------------------------------------------------------------
  height |   1563.138   133.4476     11.713   0.000      1301.355     1824.92
   _cons |  -84078.32   8901.098     -9.446   0.000     -101539.5   -66617.15
------------------------------------------------------------------------------
. graph earn yhat height, connect(.s) symbol(Oi) xlabel ylabel
```

Fig. 4.4 Log file from Stata showing the steps we had to go through to get the height and earnings information from the public data file. Stata commands (in `typewriter font` and preceded by the "." prompt) and output (in `typewriter font`) are interspersed with our comments (in *italics*). We give this to the students to give them an idea of the practical difficulties of statistical analysis.

the log (base 10) populations have mean $\bar{y} = 8.92$, and the regression slope is $b = 0.000\,687$, with a residual standard deviation of 0.265. We plot this line on the blackboard (where the data are displayed, having been projected from a transparency with graph paper, as described in Section 3.1) and then spend some time discussing the interpretation of the least-squares line (see Fig. 3.13b).

First, the regression line goes through $\bar{x}$ and $\bar{y}$, which means that, according to the line, in the year 1406 the log population was 8.92, so the population was $10^{8.92} = \ldots$ somewhere between 10^8 and 10^9 ... that is, between 100 million and 1 billion ... checking on the calculator, it is 831 million. The slope of the line is 0.000 687, which means that in every year, the log population increases by 0.000 687, which means the population increases by a factor of $10^{0.000\,687} = 1.001\,58$. That's a pretty small amount. What about every 100 years? In 100 years, the regression line increases by 0.0687, so the population would be multiplied by $10^{0.0687} = 1.17$, an increase of 17%. In 200 years, this becomes a factor of $10^{0.0687 \cdot 2} = 1.17^2 = 1.37$, that is, an increase of 37%, and so forth. The residual standard deviation implies that we can expect the predictions to be off by about 0.265, and a range of ± 0.265 on the log scale corresponds to a range of $[10^{-0.265}, 10^{0.265}] = [0.54, 1.84]$ in the scale of population. That is, we expect the true value to be between 54% and 184% of our prediction.

We conclude with an examination of the residuals themselves, which show a pattern indicating a changing slope over time, with a higher rate of increase in recent years.

4.2 Correlation

We cover correlation immediately after linear regression, and we define the correlation coefficient between x and y as the regression slope, after both variables have been scaled to have standard deviations of 1. This allows us to continue to develop students' intuitions about regression effects. We illustrate correlation with data collected from students: physical measurements, grades, and simple quizzes or memory tasks that can be performed in class. This builds upon our earlier use of correlation between exam scores (see Fig. 3.11) to illustrate the concept of a bivariate distribution.

4.2.1 Correlations of body measurements

It is traditional to look at correlations of measurements of body parts in individuals or compared to relatives or spouses. We have fun collecting measurements of unusual body parts on students in our classes. For example, we pass out strips of paper marked with a centimeter rule and have students measure the span of their right hand (distance from thumb to little finger when the fingers are spread apart), the length of their left foot, or the length of their left forearm. Or if we have a small class, we have students stand at the blackboard and measure their arm span. We pair this new body measurement up with the student's height to discuss correlation (see Fig. 4.5).

In our discussion, we compare our data with those studied by Karl Pearson at the beginning of the twentieth century. These data were also collected from

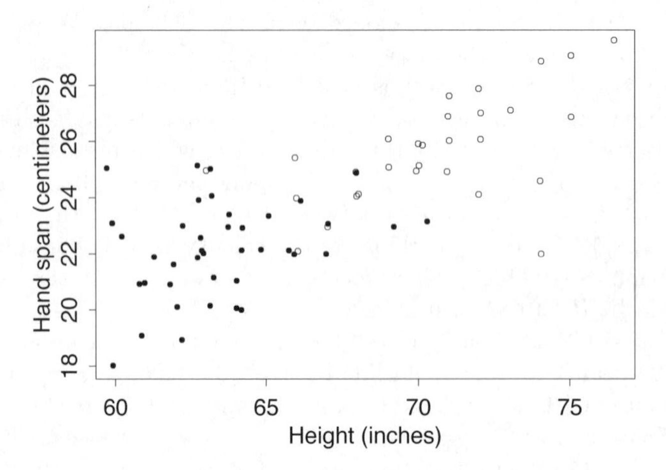

Fig. 4.5 Height and span of right hand for students in a statistics course. (The points are jittered slightly so that they do not overlap.) The average height is 66.5 inches with a standard deviation of 4.5 inches, the average hand span is 23.5 centimeters with a standard deviation of 2.5 centimeters, and the correlation is 0.74. This scatterplot combines men (empty circles) and women (solid circles). When considered separately, the 41 women had an average height of 63.4 inches with standard deviation 2.5 inches, average hand span of 22.2 centimeters with standard deviation 1.7 centimeters, and a correlation of 0.29. The 29 men had an average height of 70.5 inches with standard deviation 3.2 inches, average hand span of 25.5 centimeters with standard deviation 2.0 centimeters, and a correlation of 0.57.

university students. Today's students are taller than students 100 years ago, but the correlations between lengths of various body parts has remained about the same.

Sometimes, we surreptitiously lump data for men and women together—the correlation then becomes much higher than for each group alone—and we ask our students why they think the correlation is stronger now than it was 100 years ago. After they come up with a few plausible (but false) explanations, we reveal to them that we have combined the data—and then we discuss, using a diagram of a scatterplot, why the correlation is higher for the mixture than for either sex alone.

4.2.2 Correlation and causation in observational data

To make a different point, we recorded one year, for each student, the score on the final exam and the number of pages used by the student in the blue book to write the exam solutions. The two variables were negatively correlated. Since then, we have used these data to illustrate Simpson's paradox and the distinction between correlation and causation. A naive interpretation of the negative correlation between pages written and exam scores would suggest that students could raise their scores (on average) by writing less. But this is not so—for any given student, it would only help to write more. This is similar to the high scores of the students who finish early on exams (see Section 3.5.1): students who require the entire class period to finish their exams have lower scores, on average, than those who finish early, but, for any given student, staying on and working through the entire class period can only increase his or her score.

These examples are a natural lead-in to other discussions of correlation and causation. For example, people who own BMWs have bigger bank balances, on average, than people who own Volkswagens; but this does not mean that if you sell your Volkswagen and buy a BMW, you will have more money in the bank. For another example, baseball players with higher batting averages receive higher salaries, on average; does this mean that if a professional baseball player raises his batting average he will likely get a higher salary? Well, yes ..., but why is that? Obviously the correlation alone is not enough to convince us.

4.3 Regression to the mean

We use a series of examples and class activities to demonstrate the ubiquity of regression to the mean. This is a notoriously difficult topic to understand, and we hope that we give students a chance to see the connections to correlation and regression by involving them in several examples.

4.3.1 Mini-quizzes

We introduce the concept of "regression to the mean" with a demonstration in which students are given two short tests separated by a few minutes. The students can be tested on just about anything; it is important, however, that the test have a mix of skill and luck so that scores are neither completely random nor completely predictable. One thing we have tried, which works but is not as exciting as we might want, is memory testing: we read aloud a list of fifteen words, then wait 15 seconds, then ask each student to write all of these words he or she remembers. We explain the task to the students ahead of time (so that they know to try to memorize the words), and they typically can remember 5 to 10 of the words. We use different sets of 15 words for the two tests.

After the first test, we tally their scores and ask each student what score they expect to get on the second test. Then we compare these guesses to what actually happens. Typically, students expect to do about the same on the second test as the first, but actually the students who do worst on the first test tend to improve, and the students who do best tend to decline. For example, Fig. 4.6 shows results from a class of 25 students.

We then ask the students why these improvements and declines occurred. They generally give explanations specific to the example: the students who did the best relaxed on the second quiz, the students who did poorly tried harder, and so forth.

Fundamentally, these explanations are not correct. It would be more accurate to say that some people do better on this test than others, but the very best scores on the first test are from people who are both good *and* lucky. They will

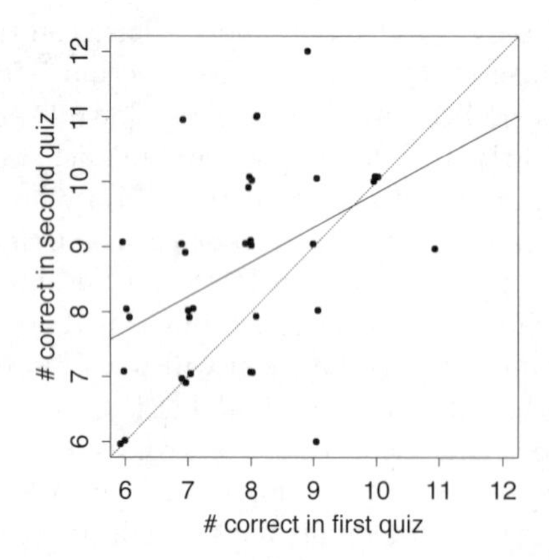

Fig. 4.6 Number of words successfully remembered (out of 15) in each of two successive memory quizzes: each dot indicates a student in the class. The solid line shows the regression line, which can be compared to the 45° line showing $y = x$. Students who perform poorly on the first test tend to improve, and high performers tend to decline (or, at least, not increase as much as the students who performed poorly the first time), an instance of the regression effect.

still probably do well on the second test, but most of them will not be so lucky the second time. However, we defer the discussion on this until we have considered some more examples.

4.3.2 Exam scores, heights, and the general principle

We next remind students of the data on scores in two successive midterm exams (Section 3.6.3). The students who performed the best on the first exam tended to decline on the second exam, and those who performed the worst tended to improve. This pattern is clear in the regression line (see Fig. 3.11). As with the memory quizzes described above, the students give specific explanations tied to the psychology of exam-taking.

We now bring in the killer example, which is also the standard example of regression to the mean: heights of parents and children. For simplicity, we consider heights of women and their adult daughters, so that we can ignore the complications of sex differences. Sometimes we gather data from the students (simply asking the women in the class to write the heights of themselves and their mothers); other times we simply present a scatterplot of already-recorded data. In either case, the data and the regression line show that the tallest women have daughters who are tall, on average, but closer to the mean in height; and similarly for the shortest women. We personalize it by considering individual examples: for example, Janet Reno is 6′2″ tall; if she has a daughter, she would probably grow up to be tall, but probably quite a bit shorter than her mother.

What is the explanation for this "regression to the mean"? We point out to the class that, mathematically, these data have the same form as the two memory quizzes or the two exam scores, but it is ridiculous to suggest, for example, that because Janet Reno is so tall, her daughter will not try so hard and hence will "regress" toward mediocrity.

The students are finally prepared for the general point, which is such patterns occur in *any* pair of variables with correlation less than 1. We remind the class of the identity $r = bs_x/s_y$; hence, the correlation r equals the regression coefficient b of the standardized scores (for which $s_x = s_y$). For example, in the memory quizzes and the midterm exam scores, even if everybody is trying equally hard on both exams, the correlation will not be perfect on the two exams, hence the regression slope of the standardized scores will be less than 1, hence the people with the highest scores on the first test will decline (relative to the population) and those with the lowest scores will improve, on average.

We also point out that regression to the mean, like correlation, is symmetric in the x and y variables. Thus, for example, students who scored best on the second exam probably were closer to the mean, on average, for the first exam. And daughters who are much taller than the mean will probably have mothers who are not so tall. We can confirm this by asking individual short and tall women in the class what their heights and their mothers' heights are.

5

Data collection

One way students learn about data collection is actually to collect some data. This teaches some of the principles of experimental design and also gives the students a feel for the practical struggles and small decisions needed in real data gathering. This chapter includes several classroom demonstrations and examples to illustrate key ideas, as well as examples of instructions for longer projects.

5.1 Sample surveys

It is fun and instructive for students to learn about sampling by conducting simple classroom or homework assignments in which they use random numbers to sample from actual populations. Unless it is part of a major class project, we don't recommend assignments with personal interviewing—this is too hard, and the difficulty of contacting people, although providing an important lesson in itself, is a distraction from the mathematical concepts of sampling.

There are all sorts of examples that don't involve personal interviews, such as sampling books from the library, pages from books, and words or letters from pages.

For actually generating random numbers, we prefer using dice or personalized sets of random digits rather than computers (can be awkward to use in class) or random number tables (which are tricky if you want different numbers for different students); see Section 7.2 for details.

5.1.1 Sampling from the telephone book

Random sampling of telephone book entries is a surprisingly subtle task, as we show in a demonstration that involves the whole class. The students divide into pairs (as usual), and we show them a local telephone book, from which we tear off several pages that we have previously sampled at random. We pass out one page for each pair of students. We then explain our plan: we would like each pair to sample 10 entries at random from their phone book page and record the phone numbers and addresses. (We later use these to demonstrate Benford's law as described in Section 5.1.2.)

We ask the students to spend five minutes coming up with a formal sampling plan for obtaining 10 entries at random from their page. They must use their random digits or dice (which yield random digits between 0 and 9; see Section 7.2) to do the sampling. We used the Manhattan phone directory, which has five

	Page	Column	Entry	Address #	Telephone #
1					
2					
3					
4					
5					
6					
7					
8					
9					
10					

Fig. 5.1 Form for pairs of students to record their samples from the telephone book, for the demonstration described in Section 5.1.1.

columns per page with 126 lines in each column. It is thus reasonable for students to first pick a column and a page (recall that their telephone book pages are two-sided) and then a line within a column. They can repeat this procedure 10 times to get their 10 addresses and telephone numbers. (Before class, we prepare a special "ruler" with a list of numbers from 1 to 126 photo-reduced so that the numbers exactly line up with a column of the phone book. We hand out copies of this ruler to all the pairs of students so that, once they have sampled the entry number, they can easily read off the address and phone number.)

Picking a column and a page is easy (simply roll the die and assign the numbers 0 through 9 to the 10 columns on the two pages), but getting a random number between 1 and 126 is more difficult. The simplest approach is to roll the die three times, once for each digit, and discard if the number is not between 1 and 126. However, this method would require a huge number of die rolls. A more efficient procedure that some students came up with is to restrict the sampling from 000 to 199:

1. Roll a die to pick the first digit: select 0 if the die roll is even or 1 if it is odd;
2. Roll two dice to pick the second and third digits, thus yielding a number between 000 and 199;
3. If the number was not between 001 and 126, go back to step 1 and reselect the number.

Of course there are many ways to select an entry at random. Rather than simply telling them the above procedure, we see what the students come up with and then lead the class in a discussion. We raise questions such as, *What do you do when the last two rolls gives us too big a number?* or *What if you sample a blank line?* (In the first case, the answer is to sample a new entry from the column; in the second, one must sample a new page and column as well. In the second case, the natural strategy of sampling the next non-blank line would *not* work because it would oversample items that fall immediately after blank lines.)

KASSOMBOLA—KATZ 509

KATOPIS Theodore 120 E 82	212 249-3047
KATOVITZ Michael 299 W 12	212 929-9511
KATOWSKY Marc 215 E 95	212 706-2855
KATRAGADDA Sireesha	
31 E 31	212 532-6457
KATRANCI Elif 155 E 99	212 722-1951
KATRI Edmond 160 E 48	212 588-0118
KATRITSIS A	212 741-0174
KATROV Marat P 747 10 Av	212 757-4845
KATS Amir 531 W 48	212 333-5811
Ester 15 Willett	212 477-2490
Guyora 230 W 82	212 362-5351
I	212 588-1244
Inna 1277 3 Av	212 288-7739
Michael 345 E 93	212 987-2902
Victor 75 West St	212 385-1686
KATSAMAKIS Basil 315 E 69	212 628-9512
Basil 530 E 72	212 628-0312
KATSANOS Andrew 321 E 71	212 717-9393
Christina 417 W 47	212 459-2304
KATTULA Jennafer 409 E 69	212 327-2845
KATUN Mosammat 316 W 95	212 666-4817
KATUS B 210 W 89	212 362-9715
KATUSAK F J 176 E 77	212 737-8955
KATVAN Moshe 40 W 17	212 627-2169
Moshe 40 W 17	212 627-4362
Moshe 40 W 17	212 627-5035
Moshe & Rivka 117 W 17	212 627-5034
KATWAROO Dianna 434 W 163	212 568-0636
Errol 434 W 163	212 568-3629
KATYAL Monica 617 W 115	212 222-3669
KATYANG Keo 104 W 96	212 749-8386
KATZ A	212 721-3504
A	212 725-6758
A 268 E Bway	212 982-8619
A 737 Park Av	212 517-8897
A 2 5 Av	212 533-9692
A 148 10 Av	212 366-6487
A 315 E 86	212 831-7554
A D 433 W 21	212 255-1769

Fig. 5.2 Section from the white pages of a telephone book. As discussed in Section 5.1.1, blank lines and multiline entries create potential for biases if the sampling design is not chosen carefully.

Once we agree on a protocol, the students start generating their samples and collecting data. The students write the information for each entry on a sheet that we provide (Fig. 5.1). Selecting the 10 random entries and finding the telephone numbers and addresses takes about 5 minutes. At this point, we ask if any difficulties arose. As is typical in real-life settings, the act of data collection reveals problems that were not anticipated in the planning stage.

One difficulty is what to do with listings that take two lines; see, for example, Sireesha Katragadda in the displayed phone book page in Fig. 5.2. We ask students if they encountered any such entries in their sampling (there are enough in the phone book that the answer will be Yes) and how did they handle them: for example, what would they do if they had selected line 4 or line 5 from that column. Typically, the students say that they would record Sireesha Katragadda's information in either case. We explain that such a procedure would yield a biased sample, since people with two-line listings would then be twice as likely to be sampled, and the sample as a whole would overrepresent such people. A better method is to set up a rule ahead of time, for example, to only record those phone book entries with a telephone number on the selected line (such as line 5 in this example). If a line without a telephone number (such as line 4) is picked, the students should resample, just as if they had sampled a blank line or a line number larger than 126.

Questions of duplicate listings and unequal sampling probabilities lead naturally into a brief class discussion on real-life difficulties in survey sampling. We ask the students to briefly divide into pairs and come up with possible sources of bias in telephone sampling. Then we rejoin and consider the possibilities, including households with multiple telephone numbers, unlisted numbers, non-

	Page	Column	Entry	Address #	Telephone #
1	520	5	100	15 W 53 St	586-7149
2	519	2	116	240 W 116 St	663-1076
3	519	4	087	710 West End Ave	749-2245
4	520	2	081	511 E 20 St	533-0614
5	519	4	115	2 Horatio St	206-7914
6	519	3	124	256 ...	304-2769
7	519	2	110	350 ...	308-4620
8	520	1	107	129 ...	xxx-2xxx
9	520	5	126	315 ...	xxx-2xxx
10	520	2	040	104 ...	xxx-1xxx

Fig. 5.3 Example of a filled-out form (from a pair of students given pages 519/520 of the telephone book) for the random sampling demonstration described in Section 5.1.1; for the last few entries the students just wrote the numerical address and the first digit of the telephone number suffix, since these were all that were required for the next part of the demonstration. There is evidence from these data that the students did not do the sampling correctly. Can you figure out the incorrect sampling method the students probably used? Think about this before reading on.

response, and multiple people living in a household. (In Fig. 5.2, consider Moshe Katvan, who appears to be listed at three different telephone numbers at the same address.)

Finally, we can go back and look at the students' results and diagnose points of confusion. For example, Fig. 5.3 shows an example of a form filled out by a pair of students. A quick glance at this form revealed that the sampling had been done incorrectly—it is a fun puzzle to figure out the mistake.

The problem with the filled-out form in Fig. 5.3 is that too many of the entries (7 out of 10) are in the hundreds. When sampling numbers from 1–126, only about $27/126 = 21\%$ of the numbers should be between 100 and 126. (In Section 8.5.2, we perform a formal significance test to check if a discrepancy this large could plausibly have occurred by chance.) We suspect that the students sampled in the following way:

1. Roll a die to pick the first digit: select 0 if the die roll is even or 1 if it is odd;
2. Roll two dice to pick the second and third digits, thus yielding a number between 000 and 199;
3. If the number was not between 001 and 126, go back to step 2 and reselect the second and third digits.

This method differs from the recommended approach (see page 49) only in step 3: if the number selected is outside the desired range, only step 2 is repeated, rather than steps 1 and 2. This may seem to be a more economical approach, but it does *not* sample all 126 numbers with equal probability but rather oversamples the numbers between 100 and 126. Instead, it gives a 1/2 probability that the

First digits of phone number suffixes		First digits of addresses	
0			
1		1	
2		2	
3		3	
4		4	
5		5	
6		6	
7		7	
8		8	
9		9	

Fig. 5.4 Copy this sheet onto transparency paper, then pass it around the room along with a marker so students can tally their results for the telephone book sampling. When the students are done, you can display it on a screen or blackboard. See Fig. 5.5 for an example of results.

number begins with a 0 and a 1/2 probability that it begins with a 1.

For another example, we tell the students to imagine a man, a dog, a beetle, and the following sampling scheme: pick an animal at random, then a leg at random from the animal selected. Are all twelve legs equally likely to be sampled?

5.1.2 First digits and Benford's law

The data from the telephone book sampling in Section 5.1.1 can be immediately used for a fascinating demonstration of the distribution of first digits. We ask the students to guess the answers to the following questions about the addresses and phone numbers they have sampled:

1. What proportion of the phone numbers have a 1 for the first digit of the suffix (for example, 642-1034, or 530-1200)?
2. What proportion of the addresses have a 1 for their first digit (for example, 15 Hill Road, or 124 W 89th St., Apt. 4E)?

Reasonable guesses are 1/10 for the phone numbers and 1/9 for the street addresses (which cannot start with 0).

The next step is to see what actually happened. We can circulate a marker and a sheet of transparency paper on which space is allocated for tallying the first digits of the addresses and telephone number suffixes (see Fig. 5.4). The students can pass the sheet around and tally their data while the lecture proceeds, and when they are done we can display the results on the blackboard.

Figure 5.5 shows a typical set of data: the first digits of the phone numbers are uniformly distributed (as one might expect), but the addresses are much more likely to begin with low digits. We ask the students why this occurred. The story for the phone numbers is simple: the last four digits can range from 0000 to 9999, and it is reasonable to assume a uniform distribution (except for minor

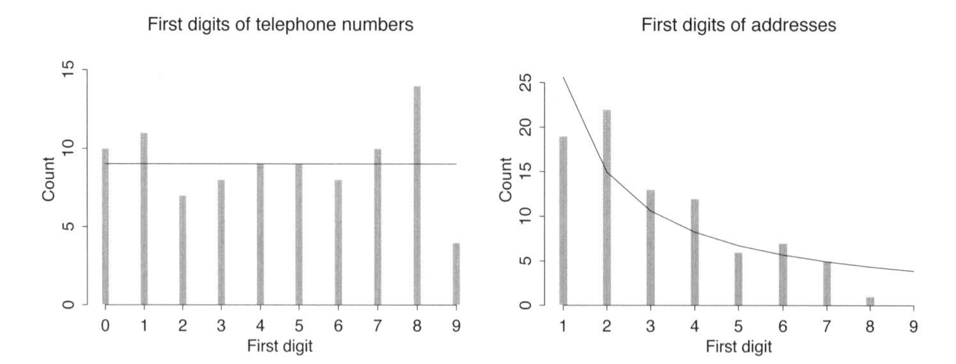

Fig. 5.5 Data from a random sample of telephone-book entries from the class demonstration described in Section 5.1.2: (a) first digits of telephone number suffixes; (b) first digits of addresses. The phone digits roughly follow a uniform distribution, but the addresses show a striking pattern of many more low than high numbers. The pattern of the first digits of addresses is characteristic of Benford's law (see Section 5.1.2). The lines on the graphs show expected frequencies under the uniform and Benford distributions, which are in fact consistent with the telephone and address digits, respectively (see Section 8.5.4 for the formal hypothesis tests).

issues such as that suffixes of 0000 are presumably more likely to be assigned to businesses and thus would not show up in pages from a residential directory).

But what about the addresses? One answer that comes up is: if a street is only numbered up to 400, say, then the leading digit is most likely to be a 1, 2, or 3. More generally, there will be some streets numbered only up to 100 (in which case each digit is roughly equally likely), some numbered up to 200 (and then about half of the addresses begin with a 1), some numbered up to 300, and so forth. Considering all of these, it makes sense that 1 is the most common leading digit, followed by 2, then 3, etc.

More formally, one can model the distribution by assuming that street addresses are approximately uniformly distributed on the logarithmic scale: that is, there should be approximately as many addresses between 100 and 200 as between 200 and 400, or 400 and 800, and so forth. In this case, the probability that the leading digit is i, for any i between 1 and 9, is $\log(i+1) - \log(i)$, which starts at 0.30 for $i = 1$ and 0.18 for $i = 2$ and continues down to 0.05 for $i = 9$. This theoretical curve is shown as the line in Fig. 5.5b, and, amazingly enough, it fits the data well. (See Section 8.5.4 for a formal check of the fit.)

We then briefly relate the history of "Benford's law," as this phenomenon is called, after Dr. Frank Benford, a physicist at General Electric Company in the 1930s who noticed that in books of logarithms, the pages corresponding to numbers with a leading digit of 1 were more worn than other pages. (Actually, this was first noticed by Simon Newcomb in 1881, but Benford rediscovered it.) Benford followed up by collecting data sets from many different sources in the natural and social sciences and found that the logarithmic pattern generally held.

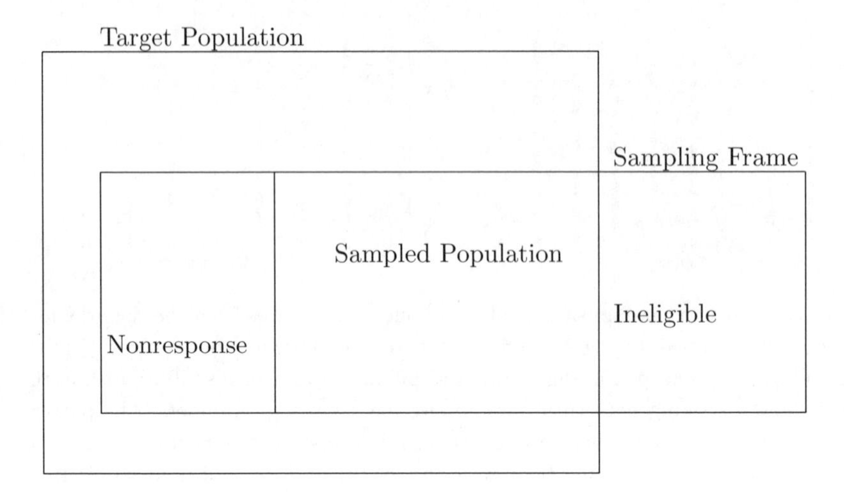

Fig. 5.6 Students fill in this template by identifying the target population, sampled population, and sampling frame, for surveys discussed in class.

More recently, Benford's law has been in the news as a method for checking for fraud in financial data.

5.1.3 Wacky surveys

It is always fun to illustrate lecture points with examples of surveys reported in the popular press. We choose news stories that report on surveys with possible sources of selection bias, measurement bias, question bias, nonresponse bias, and so forth. In class discussion, it's important to be critical without being dismissive: we lay out possible sources of bias while at the same time recognizing that these imperfect survey results might be useful as part of a comprehensive perspective on an issue. To help students understand the sampling scheme and identify biases, we hand out copies of the template shown in Fig. 5.6. Students practice distinguishing between the target population (the group we want to study), the sampled population (the group from which the sample was taken), and the sampling frame (the list used to contact the sampled population).

For example, on August 13, 1999, the Associated Press reported on an online survey conducted by ABC News that estimated there to be 11 million Americans suffering from "some form of addiction to the World Wide Web." This estimate was based on data collected from 17 251 responses to an Internet use questionnaire distributed and returned through the Web site `abcnews.com`. Here the target population consists of persons who use the Web, the sampled population are those visits (and submissions) at the ABC News Web site, and the sampling frame are the hits on that site. The sampled population is not well defined because we don't know if a person can complete the survey multiple times. In addition, sample selection probabilities are unequal: the more time someone is on the Web, the more likely he or she is to hit this particular site.

As another example (see Section 6.6.2), a telephone survey of 2000 children ages 10–16 found that 25% of the children were slapped, punched, kicked, hit, or threatened with an object in the past year by an adult, sibling, or another child. Here the target population is all children ages 10 to 16 in the United States; the sampling frame is all phone numbers in the United States; and the sampled population is children living in a home (as opposed to an institution) with at least one phone number who would be at home the time of the call and whose guardian would consent to the interview. Measurement bias is key to this survey. The definition of child abuse departs markedly from the more common definition used by the National Crime Survey. The authors of the survey chose a broader definition of abuse with the goal of raising public awareness about the violence to which children are exposed.

In another news story, "Poll Finds 1 out of 3 Americans Open to Doubt There was a Holocaust," *Los Angeles Times*, April 20, 1993, the results from a Roper poll created a big stir. But a closer look at the question asked reveals a potential problem with question bias: "Does it seem possible or does it seem impossible to you that the Nazi extermination of the Jews never happened?" The compound structure of the sentence and the use of the double negative makes the question confusing. 22% reported that it was possible that the Holocaust didn't happen, and 12% didn't know. A year later, Roper repeated the survey, keeping all other questions the same and rewording this one: "Does it seem possible to you that the Nazi extermination of the Jews never happened, or do you feel certain that it happened?" This time only 1% reported that it was possible that it didn't happen, and 8% didn't know.

Hemenway (1997) reports a more consistent error in survey responses:

> Using surveys to estimate rare events typically leads to overestimates. For example, the National Rifle Association reports 3 million dues-paying members, or about 1.5% of American adults. In national random telephone surveys, however, 4–10% of respondents claim that they are dues-paying NRA members. Similarly, although *Sports Illustrated* reports that fewer than 3% of American households purchase the magazine, in national surveys 15% of respondents claim that they are current responders.

Students can discuss possible reasons for this bias and methods of correcting for it.

5.1.4 An election exit poll

Election polling is a standard example for teaching sample surveys and provides many interesting examples of successes and failures in sampling. One example we personally encountered (but, alas, too late to fix) was an exit poll in New York City, in which pollsters selected a sample of polling places and, at each, intercepted every fifth exiting voter to ask whom they voted for, demographics (sex, ethnicity, marital status, income, and so forth), and various other questions (including political ideology, previous voting, and opinions about the news media). But a key piece of information was *not* gathered by the pollsters ... (At this point, we give the students a minute to try to think of the missing information.) If none of the students figure it out, that's ok—neither did the political scientist who conducted the poll: the pollsters should have been instructed to gather information on the *nonrespondents*, to record the sex, ethnicity, and approximate age of the people who refused to cooperate. This would allow some judgment as to the scale of nonresponse bias.

5.1.5 Simple examples of bias

Ask the students in your class to raise their hands if they love statistics. The number who raise their hands, divided by the number of students in the room, is an estimate of the proportion of people with this opinion in the general population. Now give the students a minute to work in pairs and come up with as many sources of bias as they can think of for this example. These can be listed on the blackboard and divided into categories: differences between sampled and target populations (students in a math or statistics course are not typical of all students, let alone all people, and, in addition, taking a statistics course may increase your love for the subject!), nonresponse bias (most students hesitate to raise their hands), and response bias (students may want to please the teacher and say Yes, or conversely they may be embarrassed and say No even if they do love the subject).

This mini-example can lead to a class discussion of biased survey questions: when do the news media or other organizations get misleading impressions because of question wording? How much of a difference can question wording actually make? This can be a topic for a class project, in which groups of students try to come up with loaded survey questions.

5.1.6 How large is your family?

After discussing sampling bias, with examples such as given above, we drive the point home with a demonstration. Each student is asked to tell how many children are in his or her family ("How many brothers and sisters are in your family, including yourself?"). We write the results on the blackboard as a frequency table and a histogram, and then compute the mean, which is typically around 3 (see Fig. 5.7 for an example).

We tell the students that the average number of children in families that were having children 20 years ago (about the age of the students in the class) was about 2.0. Why is the number for this class so high? Students give various suggestions such as, perhaps larger families are more likely to send children to college. After some discussion, someone will realize that if the family had zero children, they certainly did not send any to college. This is part of the explanation but not the whole story.

The 2.0 figure is the average number of children when sampling *by family*; 3.0 is the average number of children when sampling *by child*. When sampling by child, a family with n children is n times more likely to be sampled than a

Family size (# of siblings, including self)	Count
1	3
2	3
3	5
4	5
5	3
6 or more	0

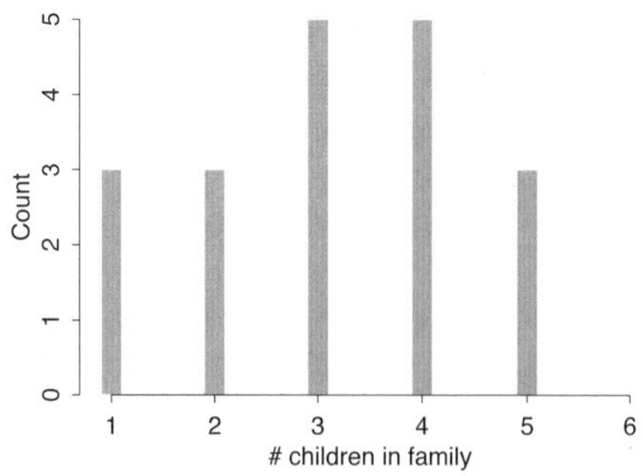

Fig. 5.7 Data on number of children in families (displayed as frequency table and histogram) from students in a typical class. The average is 3.1, which is at first a surprise, given that the average family has about two children. See Section 5.1.6 for the explanation.

family with 1 child. This illustrates the general point that it is not enough to say you sampled at random; you must also know the method of sampling. It can also be considered as an example of *sampling bias* (a topic we return to in Section 10.2).

At this point, we can get the students further involved by asking the question, how can data be gathered to estimate the average number of children per family? In discussing the problem, students can consider the relative difficulties of correcting the data on students for sampling bias, compared to the direct approach of sampling families. The former approach is tricky because it still requires some estimate of the proportion of families with zero children. The students also have to realize that, for either approach, a careful definition of "family" is required. We sometimes ask the students how many children their oldest aunt or uncle has. Leaving out those students who have no aunt or uncle, the average number of children is much smaller (in one class survey we found it to be about 2.2).

Related issues arise in telephone sampling, when you call a telephone number at random and then pick a person at random in that household to interview: Is a person with many phone lines more or less likely to be sampled than a person with one line? Is a person with many roommates more or less likely to be sampled than a person living alone?

5.2 Class projects in survey sampling

Chapter 11 discusses various sorts of projects where students can get direct experience with data collection. One approach we have tried is to have our students design a questionnaire, develop a sampling plan, collect the survey responses, clean them and enter them on the computer, analyze the survey data, and write a report of the findings. This is a big job, and we often run it as a whole-class project. The project takes the better part of a semester to complete, although the effort required from each student is limited to a few hours over one or two weeks because we break the project into smaller manageable tasks where groups of students work on these different tasks. We oversee the project by scheduling groups to report on their work to the rest of the class, and we use this reporting time to solicit input from all of the students on each aspect of the project. Each group writes a summary of its contribution, which appears in the final report as an appendix.

We introduce the project early in the semester and set deadlines for completion of the main pieces. If the class is large, more than one group works on the same task. For example, two groups may design questionnaires with the advantage being that we can merge the best parts of each into one questionnaire. Typically, one large group of about six to ten students takes charge of data collection and validation, but everyone in the class helps collect data.

We find that our students really enjoy working on these large projects. The whole-class format has many advantages. Students can choose the task for which they are best suited, and the value of the role they play in the process is evident to everyone in the class because the entire process is viewed collectively. In addition, the workload for each student is kept to a reasonable level. The instructor's involvement at each step helps secure a successful outcome, and it sets a good example of how to carry out independent research projects.

5.2.1 The steps of the project

Although only a subset of the students work on any one task, we organize class discussions on all the topics. We provide students with handouts, presented here in Figs. 5.8–5.12, to clarify each group's job and to focus our class discussions.

The sampling plan

We ask the team designing the sampling plan to come to class prepared to identify the following:

- The target population: the group we want to study
- The sampling frame: the list to use to reach the members of the target population
- The sampled population: the group from which the sample will be taken.

The students must also come up with a probability method to choose individuals for the sample, and they must specify the sample size.

After the class discussion, the group draws up a formal sampling plan and submits it for review and finalization by the instructor.

The questionnaire

The entire class first brainstorms to come up with a list of information that they want to find out about the population. Then we ask smaller groups of students to bring sample questions to class for discussion. We provide these groups with a list of examples of what not to do in order to assist them in reducing bias. The groups must try their questionnaires out on a few guinea pigs to uncover points of confusion and improve the questionnaire. We combine and edit their efforts to produce the final questionnaire.

Designing your sampling plan

You must provide a set of explicit instructions for those in charge of the data collection and verification stage of the process. This should include instructions on how to follow up hard-to-reach cases and how to avoid interviewer bias.

- Determine the population to be surveyed. Differentiate between the target population and the sampled population.
- What is your sampling frame; that is, how do you plan to reach the members of the sampled population?
- How do you plan to choose members of the population for your sample? It is not acceptable to simply use a convenience sample such as any student who walks by the library. You must employ a probability method for selecting the sample.
- How large should the sample size be? A large sample size has the advantage of increased accuracy in the estimate of the population percentage. But, keep in mind the cost of taking a large sample. Here the cost is the time that it will take us to conduct the survey—including the time to hunt down the hard-to-find students. You can expect each class member to spend two hours collecting data for the survey.

Fig. 5.8 First part of instructions for students for the design of their sampling project described in Section 5.2.1. The instructions are continued in Figs. 5.9–5.12.

Data collection

We expect each member of the class to spend two hours collecting data for the survey. The data collection team is responsible for providing explicit instructions on how to do it. They oversee the data collection process, including handling nonresponse, data entry, and cleaning. We ask the data collection team to give frequent progress reports in class in order to keep the data collection on track and so we can step in and help out when problems arise.

Here is an excerpt from one student's report from the data collection group.

I was assigned to the last discussion section for Stat 21, Room 344 Evans from 3–4 on Tuesday Nov 15. I arrived there early to inform the TA of the purpose of my visit. We decided to schedule my presentation about halfway into the class; that way tardy students would have a chance to be surveyed (if they were on my list). As I sat there, listening to the TA's lecture, I was going over in my mind the presentation. How can I make this presentation interesting and effective? I have taken questionnaires before, and I know that if the questionnaire is not interesting or if the questionnaire does not engage me in some way then my response is lukewarm and cursory. I have a plan.

I broke the ice by asking about how they feel about the labs in Stat 21, especially after the TA's announcement of a lab due next week. I told them my aversion to labs and how I often postponed them until the last minute. (Most of them smiled at this point.) I seized the opportunity by telling them about the Stat department's idea of redesigning the lab format to a video-game-like format. In order to give them some concrete ideas about some of the types of possible games, I spoke of kicking Mr. SD's ass (excuse my language), philosophizing with Mr. Probability about life, love and relationships, and playing SimStat. (Their interest was quite high at this point.) Finally, after telling them that their anonymity will be preserved and about the randomness of the selection process, I asked them for their input. I decided that, instead of giving

Writing your questionnaire

The end result of this task is to create a questionnaire, including instructions on how to complete the questionnaire and how to administer it.

- List the information that you wish to find out about the students. Some questions may be sensitive or otherwise hard to collect. For example, students may be unwilling to answer questions about their grades or test scores.

- Create questions for obtaining the information on your list. The way a question is worded can bias the response.

  Confusing, long questions may be left unanswered or misread. For example, "Do you not believe that video games could possibly not be educational?" is confusing because of the double negative.

  For some questions, it may be difficult to get honest replies. For example, "Do you use illegally copied video games on your PC?"

  For other questions, it may be difficult to get reliable responses. Of the following two questions, the second will produce more reliable responses: "How many hours do you spend a week on average playing video games?" "How many hours did you spend last week playing video games?"

  Also, consider the difference between these two questions. "Do you like math?" "Do you hate math?" Students who like math may be unwilling to admit that they do, but willing to say that they don't hate it.

- Organize your questions into a questionnaire. The questionnaire should be easy to read and easy to answer. It should appear organized and user-friendly. Asking for personal data at the beginning may turn a respondent off to the entire questionnaire. Keep your questionnaire short and to the point, and keep in mind the overall objectives of the study.

- Try your questions out on a few volunteers. These "guinea pigs" may uncover points of confusion and other problems with the questionnaire. For example, when asked where they usually play video games—on a home computer, a home system, or in the arcade—some students checked more than one response. A different question design might avoid this problem.

- Use your findings from the pilot test to improve your questionnaire. Clean up confusing or leading questions. Eliminate unnecessary ones. Finalize your questionnaire.

Fig. 5.9 Second part of instructions for students for the writing their survey form for their sampling project described in Section 5.2.1. See also Figs. 5.8 and 5.10–5.12.

them rewards after taking the survey, I would give them candies while they were filling out the survey to keep them thinking about the survey while being satisfied and to reduce the possibility of doing the survey in a superficial, hurried fashion.

I found two problems with the questionnaire. One is that the respondents chose multiple answers to questions that were designed for a single answer. For example, question 5 asks for the usual place that the respondent plays the video/computer game. One respondent checked both arcade and home on a system. Another example is question 17, where, for example, one respondent checked both an A and a B for his expected course grade. The second problem is missing data. A few respondents chose not to

Collecting and validating your data

The goal of this task is to collect accurately the information from the students selected for the sample. This part of the project requires a lot of organization.

- Make a plan for data collection and a procedure for monitoring the progress of the data collection. Frequent updates and progress reports will ensure fast and accurate collection. Include in the plan ways to follow up on the nonrespondents.

  Be careful that your methods preserve the anonymity of the respondents. Respondents' names should not appear on the questionnaire. A numbering system can be used to keep track of who has responded. For example, in the video game survey, a manila envelope was provided for each section of the class, with a list of the selected students pasted on the outside of the envelope. Students' names were crossed off this list when they returned the questionnaires. Completed questionnaires were turned in for data entry without any personal identification on them.

- Develop a training program for the data collectors. They should briefly inform the respondents about the purpose of the survey. You may want to prepare a short script for them to read to the respondent before administering the questionnaire.

- Carry out your plan to collect the data. Document your efforts to obtain the survey results. Be as specific as possible. Keep track of the times and places the data were collected. Describe your efforts in follow-up. How successful were they? What problems did you or the respondents encounter?

- Enter your data into the computer and check it for accuracy. Does a preliminary analysis of the data uncover any problems with the survey design? What was your response rate? Do the respondents differ from the nonrespondents in some way that may effect our conclusions?

Fig. 5.10 Third part of instructions for students for the data collection and cleaning for their sampling project described in Section 5.2.1. See also Figs. 5.8–5.9 and 5.11–5.12.

answer some questions. Overall, however, we had a very high response rate: 92 of the 95 students surveyed completed the questionnaire.

Data analysis

The appropriate analysis of the data depends on the questions asked. At first we have the analysis team supply the class with preliminary findings, including basic tabulations and graphical summaries of the responses to each of the questions. With these summaries in hand, we brainstorm with the class about what further analyses should be prepared. We emphasize that it is important to provide a clear and complete picture of those surveyed, and this includes looking at the relationships between variables through say cross-tabulations and scatterplots. The analysis team supplies the report-writing team with a summary of its findings as well as detailed numerical and graphical results.

The write-up

We always have a target audience in mind when summarizing the findings of the survey. For example, we ask the report team to write a memo to the com-

Analyzing your data

The appropriate analyses of the data will depend on the specific questions asked of the respondents. In general, the objective of the analysis is to summarize the findings of the study. It is important to provide as clear and complete a picture as possible of those surveyed. Be careful not to include too many numbers, charts, and graphs, but also take care in providing backup information that might not appear in a final report. It is useful to give a brief "executive summary" of your findings and then more detailed numerical and graphical results later in the report.

Here are some pointers about what to include:

- Basic tabulations and graphical summaries of the responses to the individual questions.
- Confidence intervals for population percentages, such as the percentage of students that played a video game last week.
- Information on the relationships between variables. For example cross-tabulations of two or three variable may be of interest.

Fig. 5.11 Fourth part of instructions for students for data analysis for their sampling project described in Section 5.2.1. See also Figs. 5.8–5.10 and 5.12.

Writing up your results

Summarize the findings of the survey. Have a target audience in mind when writing our report. The data analysis group should have provided a broad overview of the results and list a few interesting phenomena. Not all of their numbers will appear in your report. It is up to you to determine which analyses are most informative and most compelling. Make sure that the final report includes the following information:

- Purpose of the survey.
- Summary of findings and conclusions.
- Brief description of the survey methodology.
- Detailed description of your findings with supporting numerical and graphical statistics. If possible put the results in context with other findings in the area.
- Discussion of problems encountered, including nonresponse.
- Conclusions and suggestions.
- Appendixes: group reports.

Fig. 5.12 Final part of instructions for students for writing up their sampling project described in Section 5.2.1. The instructions are continued from Figs. 5.8–5.11.

mittee in charge of designing new computer laboratories, a press release for the student newspaper on the future plans of math majors, or a report for the Chancellor's office with recommendations on mandatory community service work for undergraduates.

Not all of the numbers, charts, and graphs provided by the analysis group make it into the body of the final report. The writing team selects those that succinctly and accurately describe the population of interest. To help shape the final report, the writing team presents a rough draft to the class for critique.

We ask that the final report includes the following information: the purpose of the survey, a summary of findings and conclusions, a brief description of the survey methodology, and a detailed description of the results with supporting numerical and graphical statistics. We also ask that the report authors put the results in context with other findings in the area, discuss the problems encountered, including nonresponse and other potential sources of survey bias, and include as appendixes the reports from the other groups.

5.2.2 Topics for student surveys

We often choose the topic of study before the semester starts. We collect background materials (newspaper clippings, research papers, and reports) to frame the problem, contact authorities on the topic for advice, and arrange guest lectures to discuss the topic with our class. We choose a survey topic according to current events and political issues on campus, and the group that we study is usually some subset of the student population. We present three class projects here to give a flavor for the variety of issues that can be addressed.

Video games

The goal of this survey (Fig. 5.13) was to inform a committee of statistics faculty, who were designing new interactive computer labs for statistics classes, about students' preferences and attitudes toward video and computer games. The target population was students who would be using the computer labs in statistics courses, the sampled population was all undergraduates enrolled in Statistics 21, Fall, 1994, at the university, and the sampling frame was the list of students who took the midterm exam the week before the survey was conducted. A simple random sample of 95 of the 310 students were selected for the sample. Several steps were taken to reduce nonresponse:

1. Each section of the class (a section has 30 students and meets twice a week with a teaching assistant) was visited on the day that the exams were returned to the students. Before attending section, the teaching assistants had been informed of the project, and the instructor for the course asked that they accommodate the data collectors. The surveys were handed out and collected in section.
2. The data collectors returned to the same section later in the week to reach any students who had not come to the earlier section meeting.
3. Finally, the names of the students who had not attended either of the two section meetings were written on the blackboard in the main lecture hall. The instructor made an announcement at the beginning of class for the listed students to raise their hands, and questionnaires were distributed to them and collected in class.

Ultimately, 92 of the 95 students in the sample responded.

1. How much time did you spend last week playing video and/or computer games? (C'mon, be honest, this is confidential) ______ Hours
2. Do you like to play video and/or computer games?
☐ Never played→ Question 9
☐ Very Much ☐ Somewhat ☐ Not Really ☐ Not at all→ Question 9
3. What types of games do you play? (Check all that apply)
☐ Action (Doom, Street Fighter)
☐ Adventure (King's Quest, Myst, Return to Zork, Ultima)
☐ Simulation (Flight Simulator, Rebel Assault)
☐ Sports (NBA Jam, Ken Griffey's MLB, NHL '94)
☐ Strategy/Puzzle (Sim City, Tetris)
4. Why do you play the games you checked above? (Choose at most 3)
☐ I like the graphics/realism
☐ relaxation/recreation/escapism
☐ It improves my hand–eye coordination
☐ It challenges my mind
☐ It's such a great feeling to master or finish a game
☐ I'll play anything when I'm bored
☐ Other (Please specify) ______________________________
5. Where do you usually play video/computer games?
☐ Arcade
☐ Home ☐ on a system (Sega, Nintendo, etc.)
☐ on a computer (IBM, MAC, etc.)
6. How often do you play?
☐ Daily ☐ Weekly ☐ Monthly ☐ Semesterly
7. Do you still find time to play when you're busy (for example, during midterms)?
☐ Yes (can't stay away) ☐ No (school comes first!)
8. Do you think video games are educational?
☐ Yes (or else all my years of playing have gone to waste)
☐ No (I should have been reading books all those years)
9. What don't you like about video game playing? Choose at most 3.
☐ It takes up too much time ☐ It costs too much
☐ It's frustrating ☐ It's boring
☐ It's lonely ☐ My friends don't play
☐ Too many rules to learn ☐ It's pointless
☐ Other (Please specify) __________________
10. Sex: ☐ Male ☐ Female
11. Age: ______
12. When you were in high school was there a computer in your home?
☐ Yes ☐ No
13. What do you think of math? ☐ Hate it ☐ Don't hate it
14. How many hours a week do you work for pay? ___________
15. Do you own a PC? ☐ Yes ☐ No
Does it have a CD-Rom? ☐ Yes ☐ No
16. Do you have an e-mail account? ☐ Yes ☐ No
17. What grade do you expect in this class? ☐A ☐B ☐C ☐D ☐F

Fig. 5.13 Questionnaire developed by students for the video game survey described in Section 5.2.2.

1. How old were you when you first realized you wanted to study math seriously?
□ Ever since I could add! □ Elementary school
□ Junior high school □ High school □ College
2. Do you have a parent or close relative who is a mathematician or works professionally in a math-related area? □ Yes □ No
If yes please specify: □ Father □ Mother □ Brother □ Sister □ Other
3. In high school did you take:
Calculus or AP Math □ Yes □ No
Chemistry □ Yes □ No
Physics □ Yes □ No
4. In high school did you have a group of peers to talk about math with?
□ Yes □ No
5. What has been important to your success in math so far? Choose at most 3.
□ Quality of high school education □ A teacher showed an interest in me
□ Friends support and encouragement □ Undergraduate research experience
□ Math comes naturally for me □ Interesting and talented teachers
□ Family support and encouragement □ Hard work
□ Job experience □ Other ________________________
6. Why is math a good major for you? Choose at most 3.
□ I'm good in mathematics □ Lively intellectual community
□ Challenging subject matter □ Math is fun to me
□ Good job opportunities □ I know when I have the correct answer
□ Aesthetic appeal □ I like being able to work alone
□ Other ________________________
7. If given another chance, would you major in math again? □ Yes □ No
8. Here at Cal, do you have a group of peers to talk about math with? □ Yes □ No
9. Do you like math at Cal better than high school math? □ Yes □ No
10. Do you learn from reading required texts? □ Yes □ No
11. Most of the time, do you first approach a difficult problem by:
□ Writing a formula □ drawing a picture
12. How often do you go to your math professors' office hours?
□ Rarely–like for a letter of recommendation
□ Occasionally, when I'm really stuck on something
□ Regularly to talk about homework
13. Have you ever talked to a professor about graduate school? □ Yes □ No
14. Do you plan to pursue a graduate degree in mathematical science?
Masters: □ definitely □ possibly □ probably not □ no way
PhD: □ definitely □ possibly □ probably not □ no way
15. How often do you question your ability to succeed in your field?
□ Always □ Often □ Sometimes □ Rarely □ Never
16. Major: □ Math □ Applied Math
17. Are you a double major? □ yes in ________________ □ no
18. So far at Cal, how many units have you completed?
Honors math: _____ units Other math: _____ units
Total: _____ units
19. Cumulative GPA:______ Math GPA:________
20. Sisters: ______ older ______ younger Brothers: ______ older ______ younger
21. Father's highest degree: ______ Mother's highest degree: ______
22. MSAT ______ VSAT ______
23. Sex: □ Male □ Female
24. Age: ______

Fig. 5.14 Questionnaire developed by students for the math major survey described in Section 5.2.2.

Math majors

The goal of this survey (Fig. 5.14) was to study the differences between male and female majors in regards to their course work, self-confidence, and future plans in mathematics. The faculty and staff of the Mathematics Department gave input into the design of the questionnaire. For motivation, we provided the class with related findings from studies of students in mathematics and science. The target population was all math majors at the university. The sampling frame was the list of declared and registered mathematics majors in January, 1995. A complete census was taken because there were only 115 majors. We obtained permission from the instructors of math courses to approach the majors at the beginning of the class to complete the survey. For the few majors who were not enrolled in a math course that semester, we found them outside the meeting places of their other classes.

Community service

In 1999, California Governor Gray Davis proposed a community service graduation requirement for all college students in order to instill in them a "sense of obligation to the future." A campus task force was formed to prepare a response to the Governor's proposal. To help the task force determine the current level of student participation in service activities, and the general attitude of students toward community service, we volunteered to work with the task force to design a survey on the topic.

For background material, we read current news stories such as, "Educators, Politicians Ponder Forcing Students to Volunteer," *San Francisco Chronicle*, May 18, 1999, and the report, "Combining Service and Learning in Higher Education," from the Rand Corporation. We met with members of the task force to find out about existing avenues at the university for students to perform community service.

The target population was all undergraduates at the university, and the sampling frame was the Registrar's list of students enrolled in the Spring, 2000, semester. The questionnaire did not ask for personal characteristics because this information was available from other sources. The class fine-tuned the questionnaire in two series of pilot studies. Due to confidentiality, our data collection and subsequent analysis were limited to the pilot studies. We forwarded the findings to the committee as examples of how they could summarize the final results.

5.3 Experiments

5.3.1 An experiment that looks like a survey

At the point during the term when we are discussing designs of surveys and experiments, we hand out a folded survey form to each student in the class. We ask the students to read the forms, answer the questions independently (without discussing it with their neighbors), and then fold them back up and return them to the instructor. The form is shown in Fig. 5.15. (Because each student must separately fill out his or her survey form, this is one of the few demonstrations in which the students do *not* work in pairs or groups.)

We chose (by computer) a random number between 0 and 100.

The number selected and assigned to you is $X =$ ____ .

1. Do you think the *percentage* of countries, among all those in the United Nations, that are in Africa is **higher** or **lower** than X?

2. Give your best estimate of the *percentage* of countries, among all those in the United Nations, that are in Africa.

Fig. 5.15 Handout for the experiment that looks like a survey, described in Section 5.3.1. Make copies of this form for all students in the class. Then write, by hand, "10" in the blank space for half the forms and "65" on the others. Fold each form, shuffle them, and hand them out to the students, telling them to answer the survey questions individually, without discussing with their neighbors.

After the students have answered the questions and returned the forms, we explain to the students that only two values of X were actually assigned and that we were interested in finding the relation between X and the students' responses on the second question. If the experiment were actually performed with random numbers from 0 to 100, the analysis would be somewhat more complicated, because there would then be a continuous range of possible treatments. We then ask the students what kind of data collection was happening: survey, experiment, or observational study. In statistical terminology, the units are the students, the treatments are the hand-written values of X (10 or 65), and the outcome of interest is the response to the second question.

This experiment is adapted from a published study that reports the median responses to the second question as 25 and 45, given $X = 10$ and 65, respectively. This result is described as an example of the "anchoring heuristic," in which an estimate of an unknown quantity is influenced by a previously supplied starting point. In this example, the value of X should not affect the outcome (after all, the students were told that X was randomly generated), yet it does!

Now is a good time to discuss the principles of randomization and blindness in experimentation, now that the students have been subjects in an experiment. Incidentally, the actual value of the unknown quantity (see page 268) is irrelevant for this example—we are only studying the differences in responses between the two groups of students.

In performing this example in our classes, we have replicated the anchoring effect, although its magnitude has not been so dramatic as in the published literature; for example, histograms of responses for a class of 43 students are given in Fig. 5.16. When the data have been collected, the two groups can immediately be compared graphically using histograms; typically, as in Fig. 5.16, the histograms overlap considerably but clearly differ. The dataset can be used

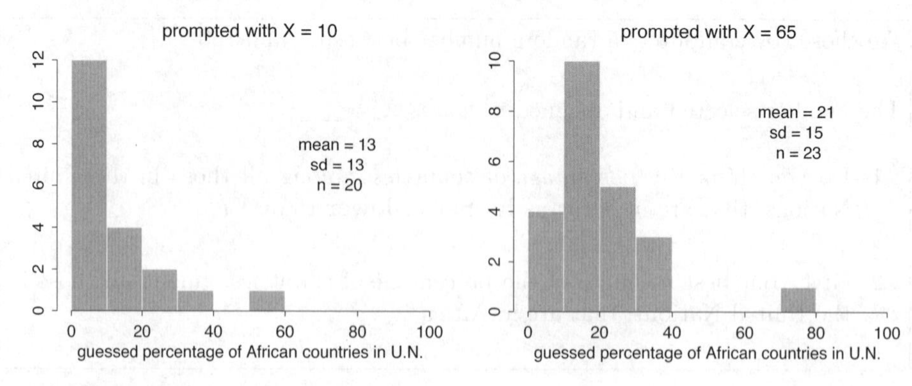

Fig. 5.16 Responses of students in a small introductory probability and statistics class to the question, "Give your best estimate of the *percentage* of countries, among all those in the United Nations, that are in Africa." The students were previously asked to compare this percentage with a specified value X; histogram (a) displays responses for students given $X = 10$, and histogram (b) displays responses for students given $X = 65$. The students were told that the value of X was chosen at random, and yet it has an effect (on average) on their responses. Similar results were obtained when the experiment was repeated in other classes.

throughout the course to illustrate various points, including the mean and median in skewed distributions, methods of testing the statistical significance of observed differences between groups, and the distinction between practical and statistical significance. With classes of about 50 students, we have found the anchoring effect in this experiment to be on the border of statistical significance as measured by a t test.

Other examples in which an experiment is embedded in a survey include studies of question wording, question ordering, and other causes of survey response bias. In addition, the experiment given here could be made more complex by randomly ordering the options "higher" and "lower" in the first question of the survey, thus giving two factors and four possible experimental conditions.

5.3.2 Randomizing the order of exam questions

One very direct experiment we have tried goes as follows. Without the knowledge of the students in the class, we prepare two versions of the midterm examination, identical in all respects except that the order of the questions is reversed. We prepare equal numbers of the two versions and mix them randomly before handing out one to each student for the exam. (We put each exam question on a separate page and grade the questions one at a time, so the grading is not influenced by the order of the questions.) We record the grades achieved by the two groups of students.

After returning the graded exams to the students, we reveal that there were two forms of the exam and present the aggregate results; for example, the average score was 65 for exam A and 71 for exam B. Should we adjust the scores of the "exam A" students upward (and the "exam B" students downward) to reflect that exam A seems more difficult, in retrospect? A student who took exam B objects, noting that the two exams had identical questions—just the order was different. But the order could have an effect, right? What if the two forms had been randomly given to 1000 students and this difference had been observed—would it be "real" then? The goal here is to get the "exam A" students and "exam B" students all fired up and holding opposite positions.

How can we address the question of whether the observed difference is due to the exams or just because, say, the better students happened to take exam B? We can consider this as an experiment designed to measure the difference in exam difficulties and use the standard methods to obtain an estimate and standard error. This demonstration motivates formal hypothesis testing: is the observed difference statistically significant? Should we adjust the scores and, if so, how much? (We return to this topic in Section 9.2.1.) To round out the discussion, we ask: What if the exams differed not just in their ordering, but in the questions themselves? How would/should our statistical methods change? This is of course a subtle question with no easy answers. Students have also raised ethical questions about basing grades on different forms of the exam.

In practice, the quality of the discussion is highly influenced by the observed difference between groups, which cannot be predicted ahead of time. If the true difference between the exams is approximately zero and the standard deviation of exam scores is 15, then with a class of 40 students the observed difference in means has an expected value of zero and a standard deviation of $\sqrt{15^2/20 + 15^2/20} = 4.7$. The actual difference might then be in the range of 5–10 points, which is enough for students to care about, or it might be less than 5 points, in which case students might be not be so interested in the result.

We connect to other course material by discussing with the students how we ensured double-blindness in this experiment. We return to this example at the end of the course when covering multiple regression (see Section 9.2).

5.3.3 Taste tests

We model an in-class demonstration and a directed project on the famous (to statisticians) experiment conducted by Ronald Fisher in 1919 when he concocted an impromptu test of a lady's ability to taste the difference between two preparations of tea. This taste test illustrates the principles of randomized experiments and hypothesis testing.

According to Fisher's daughter, one afternoon at tea-time in Rothamsted Field Station in England, a lady proclaimed that she preferred her tea with the milk poured into the cup after the tea, rather than poured into the cup before the tea. Fisher challenged the lady, and presented her with eight cups of tea, four made the way she preferred and four made the other way. She was told that there were four of each kind, and asked to determine which four were prepared properly. Unfortunately we do not know how well the lady fared, but Fisher subsequently used this experiment to illustrate the basic issues in experimentation.

Soda tasting demonstration

We demonstrate Fisher's experiment in class where one student attempts to distinguish between two kinds of soda. We bring to class two cans of chilled soda (for example, one diet and one regular cola), a bottle of water, eight small paper cups, paper towels for spills, and eight marbles in a sack, four of one color and four of another. We also have ready the hypergeometric probabilities for $n = 4$, $M = 4$, $N = 8$: they are $1/70, 16/70, 36/70, 16/70, 1/70$, respectively.

We begin by asking if any students can tell the difference between the two types of soda. We choose one of the "experts" to demonstrate his or her abilities, and we request assistance from three other students.

Before setting up the demonstration, we describe Fisher's design for testing the expert, and sometimes we extract other possible designs from the class. For example, we could flip a coin eight times where each time the coin lands heads a cup is filled with one kind of soda and each time the coin lands tails the cup is filled with the other kind. (A disadvantage of this design is the possibility that all eight cups will contain the same kind of soda.) After this discussion, we have the student-expert leave the room with one of the volunteers whose job is to keep watch over the expert while we set up the experiment. The other two volunteers divide the eight cups into two sets of four. They mark each of the four cups in one set in an inconspicuous place, and fill the cups with the same kind of soda. The unmarked cups are filled with the other kind of soda. The marbles are used to determine the order in which the cups are put into a line.

When the student-expert returns, he or she sips the soda, separates the cups into two groups of four, and tells us which group is which. The water is to help clear the palate between sips of soda. If the student makes no mistakes, we declare him or her an expert, and with one or more mistakes we say that the student did as well as one might expect someone who was just guessing.

Coffee tasting project

After our demonstration, we have students work in groups of two or three to conduct their own taste test in a directed project outside of class. If all of the groups perform the same taste test, then their results can be compared. Alternatively, the students in each group select their own food for testing. Typically students compare two brands of food, such as peanut butter, chocolate, or soft drinks. We encourage them to be creative and conduct an experiment that they care about; not to simply settle for the milk test (2% versus fat-free) because it is easily available in the dining hall. One year, a group of students performed a rum-and-coke taste test. Since then, we make sure we discuss the ethics of experimentation on human subjects before letting them loose on the project.

We hand out a set of instructions to help them design and carry out their experiments. We chose to illustrate these instructions using an experiment that tests a subject's ability to discriminate between decaffeinated and regular coffee because it seemed the modern American version of tea tasting. (Cathy, the manager of Peet's Coffee shop on Domingo Avenue in Berkeley, California, was very helpful in providing information on brewing and decaffeinating coffee.) These in-

In this lab, you will design, conduct, and analyze results from your own experiment to test the sensory discrimination of a person who claims to be able to taste the difference between regular and decaffeinated coffee. You will need to find such a person for your experiment.

1. Determine a protocol for your experiment.
   (a) *What should be done about random variations in the temperature, sweetness, and so on?* Before conducting your experiment, carefully lay out the procedures for making the cups of coffee. Keep in mind that coffee can be kept warm for about 20 minutes before it starts to turn bitter. Do not reheat coffee.
   Ideally you would want to make all cups of coffee identical, except for the caffeine. But it is never possible to control all of the ways in which the cups of coffee can differ from each other. Some ways must always be dealt with by randomization.
   (b) *How many cups should be used in the test? Should they be paired? In what order should the cups be presented?* Fisher suggests that the experiment should be designed such that, "if discrimination of the kind under test is absent, the result of the experiment will be wholly governed by the laws of chance." Also keep in mind that the number and ordering of the cups should allow a subject ample opportunity to prove his or her abilities and keep a fraud from easily succeeding at correctly discriminating the type of coffee in all the cups served.
   (c) *What conclusion could be drawn from a perfect score or from a test with one or more errors?* For the design you are considering, list all possible results of the experiment. For each possible result, decide in advance what conclusion you will make if it occurs. In determining this mapping of outcomes to conclusions, consider the probability that someone with no powers of discrimination could wind up with each outcome. You may want to make adjustments in your design to increase the sensitivity of your experiment. For example, if someone can't distinguish decaffeinated coffee from regular, then by just guessing, the guesser should have a small chance of correctly determining which cups are which. On the other hand, if the taster possesses some skill at differentiating between the two kinds of coffee, then he or she may make a few mistakes.
2. Write out an instruction sheet for your experimental process. Conduct a "dress rehearsal" to work out the kinks in the process. After your practice run, you may want to make changes in your instruction sheet to address any problems that arose.
3. You should now be ready to conduct your experiment. Record your results carefully, and note any unusual occurrences in the experimental process. It may be a good idea to keep track of the order in which the cups were prepared, and the order in which they were served to the subject.
4. Summarize your results numerically. Do they support or contradict the claim that the subject possesses no sensory discrimination? Use your list of all possible events and subsequent actions to come to a conclusion. Discuss the reasons behind the decision that you have made.
5. What changes would you make to your experimental process if you had the opportunity to do it over again?

Fig. 5.17 Instruction sheet for students for the coffee-tasting project described in Section 5.3.3.

structions appear in Fig. 5.17. To help students design their experiment, we also provide a list of pitfalls that some students have discovered in their experimental procedures:

The subject simply didn't like the brand of coffee that I bought. I chose the brand that I did because it had a decaffeinated and a caffeinated version. If I were to do the experiment over, I might try to buy a kind of coffee that I knew the expert would enjoy. This has its own problems because if you choose the expert's favorite coffee, he would probably be able to identify the version that he normally drinks (whether it be caffeinated or decaffeinated) and deduce the other one.

The expert was used to drinking coffee with sugar, and I served the coffee black. Whether this is an important factor is debatable. I think that if I were to do the experiment over again, I would sugar the coffee, to simulate as closely as possible the expert's usual experience.

Another problem was that after six or seven sips of two different coffees, the expert's ability to differentiate was diminished. Both "mistakes" made by our expert were made in the second half of the tastings. If I were to re-design the experiment, I would leave some time between each tasting. In order to make sure each cup of coffee was the same, I could make a single pot and use a thermos to keep it hot throughout the tasting period.

If we had a chance to do it over again, we would have used saltine crackers rather than water as a way for the taster to remove residual tastes. Water turned out to be less effective than we had hoped.

We would choose to blindfold the taste tester. She may have been able to differentiate the cups of soda by the amount of carbon dioxide in the soda.

With the 20 samples that we used, she had a hard time eating all of it. It was the first time I had ever seen her full of chocolate.

5.4 Observational studies

In an introductory statistics course, we usually do not teach students how to conduct observational studies, but we certainly warn of potential pitfalls in their interpretation. The in-class demonstrations of regression to the mean (Section 4.3) and adjusting for age as a lurking variable (Section 3.7.3) show some of the pitfalls of interpreting observational studies. When discussing data collection, we remind students of these earlier demonstrations to relate data collection to statistical methods of scatterplots and regressions. These ideas return at the end of the semester when we discuss multiple regression as a method of adjusting for lurking variables (see Chapter 9).

Our class discussion of observational studies is grounded in demonstrations and examples we have already considered in class. An example of an observational study we have considered so far is the regression of height on income in Section 4.1.2. This mini-study showed that tall people had higher incomes, but this is hard to interpret because sex is a potential lurking variable: men are taller than women on average, and they also have higher average incomes. We return to this example, using multiple regression to control for the lurking variable, in Section 9.1.

For another example, the test scores of schoolchildren in Texas improved dramatically in the 1990s. Is it reasonable to attribute this to the state's education reforms (see Section 1.4)? What possible lurking variables could explain the difference? How could a study be designed to learn more about the effects of school reforms?

In general, any observational study can have lurking variables, and we discuss with the students what can be done about this potential problem. Possible strategies include *restricting* the lurking variable (for example, studying only women and looking at the relation of height and income), *controlling* for the lurking variable (analyzing income and height separately for men and women), and *balancing* (comparing groups of tall and short people, with the two groups chosen to be the same with respect to their proportions of men and women.

Randomized experiments ensure balance, at least on average: for example, in the "United Nations" experiment in Section 5.3.1, we shuffled the survey forms before handing them out to the students. We ask the students how we could have performed this survey in a less balanced way, and they come up with options such as giving one form to all the women and the other to all the men, or passing out one form to the front half of the class and the other to the back.

Many topics can be studied either by experiments or observational studies, and we can discuss in class the advantages and disadvantages of each form of investigation. The rest of this section gives examples that we have used as discussion topics. Another example, in which we focus on ethical issues, appears in Section 10.5.4.

5.4.1 The Surgeon General's report on smoking

Evidence from dozens of observational studies formed an essential part of the Surgeon General's report in 1964 where it was declared that smoking causes lung cancer. For our students this seems common knowledge, but that was not the case forty years ago when nearly half of the adults in the United States smoked. The early chapters of the Surgeon General's report describe the criteria for judging whether smoking is injurious to health, and we like to distribute excerpts from the report (Fig. 5.18) and discuss the importance of large population studies in this judgment. Then when we discuss other population studies reported in the press, the students have a perspective on how to interpret the evidence.

5.4.2 Large population studies

Chapter 6 gives examples of projects and homework assignments involving detailed analyses of observational studies. In addition, we bring news clippings to class and organize brief in-class activities where students work in pairs to identify five things about a study: the subjects, the outcome or response measured on each subject, the comparison groups (or variable), a possible lurking variable, and the potential problem with the lurking variable. For example, in "Pregnant Women Warned of Chlorinated Tap Water: Miscarriage Risk—But More Study Urged," *San Francisco Chronicle*, February 2, 1998, the subjects were 5144 pregnant women enrolled in the Kaiser health plan, the outcome was

. . . Clinical, pathological and experimental evidence was thoroughly considered and often served to suggest an hypothesis or confirm or contradict other findings. When coupled with the other data, results of epidemiologic studies can provide the basis upon which judgments of causality may be made. . . .
Statistical methods cannot establish proof of a causal relationship in an association. The causal significance of an association is a matter of judgment which goes beyond any statement of statistical probability. To judge or evaluate the causal significance of the association between the attribute or agent and disease, or effect upon health, a number of criteria must be utilized, no one of which is an all-sufficient basis for judgment. These criteria include:
a) The consistency of the association
b) The strength of the association
c) The specificity of the association
d) The temporal relationship of the association
e) The coherence of the association
. . .
In order to judge whether smoking and other tobacco uses are injurious to health or related to specific diseases, the Committee evaluated three main kinds of scientific evidence:
1. *Animal experiments.*—In numerous studies, animals have been exposed to tobacco smoke and tars, and to the various chemical compounds they contain. Seven of these compounds have been established as cancer-producing . . .
2. *Clinical and autopsy studies.*—Observations of thousands of patients and autopsy studies of smokers and non-smokers show that many kinds of damage to body functions and to organs, cells, and tissues occur more frequently and severely in smokers. . . .
3. *Population studies.* . . . In retrospective studies, the smoking histories of persons with a specified disease (for example, lung cancer) are compared with those of appropriate control groups without the disease. For lung cancer alone, 29 such retrospective studies have been made in recent years. Despite many variations in design and method, all but one (which dealt with females) showed that proportionately more cigarette smokers are found among the lung cancer patients than in the control populations without lung cancer. . . .
Another type of epidemiological evidence on the relation of smoking and mortality comes from seven prospective studies which have been conducted since 1951. In these studies, large numbers of men answered questions about their smoking or non-smoking habits. Death certificates have been obtained for those who died since entering the studies, permitting total death rates and death rates by cause to be computed for smokers of various types as well as for non-smokers. The prospective studies thus add several important dimensions to information on the smoking-health problem. Their data permit direct comparisons of the death rates of smokers and non-smokers, both overall and for individual causes of death, and indicate the strength of the association between smoking and specific diseases . . .

Fig. 5.18 Excerpt from Chapter 3, Criteria for Judgment; and Chapter 4 Section A, Kinds of Evidence; of the 1964 Surgeon General's report on smoking. Hand this out and discuss issues such as what is meant by "consistency," "strength," "specificity," and so forth.

	Before	After
Control group	•	
Treated group		•

Fig. 5.19 Diagram of available data in a before/after study such as the comparison of SAT scores before and after coaching as described in Section 5.4.3. The appropriate comparison would be treated vs. control in the "after" position. In this study, however, the "before" group is considered as a control, and there is no separate control group. As a result, it is impossible to separate treatment effects from changes that would have occurred even without the treatment.

whether the woman had a miscarriage in the first trimester of pregnancy, and the comparison groups were those who drank 5 or more glasses a day of highly chlorinated tap water, compared to those who drank less water or who drank water with less chlorine. A possible lurking variable is income because those drinking "better" water, such as bottled water, may tend to have higher incomes, and higher-income women tend to have healthier pregnancies.

5.4.3 Coaching for the SAT

For an illustration of the subtleties of observational studies on a topic familiar to students, we discuss the claims made by some test-preparation organizations that, with their classes, students gain an average of 100 or more points on the Scholastic Assessment Test (SAT). We ask the students: do you think that taking one of these classes would give you (or any other student) 100 extra points? We divide them into pairs and give them two minutes to think up reasons why this might be a biased estimate of the effect of test preparation.

The before–after comparison has many problems, most obviously that students generally improve their scores even without test preparation, partly because they are more familiar with the test and partly because the second test is taken later, and they have learned more material in school. In addition, students who do relatively poorly are more likely to do test-preparation: if you did well the first time, you would not bother doing extra studying. This is related to the regression effect discussed in Section 4.3. Also, the very fact that a student is taking a test preparation class implies some level of motivation, and maybe such motivated students would improve a lot even without the formal class.

The fundamental problem is that the "after" scores should be compared, not to the "before" scores, but to the "after" scores that would have occurred with no treatment (see Fig. 5.19). Other examples of bias in observational studies appear in Section 10.1.

For the SAT example, we tell the students about a more careful study that compared students with and without test preparation: the estimated effect was about 20 points: a positive effect, but not nearly as large as reported before–after changes.

6 Statistical literacy and the news media

An important theme in an introductory statistics course is the connection between statistics and the outside world. We describe here some assignments that have been useful in getting students to learn how to gather and process information presented in the newspaper articles and scientific reports they read. We developed these assignments with the help of several undergraduate research assistants; we devote an entire chapter to this topic in order to give a sense of the details that are needed to get students on track with this sort of long-form assignment. Tracking down reports from newspaper articles can be difficult, and this assignment seems to work well only when students have direction about how to do this kind of research.

We discuss three related assignments, which are described in Sections 6.2 and 6.3. For the first kind of assignment, students work through prepared instructional packets. Each packet contains a newspaper article that reports on a scientific study or statistical analysis, the original report on which the article was based, a worksheet with guidelines for summarizing the reported study, and a series of questions. For the second kind of assignment, we scale back by reducing the original report to one to two pages of excerpts that focus mostly on the study protocol. In the third kind of assignment, each student is required to find a newspaper article themselves, track down the original report, summarize the study using our guidelines, and write a critique of the article. Here, we describe the guidelines we developed to help the student in reading the newspaper article and original source, and the procedures we used for each type of assignment.

Figure 6.2 provides instructions given to the students for the individual projects. We also include in Section 6.6 the material from four of the packets, complete with questions and answers written by students.

6.1 Introduction

Every year we teach one-semester introductions to probability and statistics. One thing we like to do is hand out clippings of newspaper articles such as "Yes, People are Right. Caffeine is Addictive," that describe the results of scientific studies or statistical analyses. These articles often lead to interesting class discussions, which led us to wonder (1) How good are the scientific studies that are reported in the press? and (2) How comprehensively and accurately do the newspaper articles summarize the findings and implications of the studies?

We embarked on a project, with the help of three students, of clipping newspaper articles and tracking down the scientific reports on which they were based. In the process, we created a set of guidelines to help the students summarize relevant information on a scientific study, and using these guidelines as a base, we developed classroom assignments that focus on understanding the issues involved in presenting quantitative findings.

Through these assignments, students connect in-class statistical knowledge to current events, and they learn how to think critically about the information found in the newspaper. An important component of these assignments, which sets them apart from other projects that use current newspaper articles, is the inclusion of the original source in the analysis. We have found that the students are better able to evaluate the merits of the study and the quality of the reporting when given the original reports, even though the reports are often quite technical. They also feel a great sense of accomplishment when they are able to apply their statistical knowledge to the primary source.

This type of assignment complements a wide range of statistics courses including those using the increasingly popular teaching tools based on hands-on work with data and news clippings.

6.2 Assignment based on instructional packets

From September 12 through November 13, 1994, three undergraduate research assistants read the *New York Times* and the *San Francisco Examiner* and clipped out every article that reported a scientific study or a statistical analysis, including medical and health studies, economic analyses ("Feds, Wilson Dispute Illegal Immigration Costs") and social statistics ("Alarming Report on Worldwide Smoking Total"), but excluding highly technical studies, such as reports of a new gene being located. Newspapers report a large number of opinion polls; we selected only those with unusual methodological features. For example, "Poll: More Lawyers See O.J. Walking," reports on a survey conducted by the National Law Journal, rather than by one of the leading polling firms.

For each article, we attempted to track down the report or reports on which the article was based. Of the articles for which we had gathered reports, we selected a subset of eleven to use for instructional packets. The articles were chosen for their diversity in statistical methods and subject matter (see Table 6.1). For each study, one of our research assistants prepared a summary, a list of questions, and answers to the questions. Each of these was reviewed by us and revised by the student until it was in acceptable shape, while still keeping the essence of the student's original input. This procedure resulted in the creation of the guidelines for summaries presented at the end of this section.

The eleven instructional packets can be used in the classroom in a variety of ways. Most simply, the students can be given an article and the associated original report (or can choose an article/report of interest to them from among our eleven packets), and then be required to summarize the report and article

Table 6.1 Articles from the *New York Times* (NYT) and *San Francisco Examiner* (SFE) for which we created course packets. In selecting these from an original set of 107 articles clipped out of the newspapers over a two-month period in 1994, we sought a diversity of kinds of studies (experiments, surveys, economic analyses, etc.) and also of topics (that is, not just medical studies). Course packet materials for the first four articles appear in Section 6.6.

Newspaper article	Source	Kind of study
Giving IV fluids to trauma victims found harmful (NYT)	*New Eng. J. Medicine*	nonrandomized experiment
1 in 4 youths abused, survey finds (SFE)	*Pediatrics*	sample survey
Monster in the crib (SFE)	*Science News*; *Pediatrics*	observational study
Illegal aliens put uneven load on states (NYT); Feds, Wilson dispute illegal immigrant costs (SFE)	Office of Governor of California; Urban Institute	model-based analyses
Yes, people are right. Caffeine is addictive. (NYT)	*J. Amer. Med. Assoc.*	observational study; randomized expt.
Over-control of eating leads to fat children (NYT)	*Pediatrics*	randomized expt.
Runners are far ahead in aging healthfully (NYT)	*Ann. Internal Medicine*	observational study
Surgeons may operate better with music (SFE)	*J. Amer. Med. Assoc.*	randomized expt.
Walnuts add a happy crunch to life (NYT)	*New Eng. J. Medicine*	randomized expt.
Panel finds no major risk from 'yo-yo' dieting (NYT)	*J. Amer. Med. Assoc.*	meta-analysis
Working women say bias persists (NYT)	U.S. Dept. of Labor	sample survey

using the guidelines displayed in Fig. 6.1 and to answer additional questions in the packet (see Section 6.6 for the questions supplied for a sample packet). If the students need additional help, they can be given a complete packet, with the worksheet filled in and the questions answered, as an example.

For a slight variation on this assignment, we cut and paste parts of the original report into a one- to two- page technical summary (see Section 6.6). We use this alternative for a shorter assignment that focuses more on the design of the study than on the statistical analysis.

Guidelines for your summary

Source of newspaper article wire-service, in-depth article, feature story, etc.
Kind of report: medical/technical journal, press release, book, etc.
Summary of article: One-sentence summary of the newspaper article.
Background as taken from the original report.
Objective of the study, as taken from the original report.
Type of study: (a) *randomized experiment*: treatments under the control of the experimenter and assigned randomly; (b) *nonrandomized experiment*: treatments under the control of the experimenter and assigned non-randomly (for example, the healthy-looking patients get treatment A and the sick-looking patients get treatment B); (c) *observational study*: treatments not under the control of the experimenter (for example, the patients choose whether to smoke); (d) *sample survey*; (e) *model-based analysis* (for example, an estimate of the economic effects of immigration); (f) *meta-analysis*: an examination of several earlier studies.
Study protocol. This should include: the setting of the study, a description of the experimental subjects or survey participants, the population to which the results are generalized, the outcome measurements or responses, control variables, summary of nonresponse.
Also, as appropriate, include: treatments used and the method of treatment assignment (if experiment, observational study, or meta-analysis), blindness (a study is blind if the subjects do not know which treatment they are receiving and double-blind if the experimenters do not know which subjects receive which treatments), how the survey was conducted (if a sample survey).
Statistical methods. These should include graphs and tables (which the students should be expected to understand) as well as more formal methods such as t-tests and regressions (which the students may just have to mention without understanding).
Stated conclusions. Taken from the original report.
Generalization. Difficulties of generalizing to the real world and other problems and potential problems of the study. How do the results relate to the rest of the scientific literature? These questions are generally addressed in the report itself, but the students are encouraged to use their own critical thinking to discuss problems not mentioned in the report.
Discussion. The students are asked to consider questions such as, How accurately did the newspaper article summarize the report? Did the newspaper article overstate the conclusions of the report? Did the newspaper article point out potential flaws in the study not noted in the report?

Fig. 6.1 Guidelines for student summaries for the statistical literacy assignments.

6.3 Assignment where students find their own articles

A longer assignment, which takes more careful management but can be more rewarding, has each student working with a newspaper article of his or her own choosing. Below, we describe the steps involved in the assignment. The handout for the assignment appears in Fig. 6.2.

The newspaper article

Each student must find his or her own newspaper article; two students may not use the same article. We remind the students that scientific and statistical studies on health, public policy, and lifestyle issues appear in all sections of the newspaper and are not just covered by science reporters. Some students use the Web to search for news clippings that interest them; we also save copies of two

The assignment. You are to find a newspaper article that reports some scientific finding of interest to you. Then track down the primary source for the article and write a synopsis of the study including information on the kind of study performed, the scientific protocol, general applicability and limitations of the study, and a summary of the findings. In addition, you are to revise the clipped newspaper article to correct any inaccuracies in reporting and add findings or other information which you thought relevant to the story.

Finding an article. Read the newspaper and clip out an article that reports a scientific study or statistical analysis. Scientific and statistical studies on health, public policy, and lifestyle issues appear in all sections of the newspaper and are not just covered by science reporters. The article may be a medical study of a clinical treatment, an observational study, an economic analysis, or social statistics. Do not clip out reports of highly technical studies. Also skip opinion polls, unless they have some unusual methodological feature.

Finding the report. Determine the source of the scientific or statistical study for your article, and track down the report or reports on which the article is based. Science-related newspaper articles come from a variety of sources, ranging from in-depth studies by local reporters featuring many references and interviews, to ten-paragraph summaries of just-released scientific studies in publications (which send advance copies of newsworthy findings to major newspapers), to press releases and interviews. Some of the reports cited in newspaper articles are in widely circulating journals and can be found in the library. In other cases, articles in other news publications may be cited, and you will have to go further to track down the scientific reports. For some, you may need to send away for a report, or telephone a person cited as a source in the newspaper article. The goal is to find a primary source that has technical details on how the study and data analysis were performed, along with numerical summaries (tables and graphs) of the results.

The analysis. Carefully read the primary source, and use the guidelines provided to make a worksheet summarizing the scientific findings. The examples in the handout can be used as a model for your summary.

The write-up. Use our synopsis as a guide in writing a one- to two-page description of the study. Next write a one-page critique of the news story. Discuss which information was reported on in the newspaper article (or could be inferred from the information provided in the article) and which was solely available from the primary source. Address the following questions in your critique.

- How accurately does the title reflect the findings of the study?
- If you could change one sentence in the article, which sentence would you change, how would you change it, and what are your reasons for changing it?
- If you could add one paragraph to the newspaper article to give more information about the study, what would you say? Explain your reasons for choosing the information you did.
- If there were any graphs of figures, comment on their usefulness. Do you think additional figures or graphs in the article would be helpful? If so, what kind?

Fig. 6.2 Handout for students describing the long-form assignment in statistical literacy described in Section 6.3.

newspapers every day for a month before starting the assignment and then bring them to class for the students.

Often it is hard for students to judge the appropriateness of an article. For example, some students choose articles that contain only a single summary statistic with no indication of a more in-depth study underlying the news report. To help students select an appropriate article, we require them to obtain instructor approval of their article before proceeding with their search for the original source.

The original source

When the students choose an article that interests them, they can be very resourceful in searching for the background material and have less difficulty reading the source (in comparison to reading on a subject not of their choosing). Nonetheless, it is useful to provide some guidelines about what kinds of reports are relatively easy to obtain, along with strategies for tracking down hard-to-find sources. We discuss these further in Section 6.4.

Some of the reports cited in newspaper articles are in widely circulating journals (mostly in medicine and public health) and can be found in the library or a reprint can be obtained from the author. In other cases, articles in other news publications are cited, and the student will need to go further to track down the scientific reports. To obtain such reports, the students have found it best to make telephone calls to the persons cited as sources in the newspaper articles. Phone numbers can be obtained from the Web, or from directory information. Most reports will be 5 to 20 pages long, but in a few cases, they are book-length studies.

At the time the student receives approval for the article, we outline a course of action for obtaining the original source. We may help identify the journal that contains the source and the library where it can be found, or we may help look up a phone number and make contact with the person cited in the article.

The write-up

The student prepares a summary of the study using the general guidelines in Fig. 6.1. Students are asked to keep track of the information on the study that was available from the news article alone. This helps them learn how to carefully read the newspaper. Students are often misled by claims made in the paper and are surprised when a closer reading shows that the journalist has been quite careful in his or her choice of wording. Using the summary, the student writes a two- to three-page report that describes the study and critiques the newspaper article.

The grade

The student turns in the newspaper article, background source, the summary of the study, and two- to three-page report for grading. We look for a thorough summary of the study, an accurate description of which study details were reported in the newspaper article, and a responsible critique of the news article. We try not to confuse the quality of the researcher's study with the quality of the student's report. For example, articles in the *Journal of the American Medical Association* are structured so that a student can concisely summarize the study from the abstract alone, and we look to see that the student has described the study in his or her own words and has included details not available in the abstract. Additionally, students do not get full credit if they merely try to punch holes in the study or the newspaper report without appreciating its strengths.

Because of the time it takes to track down the source, we allow six to eight weeks to complete this project, and we set intermediate deadlines for selecting the newspaper article and obtaining the original source. The students also find it helpful, if we review their summaries in advance of the final due date.

These projects serve as great sources for class examples, review problems, and exam questions. The students like seeing their projects used this way in the course, and they enjoy bringing their expert knowledge of the study to the class discussion.

6.4 Guidelines for finding and evaluating sources

Here, we present some of the difficulties we had in finding the sources of newspaper articles; students find these accounts very useful when tracking down their own studies.

Science-related newspaper articles come from a variety of sources, ranging from in-depth studies by local reporters featuring many references and interviews, to ten-paragraph summaries of just-released scientific studies in publications such as the *Journal of the American Medical Association* (which send advance copies of newsworthy findings to major newspapers), to press releases and interviews. Nearly all newspaper articles identify the organization sponsoring the original study and the publication in which it appeared, or the name and affiliation of a contact person, such as the author of the study.

The studies we managed to track down varied greatly in quality. Generally we found the studies in the *Journal of the American Medical Association* and the *New England Journal of Medicine* to be thorough and convincing. Articles in other scientific journals were mostly excellent but varied somewhat in quality. The press releases also varied in quality: for example, "Working Women Count," put out by the U.S. Department of Labor, presents the results of two surveys along with fairly comprehensive details of sampling design (and only a little bit of that overly dramatic press-release style of writing). At the other extreme, the information provided by the A. C. Nielsen Company about their procedures for sampling television viewers was a brief press release that had too few details to determine how the sample was chosen.

An example of an informative private-industry report is "Potential Impact of a Nationwide Workplace Smoking Ban on International Travel to the U.S.," by Price Waterhouse LLP. This was used as source material for the *San Francisco Examiner* article, "Report: Smoking Curbs May Drive Visitors Away." This report, sponsored by the San Francisco Hotel Association, is an economic analysis of what might happen to tourism in San Francisco in the event of a ban on smoking in hotels and restaurants. The report goes into great detail about the economic analysis and its assumptions. Unfortunately, the assumptions seem simplistic to the extent of discrediting the conclusions. Claims of the potential impact on tourism are based on an opinion poll of travel agencies and tour operators throughout the United States, who were asked to estimate by what percent a smoking ban would increase or decrease the number of foreign visitors to the U.S. It seems questionable to base an economic analysis (claiming billions of dollars in lost sales in San Francisco) on the results of these sorts of speculations.

Other times, we could only obtain a press release or an oral description, without enough details to fully understand the studies or their results. For example, the newspaper article, "O.J. will be Cleared, Lawyers Tell Poll," referred to the National Law Journal, where we found the article, "O.J. Will Walk, Says 61 Percent in Lawyers' Poll," which presented the results of a "poll of 311 attorneys, conducted Sept. 23–26 [1994] by Penn + Schoen Associates Inc, a New York polling company." We called both the National Law Journal and Penn + Schoen Associates, but neither would provide information on how the poll was conducted.

In some cases, the persons who performed the studies were very helpful, but the studies were too idiosyncratic to be useful for inclusion in this project. For example, to track down the source for "Restaurants in New York Show Signs of a Boom," we called Fred Sampson, president of the New York State Restaurant Association. We obtained the phone number of the association from 1-800-CALL-INFO. After playing phone tag for a while, we finally made contact and talked about the numbers quoted in a recent article about the rise in the restaurant business in New York. Asked how he obtained those numbers, Sampson explained that he got those figures informally through everyday conversations with members of his association. He said he typically receives about 50 calls a day (which we believe, since he was always on another line when we kept trying to reach him) from his members. Whenever he talks to them, he casually asks how well their businesses are doing and jots down the responses, usually percentages, on a little note pad. Often approached by the media, he then, from his note pad, estimates the status of the restaurant business in New York. No explicit analysis or formal surveys were done, although he answered this question with some advanced statistical terminology. Therefore, all the numbers reported in the article came from his own judgment.

In summary, articles from scientific journals are by far the most reliable sources: generally easy to track down, and with comprehensive descriptions of the studies. However, some of the most interesting reports come from other sources—here, more effort must be put into tracking down the source, and even then there is not always enough information to fully understand the study. Despite this difficulty, we do not recommend relying only on scientific journals, because the students who gathered material from alternative sources found it quite educational and because these reports offer a greater variety of topics.

6.5 Discussion and student reactions

We asked the three students who prepared the instructional packets what they had learned. Their responses are similar to the reactions of students in a class to the assignments.

They felt that all of the studies used in this project had good objectives, but some were so flawed that their results became unbelievable and misleading. Many of the erroneous studies were actually harder to obtain because of the reluctance of the researchers and sponsors to provide information. The students also noted that the results of studies can depend strongly on the definitions of what is being counted or observed, as in the illegal immigration study (see Section 6.6.4) and the abused children study (Section 6.6.2).

About what gets reported, one student noted that, looking through a journal's table of contents, there are many studies that seem as interesting as those reported on in papers, yet they aren't given the same press coverage. It appears as if the studies reported on in newspapers are selected haphazardly. This can be seen to an extent from the fact that most studies reported on in the *New York Times* were not reported on in the *San Francisco Examiner* (and vice versa).

The students were generally impressed with the quality of the newspaper reporting, given the shortness of most of the articles. However, they commented that the newspapers tend to make the research findings more convincing than they actually are. The newspaper article often omits potential flaws or limitations of the study that are actually mentioned in the original report. Newspaper reporters have the difficult task of reading a report and writing an article that retains enough important facts for the reader to understand the significance of the study, under the constraints that it be readable and the right length for the space provided.

On the whole, we have found that these assignments provide a framework for the students in an introductory statistics class to combine their statistical knowledge with their general powers of critical thinking. This is more important than just criticizing newspaper articles; students learn some of the limitations of scientific studies and the role of statistics in making conclusions and presenting them. In these assignments, the students learn much more about issues of data collection and design, and also about the relevance of statistical methods to real-world problems.

6.6 Examples of course packets

We illustrate with four packets, corresponding to an experiment, a survey, an observational study, and an economic analysis. These packets are complete except for the cited reports.

The summaries, questions, and answers were written by students, but we discussed and revised them together. So what you see here is more of an idealized than an expected student assignment. The course packet can be used as teaching material in two ways: give students the entire packet to read as preparation for the classroom assignment (see Fig. 6.2); or give students the article, report (or excerpts from it), and questions, and require them to make a summary and

THE NEW YORK TIMES **NATIONAL** FRIDAY, OCTOBER 28,

Giving IV Fluids to Trauma Victims Found Harmful

By WARREN E. LEARY

Special to The New York Times

WASHINGTON, Oct. 27 — A study of seriously injured people receiving emergency trauma treatment has concluded that the common practice of immediately giving many of these patients intravenous fluids may be detrimental, researchers said today.

It has long been standard treatment to give fluids to bleeding trauma patients in hopes of keeping their blood pressure up, staving off shock and preventing damage to organs from blood deprivation.

But a study of almost 600 patients in Houston indicates that the practice of immediately giving intravenous fluids, or fluid resuscitation, before the sources of bleeding are stopped may hurt many patients more than it helps, according to a report published today in The New England Journal of Medicine.

If the results are borne out by further studies, the researchers said, they could cause trauma and emergency care services to change decades of practice. Giving intravenous fluids — most often a salt or a sugar and mineral solution — to bleeding patients has been considered so important to care that the treatment is commonly begun in ambulances before the injured reach a hospital.

The study found that severely injured patients who did not receive fluids until surgeons were ready to close their wounds had a slightly better survival rate than those who received the standard fluid regimen.

Researchers looked at hundreds of adult patients who suffered penetrating torso wounds, usually from gunshots or stabbings, and who were taken to the major trauma center for Houston, Ben Taub General Hospital, during a 37-month period. Of the study's 289 patients whose fluid resuscitation was delayed, 202 of them, or 70 percent, survived to be discharged. Of the 309 patients who received immediate fluid replacement treatment, 183, or 62 percent, survived, researchers reported.

The study also found that patients denied early fluids tended to be discharged faster and appeared less likely to suffer complications like infection, kidney failure, pneumonia and other respiratory problems, and blood-clotting difficulty immediately after surgery.

The report said animal studies indicated that raising blood pressure with fluids might be harmful because higher pressure encouraged continuing hemorrhaging in torn arteries or veins, or could dislodge fragile new blood clots that were trying to plug the leaks in those blood vessels. Giving fluids could also dilute coagulation chemicals in the blood, which help in clotting, and lower blood viscosity, or thickness, which would make it easier for blood to flow around an incomplete clot, the researchers said.

Dr. William H. Bickell, the lead investigator for the study, said the findings only applied to bleeding adult patients with penetrating wounds to the chest and abdomen, because those suffering other types of trauma, like internal injuries from falls or car accidents, were not studied. In addition, without further research, the results would not apply to children or adolescents, or to people with head wounds, who were also excluded from the study, he said.

But Dr. Bickell, a former research fellow at Ben Taub who is now director of research at St. Francis Hospital in Tulsa, Okla., said in an interview that the results were strong enough to recommend delaying fluid resuscitation in the types of patients studied. "Over all, the data indicates delaying all fluid resuscitation for these patients is best," he said, "but the door is still open for modifying this recommendation.

A study questions a common way to treat trauma cases.

"There need to be more studies to see if perhaps more limited resuscitation might be better for some patients."

In an editorial in the journal, Dr. Lenworth M. Jacobs of Hartford Hospital in Connecticut praised the study as well designed and executed, but he cautioned doctors about applying the results prematurely. Before withholding fluids from patients at the scene of accidents or in trauma centers, he wrote, studies are needed to show which groups of patients would most benefit, and under what conditions of severe bleeding and low blood pressure the denial of fluids would be most helpful.

Dr. Kenneth L. Mattox, director of trauma care at Ben Taub and an author of the report, said the study was significant because it raised questions about what was the best trauma care and made people rethink standard practices.

"It raises questions about the relevance of giving IV fluids to patients with penetrating wounds and the fact that this practice is not based on good research to begin with," Dr. Mattox said in an interview.

The researchers said the wisdom of giving fluids to increase blood pressure in trauma patients has been questioned since World War I, but people largely stopped debating the issue about 30 years ago because some research with dogs, which was not related to human injury, indicated it was helpful.

Dr. Mattox and Dr. Bickell said limited research indicated some patients suffering blood loss from injuries experienced very low blood pressure for long periods of time with no apparent lasting effects after treatment. This could mean there is a natural defense mechanism by which the body drops blood pressure to decrease blood flow through an injured vessel.

"In some cases, by giving fluids, we may be interfering with a natural defense mechanism that the body uses to protect itself from hemorrhage," Dr. Bickell said. Questions like these only can be answered with more research in trauma medicine, which has traditionally gotten little funding, researchers said.

Fig. 6.3 Article from the *New York Times* for the course packet in Section 6.6.1.

answer the question. We have included sample excerpts for three of the four packets presented here, and we marked those questions that cannot be answered from the excerpts alone (without the full report).

We have also included a student's comments on each of these packets to give some sense of how they might be received.

6.6.1 A controlled experiment: IV fluids for trauma victims

"Giving IV fluids to trauma victims found harmful," *New York Times*, October 28, 1994, p. A9 (see Fig. 6.3).

Excerpt from "Immediate versus delayed fluid resuscitation for hypotensive patients with penetrating torso injuries," W. H. Bickell et al., *New England Journal of Medicine* **331**, 1105–1109:

For the past two decades the preoperative approach to hypotensive patients with trauma in North America has included prompt intravenous infusion of isotonic fluids. The rationale for this treatment has been to sustain tissue perfusion and vital organ function while diagnostic and therapeutic procedures are performed. This approach was based largely on the demonstration in animals in the 1950s and 1960s that isotonic-fluid resuscitation was an important life-sparing component of therapy for severe hypotension due to hemorrhage.

Patients eligible for this study were adults or adolescents (age at least 16 years) with gunshot or stab wounds to the torso who had a systolic blood pressure less than or equal to 90 mm Hg, including patients with no measurable blood pressure at the time of the initial on-scene assessment by paramedics from the City of Houston Emergency Medical Services system. ... Pregnant women were not enrolled in the study. All patients within the city limits of Houston who met the entry criteria were transported directly by ground ambulance to the city's only receiving facility for patients with major trauma, Ben Taub General Hospital.

Patients included in the final study analysis were those for whom fluid-resuscitation might affect outcome. Part of the prospective study design was to exclude the following from the outcome analysis: patients with a Revised Trauma Score of zero at the scene of the injury, those who also had a fatal gunshot wound to the head, and patients with minor injuries not requiring operative intervention. The paramedics caring for the patients were not aware of these exclusion criteria and treated all hypotensive patients with penetrating torso injuries according to the protocol.

All patients enrolled in this study were assigned to one of two groups: the immediate-resuscitation group, in which intravascular fluid resuscitation was given before surgical intervention in both the prehospital and trauma-center settings, or the delayed-resuscitation group, in which intravenous fluid resuscitation was delayed until operative intervention. ... Patients injured on even numbered days of the month were enrolled in the immediate-resuscitation group, whereas those injured on odd numbered days were enrolled in the delayed-resuscitation group. The alternating 24-hour period corresponding to the 24-hour shifts worked by both the paramedics and the trauma teams. Because there were three rotating teams of paramedics and surgical house staff, assignments to the groups were alternated automatically for both prehospital and hospital staff members.

Among the 289 patients who received delayed fluid resuscitation, 207 (70%) survived and were discharged from the hospital, as compared with 193 of the 309 patients (62%) who received immediate fluid resuscitation. The mean estimated intra-operative blood loss was similar in the two groups. Among the 238 patients in the delayed-resuscitation group who survived to the postoperative period, 55 (23%) had one or more complications (adult respiratory distress syndrome, sepsis syndrome, acute renal failure, wound infection, and pneumonia), as compared with 69 of the 227 patients (30%) in the immediate resuscitation group. The duration of hospitalization was shorter in the delayed-resuscitation group.

A student's summary

One-sentence summary of the newspaper article: Bleeding trauma patients are usually given fluids in order to keep their blood pressure up, but this practice may actually be harmful.

Background: Giving IV fluids to trauma patients before their bleeding is controlled may be detrimental. This procedure has been standard for the past two decades because it was proven successful for severe hypotension due to hemorrhage.

Objective: To determine the effects of delaying fluid resuscitation until the time of operative intervention in hypotensive patients with penetrating injuries to the torso.

Kind of study: Nonrandomized experiment.

Subjects: 598 adults with penetration torso injuries with a pre-hospital systolic blood pressure less than or equal to 90 mm Hg.

Setting: City (Houston) with a centralized system of pre-hospital emergency care and a single receiving facility for patients with major trauma.

Treatments: Immediate-resuscitation group and delayed-resuscitation group. The difference between the two is that the first group received fluid resuscitation before they reached the hospital and the others didn't receive it until they reached the operating room.

Treatment assignment: Depended on whether the person was injured on an odd-numbered or even-numbered day.

Outcome measurements: Survival of patients.

Blindness: Patients knew how they were treated, but it probably didn't alter their will to survive. Physicians were not blind.

Population: People aged 16 and over with a gunshot or stab wound to the torso and with a blood pressure less than 90 mm Hg. Pregnant women were not enrolled in the study. Patients were also measured for the Revised Trauma Score to see if they were to be included in the study; those with a score of 0 were not included, as well as those with fatal gunshot wounds to the head and patients with minor injuries not requiring operative intervention.

Statistical methods used: A comparison of different characteristics to check that the two treatment groups were similar (Table 1 of the report). Comparisons between treatments were made using two-way tables and binomial p-values for comparison of proportions.

Stated conclusions: Of the 289 patients who received delayed fluid resuscitation, 70% survived and were discharged from the hospital. In the other group, 62% of 309 patients survived (this difference between the two percentages is statistically significant at $p = 0.04$). Among the 238 patients in the delayed-resuscitation group who survived to postoperative period, 23% had one or more complications. Among the 227 patients in the immediate-resuscitation group who survived to the postoperative period, 30% had complications (the difference is statistically significant at $p = 0.08$). In addition, the duration of hospitalization was shorter in the delayed-resuscitation group.

Nonresponse: None (for all the patients that met the initial eligibility requirements).

Difficulties of generalizing to the real world: This study only found the effects of specific types of wounds. Also, those with blood pressures below 40 mm Hg, who rarely survived, were not included in the study, and so this cohort should also be examined in a different study.

How does this relate to the rest of the scientific literature? These findings are similar to conclusions of studies done on animals.

Questions

1. (a) What were the two treatments in the study?
   (b) Who were the subjects? How many subjects were there?
   (c) What pre-treatment characteristics were recorded?
   (d) What were the main outcome measurements?
2. Describe the method of treatment assignment. Why do you think they did not assign treatments randomly?
3. Was the treatment assignment blind to the subjects? The doctors?
4. The report gives details on how the sample size was chosen. Describe what they did and what reasoning they used to decide the sample size.
5. Seeing the results, do you think the sample size turned out to be too small or too large?
6. Given the information in the report, do you think the article title, "Giving IV Fluids to Trauma Victims Found Harmful," is a reasonable claim?
7. Among the 238 patients in the delayed-resuscitation group who survived to the post-operative period, 55 (23%) had one or more complications. Among the 227

patients in the immediate-resuscitation group who survived to the post-operative period, 69 (30%) had complications. This comparison does not take into account the patients who died right away. How does this adjustment affect this comparison?

8. (*This question can only be answered with the full report, not merely the excerpt given above.*) If you have covered *p*-values in your statistics course, explain how the *p*-value in the first row of Table 5 was determined. What does the *p*-value of 0.04 tell you?
9. Why do you think pregnant women were excluded from the study? What about car accidents?
10. If you could add one paragraph to the newspaper article to give more information from the study, what would you say?

Answers to questions

1. (a) The two treatments were the immediate-resuscitation group (in which intravascular fluid resuscitation was given before surgical intervention in both the pre-hospital and trauma-center settings) and the delayed-resuscitation group (in which intravenous fluid resuscitation was delayed until operative intervention). (p. 1106, "Methods: Study Interventions")
   (b) The subjects were patients over 16 years of age who were transported to Ben Taub General Hospital with gunshot or stab wounds to the torso and a systolic blood pressure less than or equal to 90 mm Hg. They also did not have a Revised Trauma Score of zero at the scene of the injury, a fatal gunshot wound to the head, or minor injuries not requiring operative intervention. There were 598 patients. (pp. 1105–1106, "Methods: Study Subjects" and p. 1107, "Results: Characteristics of the Patients," second paragraph)
   (c) Pre-treatment characteristics that were recorded were blood pressure; the Revised Trauma Score (calculated from the Glasgow Coma Scale, systolic blood pressure, and respiratory rate); times at which emergency vehicles were dispatched, arrived at the scene, departed from the scene, and arrived at the trauma center; Injury Severity Score. (p. 1106, "Methods: Main Measurements and Secondary Outcome Variables")
   (d) The main outcome measurements were survival of patients and assessment of six defined postoperative complications (wound infection, adult respiratory distress syndrome, sepsis syndrome, acute renal failure, coagulopathy, and pneumonia). (p. 1106, "Methods: Main Measurements and Secondary Outcome Variables")
2. Patients injured on even-numbered days of the month were assigned to the immediate-resuscitation group, while those injured on odd-numbered days were enrolled in the delayed-resuscitation group. The researchers did not assign the treatments randomly probably because of the difficulty and confusion to choose which protocol/treatment to use on each patient where time was a factor and each patient was unique. (p. 1106, "Methods: Study Interventions")
3. The treatment assignment was blind to the subjects because they did not know that they were in an experiment. The treatment assignment was not blind to the doctors because they have to know what to give in the pre-hospital phase. (p. 1106, "Methods: Study Protocol")
4. The sample size was calculated on the assumption that death would occur in 35% of patients receiving standard preoperative fluid resuscitation for penetrating torso injuries. On the basis of experimental data and past clinical experience, an estimated 10–15% improvement in survival was predicted if fluid resuscitation was delayed until operative intervention. With an alpha value of 0.05 and a beta value of 0.2 for a hypothesis test, approximately 600 patients are needed. (p. 1106, "Methods: Statistical Analysis," first paragraph)

5. The sample size seems reasonable, but a larger sample might be good, considering that the main results are close to the cut-off p-value of 0.05. (p. 1108, Table 5)
6. The title does not seem reasonable because it implies that all trauma victims were considered and studied, whereas the study only pertains to those with penetrating torso injuries. Also, the title implies that IV fluids do not work at all, but the results show some indication that the fluids can help: not all of the immediate-resuscitation group died. (p. 1105, "Abstract: Results" and "Abstract: Conclusions")
7. The numbers may be higher if those who died right away were accounted for. In fact, the percentage for the delayed-resuscitation group could be bigger because death or complications may be the results of not getting any fluids immediately. (p. 1109, Table 6)
8. The null hypothesis is just a comparison of proportions (193/309 to 203/389). The question is if this difference occurred by chance. The observed difference is 8%. They divide this difference by the standard deviation of the difference ($\sqrt{(0.62 \cdot 0.38/309) + (0.70 \cdot 0.20/289)} = 0.035$); the result is $0.08/0.035 = 2.28$, so the difference is 2.28 standard deviations away from zero. The z-test is performed (two-sided) to give a p-value $= 0.04$. (p. 1106, "Methods: Statistical Analysis")
9. Pregnant women were excluded in the study to make the sample and the results representative and general to the whole population. Plus, there are outside complications that occur during pregnancy that might confound the study. Also moral issues could be involved (risk the life of the baby?) Those in car accidents were excluded because they are more likely not to get penetrating wounds or more likely received wounds that are not serious enough to get IV fluids. (p. 1105, "Methods: Study Subjects," first paragraph)
10. A possible paragraph: Despite indications that the delayed-resuscitation group appeared to have fewer complications, there was not enough done to study this issue. People who died immediately after operative intervention were not included in these calculations. Complications might have come from not getting fluids, which could result in death; but these patients were not accounted for in this analysis of the existence of complications. (p. 1109, Table 6)

Another student's opinions

The news article begins with the big headline, but as the reader reads on, the statement, "Giving IV fluids to trauma victims found harmful," carries less and less weight as the writer qualifies his headline. Firstly, we discover that delaying IV treatment is useful only for "bleeding adult patients with penetrating wounds to the chest and abdomen," and other types of injuries were not included, such as head wounds and victims who were children or adolescents. Furthermore, another researcher says that even though the data suggest that delaying IV treatment is best, the "door is still open for modifying this recommendation."

The journal article states that the percentage of victims who survived with IV fluids was 62%; without fluids, 70%. Furthermore, those who received immediate fluids averaged longer ICU and hospital stays suggesting that the immediate IV fluids interferes with various immune responses, as the news article explains.

There were a few interesting points that caught my attention. The first was Table 1, where they state the "probability of survival" of both groups; those who receive immediate resuscitation were 3% less likely to survive. If this is not accounted for in the final calculation of survivors, then perhaps their findings would not be statistically significant.

Overall, the researchers made a good effort to question a long-standing practice of administering IV fluids to trauma victims. Their data suggest that in some cases, doing so does more harm than it alleviates. Because of its widespread practice, more in-depth research needs to be done to evaluate delayed IV's with head traumas and injuries to children, which were not included in this study.

6.6.2 A sample survey: 1 in 4 youths abused, survey finds

"1 in 4 youths abused, survey finds," *San Francisco Examiner*, October 4, 1994 (see Fig. 2.5 on page 16 of this book).

Excerpt from "Children as victims of violence: a national survey," D. Finkelhor and J. Dziuba-Leatherman, *Pediatrics* **94**, 413–420 (1994):

The study staff interviewed by telephone a nationally representative sample of 2000 young people between the ages of 10 and 16 and their caretakers. A national sample of households was contacted and screened for the presence of appropriate age children through random digit dialing. Interviewers spoke with the primary caretaker in each household, asking him or her some questions relevant to child victimization prevention and explaining the objectives of the study. They then obtained parental permission to interview the child. Speaking to the children, the interviewers again explained the study, obtained consent and made sure that the children were alone and free to talk openly.

The participation rate was 88% of the adults approached, and 82% of the eligible children in the households of cooperating adults, despite that the study involved children, a potentially sensitive topic, a lengthy interview and required the consent of two individuals. Almost three-quarters of the child nonparticipation came from the caretakers denying permission to interview the children and the rest from the children not wishing to the interviewed. The youngest children (aged 10 and 11) had marginally lower levels of participation. An analysis of the households with child nonparticipants showed them to be demographically indistinguishable from the participants, although there was a slightly greater salience for issues of violence (more concern, more perception of threat) among the parents of participants.

Children were asked a total of 12 questions about possible victimizations. The victimization questions were followed up with more extensive questions about the details of the episode, on the basis of which the episodes were classified into one of several categories and also as attempted or completed.

A quarter of the children reported a completed victimization (excluding corporal punishment) in the previous year and over a third a completed or attempted victimization. Over half the children reported a completed or attempted victimization sometime in their lives. Nonfamily assaults were the most numerous type of victimization. Boys were over three times more likely than girls to have experienced a completed nonfamily assault the previous year, most of which were committed by known perpetrators under the age of 18. The rates did not vary by age.

Children experienced family assaults at about one-third the rate of nonfamily assaults. There were no age or gender differences. A little less than half the family assaults were committed by adults. The survey also revealed that well over a quarter of these youths were still being corporally punished by their parents. However, there was a significant decline in corporal punishment use with age (46% of 10-year-olds vs. only 15% of 16-year-olds).

Nonfamily assault screening questions:

1. Sometimes kids get hassled by other kids or older kids, who are being bullies or picking on them for some reason. Has anyone—in school, after school, at parties, or somewhere else—picked a fight with you or tried to beat you up?
2. Has anyone ever ganged up on you, you know, when a group of kids tries to hurt you or take something from you?

The definition of an attempted or completed assault included any child responding yes to either of these questions (except that any episode involving a family member perpetrator was moved from this category to the family assault category). A completed nonfamily assault was an episode that included actual punching, kicking, hitting with an object, or threatening with a weapon.

Family assault screening questions:

1. Sometimes kids get pushed around, hit, or beaten up by members of their own family, like an older brother or sister or parent. Has anyone in your family ever pushed you around, hit you, or tried to beat you up?
2. Has anyone in your family gotten so mad or out of control you thought they were really going to hurt you badly?

A completed family assault included the occurrence of actual punching, slapping, kicking, hitting with an object, or threatening with a weapon.

A student's summary

One-sentence summary of the newspaper article: According to a telephone survey of 2000 children, one in four adolescents had been physically or sexually abused within the past year.

Background: Recently there had been a great deal of public and media attention to victimized children, but this concern has been largely fragmented by focusing only on specific forms of victimization. This fragmentation has prevented a comprehensive view of the overall victimization of children.

Objective: To gain a more comprehensive perspective on the scope, variety, and consequences of child victimization.

Kind of study: Survey.

Subjects: 2000 children (1042 boys, 958 girls) aged 10–16 years.

Responses/measurements: Children were asked 12 basic questions about any occurrence of nonfamily assault, family assault, kidnapping, sexual abuse/assault, violence to genitals, and corporal punishment within the last year and in their lifetimes. Also asked were follow-up questions about the details of the act(s): attempted/attempted-and-completed, noncontact/contact/rape, family perpetrator/nonfamily perpetrator, injury/no injury.

How the survey was conducted: The study staff contacted by telephone a national sample of households through random digit dialing and screened the sample for appropriate age children. Interviewers spoke with the caretakers about victimization prevention and the study and obtained permission to speak with the children. Then the staff obtained consent from the children and, for 30–60 minutes, interviewed them alone.

Blindness: None.

Population: Children in the U.S. who were contacted through random digit dialing.

Control variables: Characteristics such as household income, race/ethnicity, region, type of metro area, and gender.

Statistical methods used: Weighted sample to correct undercounting of black and Hispanic children. Comparisons of weighted sample proportions by types of victimization and completeness of act with 95% confidence intervals. Chi-squared tests (analyses between various control factors and types of victimization). Six tables.

Stated conclusions: Within the past year, a quarter of the children had experienced a completed victimization (excluding corporal punishment), a third had encountered a completed or attempted act, one in eight had experienced an injury, and one in a hundred required medical attention as a result. Nonfamily physical assaults were the most numerous, usually with boys. Contact sexual abuse occurred to 3.2% of girls and 0.6% of boys. There were substantial numbers of incidents of attempted kidnappings and violence directed to genitals. Two-thirds of the victimizations were disclosed to someone, but only 25% to an authority. Most likely to experience victimization were black or Hispanic, lived in the Mountain and Pacific areas, and lived in large cities.

Nonresponse: (in the population) 12% of the adults approached and 18% of the eligible children with cooperating adults, due to refusal of permission from adult or child; (in sample of subjects) none.

Difficulties in generalizing to the real world: Only those between 10-16 years were studied.

Other problems: Uncertainty of effectiveness of a telephone interview (time constraints, no knowledge of body language). Exclusion of some high risk children such as those without telephones, those in juvenile correctional and mental health facilities, those with disabilities, and those who were angry/alienated to participate. Possible lack of disclosure of intimate victimizations to a stranger interviewer. Possibility of children's forgetting or repressing certain acts. Not many details obtained on every victimization due to time. Unclear control for siblings. Definitions are broad.

How this relates to the rest of the scientific literature: Some rates in this study were agreeable to past studies, but other rates were not as high or low as other past studies due to difference in categorization. This study suggests the need for better statistics and for a more comprehensive study on all forms of child victimizations.

Questions

1. The title of the short newspaper article is rather alarming. What is the definition of abuse used in the survey? Why do the investigators choose to use such a definition?
2. The survey was conducted by phone interview of children. How were the children included in the study? What percentage of the children were not included in the survey either because the parent refused permission or the child refused to talk to the interviewer?
3. What is the population of interest? How do the children surveyed compare demographically to the population of interest?
4. (*This question can only be answered with the full report, not merely the excerpt given above.*) The assault rate reported in the article is 15.6%. Find this rate in the report, and say what is it a report of.
5. (*This question can only be answered with the full report, not merely the excerpt given above.*) The results of this survey are compared to those of two other surveys in the report. What surveys are they? How do the assault rates differ from those reported by Finkelhor and Dziuba-Leatherman? Why might they differ? Which numbers does the newspaper report? What assault rate should be used in comparing this study's results with those of the National Youth Survey?
6. In the report it is stated that of all victimizations, "only 6% were reported to the police, the agency which collects crime statistics ... Obviously the scope of victimization to children cannot be assessed well through police-based statistics." Why might this percentage be so low?
7. If you could add one sentence to the newspaper article, what would it be?

Answers to questions

1. The survey asked about six kinds of abuses: nonfamily assault, family assault, kidnapping, sexual abuse, violence to genitals, and corporal punishment. Abuse was defined as: punching, kicking, hitting, or threatening with an object, kidnapping where the child is taken somewhere, touching sexual parts in a sexual way, exposing oneself to the child, asking the child "to do something sexual," penetrating or engaging in any oral-genital contact, violent contact to genitals, slapping, hitting, or spanking (Appendix, p. 419)
   The survey uses a wide definition of abuse to a child. The purpose of creating a broad definition was to measure a comprehensive view of abuse to children as opposed to other studies that only look at specific abuses such as sexual assault (p. 413, column 1)
2. The caretaker of the child was first asked permission to talk to the child. 88% of all adults asked gave permission and of that 82% of the children agreed to answer.

28% of the children contacted were not included in the study because they or their adult caretakers refused to be a part of the study. (p. 414, column 1)

3. The population of interest are children aged 10–16 years. The group surveyed had a higher rate of assault than children 12–15 years old as reported by the National Crime Survey and had a lower rate of assault than 11–17-year-olds in the National Youth Survey. (p. 415, column 2)
4. The rate is last year's completed nonfamily assaults. (p. 415, Table 1)
5. The results of this study were compared to the National Crime Survey and the National Youth Survey. The estimate from the NCS is a third of the one in the study (5.2%), and the estimate from the NYS was almost twice as much (31%) as the rate in the study. The rates may differ because of the definition they give to assault. The NYS estimate may be higher because it includes assault from siblings and other nonparents family members which the Finkelhor and Dziuba-Leatherman nonfamily assault rate does not cover. The news article quotes the NCS. In order to compare this survey with the NYS, the sum of the completed nonfamily assault rate and the completed family assault rate of this survey should be compared to the NYS rate. (pp. 415–416)
6. The percentage of children who report to the police may be low because the assault may not be seen as something that needed to be reported, such as another child hitting one child on the playground. In this situation, a child may tell an adult supervisor but not file a complaint with the police. (p. 416, Table 3)
7. Here is one possible sentence: "The study attempted to survey a more comprehensive view of child victimization and abuse in order to alert people to its extensive and diverse forms."

Another student's opinions

The newspaper article reported that 15.6% (14.0–17.2, 95% CI) of participants had been assaulted; however, it does not further qualify, leaving one unaware that this figure refers only to "completed nonfamily" assault. I believe the article should have reported the most important finding of the study: about one-quarter of all children reported completed victimization and over one-third had experienced attempted victimization. These results are stated early in the results, and in the abstract. Also, it was found that 1 in 8 children experienced physical injury and 1 in 100 required medical attention. If these data are accurate, then approximately 6 million of America's youth are suffering from completed abuse of some form, 2.8 million will be injured, and a quarter million will need medical treatment for abuse.

Due to its brevity (around 110 words!), the newspaper article omitted many findings that I found quite interesting, and disturbing. For example, in Table 5, p. 417, the percent of children victimized in the Pacific and Mountain region was 62.7% and 64.6%, whereas the average was around 52%. Also, differences were found between the social classes; the biggest difference being from under $20K/year to $20–50K/year—general victimization was 7.3% lower, and all other assaults were lower in the middle income bracket. However, there was also not much difference between the middle bracket and the highest one, over $50K/year.

If these data are accurate and representative of our family structure, then the conclusions are nothing less than disturbing and should set off further research into child abuse. Moreover, intervention should be made more readily available for children of such violence.

6.6.3 An observational study: Monster in the crib

"Monster in the crib," Keay Davidson, *San Francisco Examiner*, September 25, 1994, p. A2. "The terrible twos just got younger," *Science News* **146** (12). (See

MONSTER IN THE CRIB: Forty-two percent of mothers of children between ages 1 and 2 display "symptoms of depression," Science News reports. "About half of that group probably ranks as clinically depressed." The finding strengthens the popular belief that child

raising becomes especially difficult by, or even well before, age 2 — "the terrible twos." The report is based on research by Ardis L. Olson and Lisa A. DiBrigida of Dartmouth-Hitchcock Medical Center in Lebanon, N.H.

The terrible twos just got younger

A smile quickly crosses the face of almost anyone who sees 1- to 2-year-olds trying out their newfound walking skills or grinning at passersby from the safety of their strollers.

Children in this age group, however, appear to have a different effect on many of their mothers. A new, preliminary study of 233 mothers of children between the ages of 1 and 2 finds that 42 percent of the moms showed symptoms of depression. About half of that group probably ranks as clinically depressed, report Ardis L. Olson and Lisa A. DiBrigida of Dartmouth-Hitchcock Medical Center in Lebanon, N.H.

However, mothers were much less apt to feel depressed if they liked their role either as a stay-at-home or an employed mom, the team reports in the September PEDIATRICS.

The study participants included primarily white, married, middle-class women taking their children to a pediatrician's office for a routine checkup. Ninety-six of the mothers did not work outside the home, 79 had part-time jobs, and 58 worked full-time outside the home.

Forty-three participants — in roughly equal proportions in the three groups — reported feeling dissatisfied with their employment role. However, almost 70 percent of mothers who were not content exhibited symptoms of depression, compared to 35 percent of the satisfied group, Olson and DiBrigida report.

Moreover, they argue, results of the study suggest that "dissatisfaction with one's current work role does not simply reflect the malaise of depression."

Participants with part-time jobs had the lowest depression scores of the whole group, says Olson. On average, these women worked 20 hours a week. However, women dissatisfied with working part-time had the highest depression scores.

Part-time employment "needs to be considered as an option that may enhance mothers' mental health," Olson and DiBrigida assert.

The study's findings on part-time work confirm that having a variety of roles that boost self-esteem helps inoculate women against depression, says Ellen McGrath, a psychologist in Laguna Beach, Calif.

Other researchers have reported lower rates of maternal depression than the Dartmouth team found. Health experts often consider depression a problem of economically disadvantaged mothers. But investigations are uncovering increasing rates of depression among young adult women of all income levels, Olson and DiBrigida report.

Olson believes that their study found such a high rate of depression because they interviewed only mothers of 1- to 2-year-olds, a particularly difficult age. "I tell mothers the 'terrible twos' are the 6 months before 2," Olson says.

"The constant vigilance needed to insure physical safety as well as the child's struggle for autonomy contribute to make the toddler year particularly stressful for parents," the investigators argue.

Their questionnaire also uncovered feelings that other studies had failed to reveal, Olson says. The depressed women felt overwhelmed, alone, and full of self-doubt. "Mothers just carry this around as baggage," says Olson. "They think, 'This is how I'm supposed to feel as a mother of a toddler.'"

Robin Post of the University of Colorado Health Sciences Center in Denver applauded the study for informing pediatricians that mothers of toddlers may easily feel depressed. However, the study has shortcomings, she says. For example, the researchers do not compare their group to mothers of older or younger children. Nor do their findings necessarily apply to women from other cultural or economic backgrounds.

The researchers are again interviewing the participants, whose children are now 4 to 5 years old. They expect to find happier moms. They are also investigating how marital satisfaction affects mothers' depression. — *T. Adler*

Fig. 6.4 Articles from the *San Francisco Examiner* and *Science News* for the course packet in Section 6.6.3.

Fig. 6.4 for both articles.)

Excerpt from "Depressive symptoms and work role satisfaction in mothers of toddlers," A. L. Olson and L. A. DiBrigida, *Pediatrics* **94**, 363–367 (1994):

This study investigates maternal depressive symptoms during the second year of the child's life. To avoid over selection of problem families, we studied depressive symptoms in mothers of toddlers presenting for health supervision visits in community practices. Pediatricians currently provide advice in health supervision visits about development, sleep, and behavior problems. Maternal depression has been associated with problems in all of these areas. Thus the community pediatric practice is an important setting for investigating maternal depression.

Because most studies examine broad age ranges in childhood, we have few data of mothers when they are at most risk for depressive symptoms. The few longitudinal studies of postpartum depression show increased rates of maternal depressive symptoms and disorders in the child's second year of life. In our interviews with parents they have emphasized the year from ages 1 to 2 is the most demanding and that beyond 2 or 3 the situation improves. The constant vigilance needed to insure physical safety as well as the child's struggle for autonomy contribute to make the toddler year particularly stressful for parents.

Depression screening measures were completed by 233 mothers of toddlers (aged 12 to 24 months) at health supervision visits in two community pediatric practices in New Hampshire. Depression was evaluated with a depressive symptom screening inventory modified by Barrett, Oxford, and Gerber from the Hopkins Symptom Checklist for use in primary care population. Data were obtained on parents' socioeconomic variables, hours worked, and whether the mother was satisfied with her current role of being employed or not employed.

Depressive symptoms were present in 42% of mothers. Rates of depressive symptoms were similar in employment groups but varied significantly with work role satisfaction. When both employment and satisfaction were considered, mothers who were dissatisfied were 3.7 times more likely to be depressed. After controlling for work role satisfaction, mothers working part-time were half is likely to be depressed as mothers working full-time and not employed.

A student's summary

One-sentence summary of the newspaper article: A recent report states that 42% of mothers of children between the ages of one and two show some signs of depression, evidence that confirms the belief that raising toddlers is very difficult.

Background: Maternal depression is found in about one-third of mothers studied, with higher depression rates among mothers in poverty or those caring for handicapped children. However, recent studies have shown that high depression rates occur in women of all social strata. Because of the adverse effects of maternal depression, which include newborn irritability, parenting difficulties, severe temper tantrums, and other behavioral problems, it is important to learn which mothers are highest at risk for depression.

Objective: To determine the incidence of depressive symptoms in mothers of toddlers in community pediatric practice. The interaction of employment and work role satisfaction with depressive symptoms was also investigated.

Kind of study: Observational study

Subjects: 233 mothers of toddlers (aged 12 to 24 months) who brought their children to health supervision visits at two community pediatric practices in New Hampshire.

Setting: Depression screening measures were completed in two community pediatric practices in New Hampshire.

Outcome measurements: Depression screening measures, consisting of a depressive symptom screening inventory for use in primary care populations, were conducted using a Likert scale of satisfaction. (Inventory modified from the Hopkins Symptom Checklist by Barrett, Oxman, and Gerber)

Control variables: Parents' socioeconomic variables, hours worked, and if the mother was satisfied with her current role of either being employed or unemployed

Blindness: None

Population: Mothers who brought their one-year old children to one of two community practices in New Hampshire for a health supervision visit.

Statistical methods used: Sociodemographic characteristics of the population were examined by t-tests, one-way Anova, and chi-squared. Proportions of the population with depression were compared across categories by chi-squared. The distribution scores were examined for skewness. The interaction of significant independent variables on the total depression score was determined by univariate analysis, two-way Anova after log transformation, or nonparametric analysis of two groups. One figure, three tables.

Stated conclusions: Symptoms of depression were present in 42% of mothers, and the rate of depressive symptoms varied significantly with work role satisfaction. Mothers dissatisfied with their employment role were 3.7 times more likely to be depressed. Controlling for employment satisfaction, mothers working part-time were half as likely to be depressed as both mothers working full time and mothers not employed.

Nonresponse: The questionnaire was completed by 233 mothers, 90% of which were recruited by telephone contact. The week before the visit, 332 calls were placed, with 238 agreeing to participate. 68 of the 332 could not be reached by multiple phone calls. 209 of those who agreed to fill out the questionnaire actually did so (the other 29 had some sort of scheduling problem and had to back out), and 24 other mothers were recruited at the time of a visit to the pediatric practice.

Difficulties of generalizing to the real world: The mothers included in this study are predominantly white, married and middle class, so this study may not be applicable to the population of mothers in the U.S. who do not fit the above three categories.

Other problems: Social support, specific stresses, and marital tension were not investigated in this study, but may contribute to maternal depression. Also, the authors did not include any control groups in the study (such as women of the same age with no children); instead, they used historical controls. Also, measuring the socioeconomic values for the father and mother separately resulted in large differences in values between the two, as many more women were classified "lower-class" than were men. Since families share the money earned in the household, it would have been more accurate to use a single-family measure of socioeconomic values instead of measuring for the father and mother separately.

How this relates to the rest of the scientific literature: Past studies have only examined broad age ranges in children, so minimal information existed about when mothers are most at risk for depression. Also, earlier studies that looked at depression rates have not been updated to account for the large increase in numbers of mothers in the workforce.

Questions

1. How many subjects? How were they selected?
2. What general population are these subjects representative of?
3. 42% of the subjects in the sample display symptoms of depression. The newspaper article says, "The finding strengthens the popular belief ..." But what if 42% of the mothers of children of other ages—or women who are not mothers—display symptoms of depression? (To address this issue, the report presents results from comparison groups—comparable women who are not mothers of two-year-olds. Where in the article are comparison groups mentioned? What are the groups and what are their depression rates?)
4. If you could do a study of one more comparison group, what would that group be?
5. (*This question can only be answered with the full report, not merely the excerpt given above.*) The numbers in Table 2 of the report are presented to 3 decimal places. For practical purposes, would it be OK to use just 2 decimal places? What about 1 decimal place?
6. A big thing in the report that is not mentioned in the article is a comparison

of rates of depression among different groups of mothers of two-year-olds. If you could add another paragraph to the *Science News* article to cover this topic, what would you say?

7. If you had to remove three sentences from the *Science News* article, what would they be? Why?

Answers to questions

1. 233 subjects. They were mothers of toddlers who brought their children for health supervision visits in two pediatric primary care group practices in New Hampshire. These children were between 12 and 24 months of age. They were first recruited by phone one week before their visit but were recruited at the time of the visit if phone contact was not made. Informed consent was obtained. ("Study Design: Methods," p. 364)
2. These subjects were primarily representative of the white, married, middle-class population of mothers. ("Results: Population Characteristics," p. 364)
3. The only comparison group found was the group of adult females who were screened with the same measure as this study in another New Hampshire primary care study (first paragraph of "Discussion" in article). Other characteristics of this comparison group are unknown from this article, but their depression rate (21.5%) is half of that of the mothers in this study. Other than this group, this study did not compare groups of mothers of other ages (no control group) and did not focus on these kinds of comparisons. If 42% of mothers of children of other ages or nonmothers had symptoms of depression, then nothing can be said scientifically to justify the popular belief from this study. The popular belief cannot also be disproven because the study was not focusing on comparing different groups of mothers/nonmothers. (pp. 363, 365)
4. A possible comparison group would be mothers with children aged 2 to 4 years (ages when the children are still not in school). This would maintain a situation where the mothers are the primary caretakers of the children. (opinion)
5. It would be OK to have the numbers reported to 2 decimal places because the standard deviation is big and the variation is such that the third decimal place just gives additional but insignificant information on the variation (variation is not as big). Reporting to 1 decimal place is not as acceptable but can be used because the variation is so big. (Table 2)
6. A possible paragraph would be: "This study did not focus on comparing characteristics within the groups of mothers of two-year-olds. The subject population was very similar in terms of ethnicity, marital status, socioeconomic status, and age, and so it was assumed that this group would act similarly. Any differences would be insignificant. The goal of this study is to see if only employment status is correlated with depression."
7. One possibility is to remove the third-to-last paragraph of the *Science News* article: it is not clear that these findings are different than for non-mothers.

Another student's opinions

Child-rearing is not an easy task. In particular, women who work and are dissatisfied with work are 3.7 times more likely to be clinically depressed than women who were unemployed or were satisfied at work. Thus for women, it appears young infants are a risk factor for developing depression and that lack of job satisfaction makes them vulnerable to depression. Their main conclusion claims that the amount of time mothers spend working is the biggest factor affecting depression; the more time spent working, the greater the likelihood of becoming depressed.

Interestingly, the journal article mentions studies of postpartum depression (after childbirth) increasing during years 1–2, considered the "most demanding" and that after ages 2–3, "the situation improves." Thus, the researchers studied women who were exposed to the greatest difficulties in motherhood and saw how they coped. Instead of simply recording women's attitudes and symptoms of depression, I believe it would have been more interesting had some form of counseling/family therapy treatment been introduced as a subgroup to study. If successful, the results would highlight the need for post-natal care, and if not, the results would stress further, the potential risks of developing depression during the toddler's first two years. It is somewhat disconcerting to realize that more and more mothers are having to work more and more hours, thus having less time for young children, and also increasing incidence of depression. Studies such as this could help persuade more companies and corporations to realize the importance of adequate and proper child-rearing, and how greatly work impacts upon new mothers.

6.6.4 A model-based analysis: Illegal aliens put uneven load

"Illegal aliens put uneven load on states," *New York Times*. "Feds, Wilson dispute illegal immigrant costs," *San Francisco Examiner*, September 15, 1994, p. A4 (see Fig. 6.5).

"Illegal immigration: impact on California," Office of Governor Pete Wilson, press packet (1994). "Fiscal impacts of undocumented aliens: selected estimates for seven states," Urban Institute (1994). (These articles are not excerpted here.)

A student's summary

Background: The federal government has established a program that provides incentives for undocumented immigrants to violate U.S. immigration laws. As a result of federal mandates, states are required to provide health care and education services to illegal immigrants and their children. Further, the federal government confers citizenship to children of parents residing illegally in the state, guaranteeing the children education, welfare, and health care. The federal government has failed to fully reimburse states for costs associated with its immigration policies.

These two reports (from the Governor's office and the Urban Institute) provide two contrasting analyses of the economic costs and benefits of illegal immigration.

Summary of the Governor's office report: Illegal immigrants place a large financial burden on the state of California. The federal government mandates states to provide services to illegal immigrants at the expense of the other residents of the state.

Summary of the Urban Institute report: The costs of public school education of illegal immigrants (undocumented aliens) are estimated for seven states: California, Arizona, New York, New Jersey, Texas, and Illinois. These states account for over 85% of the undocumented population based on estimates from the Immigration and Naturalization Service. The estimated costs from the Urban Institute Study are much lower than the estimates from the Governor's office.

Objective: To estimate the economic costs of illegal immigrants on state budgets for education (and also some estimates of costs of incarceration).

Kind of study: Economic analysis.

Measurements: Estimates of number of undocumented immigrants in the state, proportion of school age, percent attending school, and per-pupil state and local costs of education.

Statistical methods used: Estimated cost is just obtained by multiplying the estimates of the measurements given above. The estimates of the number of illegal immigrants comes from the Census Bureau (updating the April 1993 estimate from the Census Bureau and the Immigration and Naturalization Service and adding a projection up to January, 1995, based on a Census Bureau estimate of an annual illegal immigrant influx of 100 000). A 1992 Los Angeles County study was used to estimate

Feds, Wilson dispute illegal immigrant costs

By Dan Freedman
EXAMINER WASHINGTON BUREAU

WASHINGTON — Gov. Wilson, who is turning Washington's reluctance to pay for illegal immigrants into a major issue in his re-election campaign, has overestimated the cost of such services to California by $1 billion, a Clinton administration-funded study says.

Wilson said the question of the exact price tag was mere quibbling and that the study backed his main point: that California still is being stuck with a substantial bill for illegal immigration.

The administration commissioned the study last winter after Wilson and six other governors — from Arizona, Florida, New Jersey, New York, Illinois and Texas — publicly pushed for federal reimbursement of the education, medical and incarceration costs tied to illegal immigration.

Five of the seven states, including California, have sued the federal government to force reimbursement for the services, and Wilson is airing a campaign ad that focuses on the expense of schooling illegal immigrant children.

According to the study, released Wednesday by the Urban Institute, a Washington-based independent think tank, California inflated the annual cost of hospital emergency care, jails and schools for the undocumented by $1 billion. The institute said California reported that it spent $2.8 billion on the services.

Wilson, who is seeking $2.3 billion in reimbursements in the state's lawsuit against the federal government, said: "The report validates my key point: The federal failure to control the border is creating a multibillion-dollar problem for California.

"We can quibble over numbers, but the Urban Institute has effectively sent the president a bill totaling $1.8 billion for California."

Although he hasn't endorsed it, Wilson has said he will vote for a measure on the state ballot that would bar undocumented immigrants from receiving public education and other government services, other than emergency health care.

The measure, Proposition 187, is favored by 62 percent of those likely to vote Nov. 8, according to a Los Angeles Times poll published Wednesday. Twenty-nine percent said they opposed it.

Clinton administration officials used the Urban Institute report to bolster their contention that the situation wasn't as dire as state officials such as Wilson were portraying it.

"I think a careful reading of the record shows we have taken responsibility (for controlling illegal immigration) and made very significant advances," said Tom Epstein, an aide in the White House political office.

Christopher Edley, associate director of the White House Office of Management and Budget, said President Clinton had dramatically raised spending on efforts to beef up border defenses. The administration's 1995 budget request for immigration spending is nearly $6 billion more than Bush administration's 1993 budget of $19.2 billion, and California soon will receive the first of 300 additional border agents.

But the officials cautioned that a strapped federal government would be unable to provide massive reimbursement to states.

"We do not have a stack of blank checks in Washington to mail to governors for purposes of fiscal relief," Edley said.

Much of the conflict between California and Washington revolves around complex data interpretations that defy all but the most hearty of number crunchers.

Using Immigration and Naturalization Service estimates for 1992, the Urban Institute said 1.44 million undocumented immigrants lived in California, 2.9 million resided in all seven states studied, and about 3.4 million lived in the entire nation. About 120,000 more immigrants have entered California each year since then.

The institute determined that the cost of imprisoning undocumented immigrants in California had come to about $368 million in March 1994.

It also found that California had spent about $1.3 billion to educate 307 000 undocumented immigrant children in the 1993-94 school year.

California officials said Urban Institute researchers had neglected to take into account cost factors unique to the state: school debt service and added payments into the teacher retirement fund, for instance.

In a separate study released Wednesday, Wilson administration officials that California would spend $474 million this fiscal year to imprison undocumented immigrants, $1.5 billion on schooling and $395 million on health care.

Illegal immigrants also will use $1 billion worth of other state services this year, including courts, roads, parks and environmental protection, the report said.

And although Wilson has previously discredited contentions that taxes paid by undocumented immigrants partially offset the cost of the services they use, his report looked at the issue and determined they would generate state and local taxes of between $528 million and $1.4 billion in the current fiscal year.

Examiner news services contributed to this report.

Fig. 6.5 Article from the *San Francisco Examiner* for the course packet in Section 6.6.4.

the school participation rate of school-age illegal immigrant children, which took into account dropout rates. Average costs per pupil come from state sources.

Two tables, 10 graphs.

Stated conclusions: According to the Urban Institute, public schooling of undocumented alien children cost California $1.3 billion in 1994. According to the Governor's office, this cost was $1.7 billion.

How this relates to the rest of the literature: The numbers from the two organizations differ because the Urban Institute estimates the number of illegal immigrant kids in school to be 307 000 (p. 9 of that report), and the Governor's office estimates it to be 392 000 (p. 3 of that report). There is a big dispute over how to estimate the number of illegal immigrants, the proportion of them who are children, and the proportion of these children who are in school.

Questions

1. The newspaper articles describe two studies that give quite different conclusions about the costs of immigration in California. Briefly summarize the estimates from the Governor of California and the Urban Institute.
2. The two reports are long. We will focus on the costs of educating illegal immigrants in the California public schools. We have included compressed versions of the two reports here. For each of the two studies, answer the following question:
   How many illegal immigrant children are estimated to be in the California public schools? On what information is this estimate based?
3. Explain why the two reports give such different answers for the number of illegal immigrant children in school. Which numbers do you believe more? Why?
4. Explain why the two reports give slightly different values for per-student expenditures.
5. Does either newspaper article give an accurate description of why the studies give different estimates for the cost of educating illegal immigrant children in California?
6. Discuss the difficulties of writing a newspaper article about an economic analysis, as compared to a survey, experiment, or observational study.

Answers to questions

1. California initially made the estimate that the annual cost of hospital emergency care, jails and schools for the undocumented was around $2.8 billion. The Urban Institute, however, estimated that total annual costs were around $1.8 billion (*SF Examiner* article, paragraph 5). After hearing about the Urban Institute's report, California released a new study, with calculations closer to the Urban Institute's findings (*NY Times* article, paragraph 7).
2. The state of California estimates that there are 392 260 illegal immigrant children being educated in California public schools (1994–95), whereas the Urban Institute believes this figure is at 307 024 (from Table 4.13 of the report). The California figure is found using the Census Bureau's April 1993 estimate of California's illegal immigrants; this number is multiplied by the percentage of 5–17 year-old immigrants; this is then multiplied by the estimated percent attending school (from "Methodology for calculating California's costs of educating illegal immigrants"). The Urban Institute uses the same multiplication process, but it starts out with the estimates from INS 1992 data. They also use the Census data to find information on recent immigrants, and use these as "proxy populations," with the goal of finding out about the characteristics of the undocumented population.
3. The two studies get such different numbers because the Urban Institute bases its figures on the INS, and California bases its figures on the Census. According to the Urban Institute (p. 24, 2.3.5), the Census Bureau estimate is flawed because at

the national level, the Special Agricultural Workers are omitted for the population estimate of legal residents. This results in an overstatement of the number of illegal immigrants in the range of 400 000 to 1 000 000 people. Because the INS is focused on immigration issues and the Census just tries to record characteristics for all people in the U.S., I would tend to think that the INS puts more care in getting accurate illegal immigration figures.

4. These values are slightly different because each agency used a different school-year for their measurements. California made measurements using a projected 1994–95 figure, while the Urban Institute used 1993–94 figures. (Urban Institute report: Section 1.2.3 and Table 4.13)
   (Also, the Urban Institute's Table 4.13 shows that they used a value under the heading "State and local per student costs," whereas the California governor's report used a value under " 'Proposition 98' per student costs." However, in California's methodology, they state that they did use the per-pupil state and local expenditure level for their estimates, so the Urban Institute may have just incorrectly stated what numbers California used for its estimates.)
5. Only the *SF Examiner* gave a description of the reasons for the different estimates, although it was a brief and incomplete one. The article stated, "Much of the conflict between California and Washington revolves around complex data interpretations that defy all but the most hearty of number crunchers" (paragraph 7, column 2). The Urban Institute used Immigration and Naturalization Service estimates for their analysis, and California officials stated that Urban Institute researchers did not take into account school debt service and added payments into the teacher retirement fund. However, the article does not state how California calculated its estimates; we only know what the Urban Institute did and why the state of California doubted the findings.
6. It is much more difficult to write an article on an economic analysis because many of these studies are based on forecasting and estimation. In surveys, experiments and observational studies, researchers know what they are looking for and they explain this explicitly. As we can see with the immigration articles, there are many different ways to estimate a value for the cost of a certain item, such as the education of children. Thus people who write newspaper articles on topics such as immigration should try to report on the concrete evidence in economic studies, but often, this evidence in uncertain. It becomes even more difficult when multiple agencies are working on the same topic. All the reporter can do is give the conclusions of each entity, and hope that the entities have collected data in similar ways.

Another student's opinions

This is a hotly contested issue in California, and there will inevitably be many points of view. One problem with any study of this nature, is that all figures are based on estimations, and are not objectively experimental. Moreover, it is possible that the data are biased against undocumented aliens, as most research is done by the U.S. government and other agencies wishing to draw attention to a potential problem that affects all U.S. citizens.

Because illegal immigration is widespread, and burdens a whole variety of social services, I believe it is best to break down services into smaller, more manageable fragments and examine them on an individual basis, and then compare them and see how this relates to the big picture. For example, when looking at public school enrollment (Table 4.9 of the Urban Institute report), I believe it is more useful to look at a breakdown of exactly which counties are most affected and how undocumented aliens are affecting class sizes, costs to schools, etc. Then compare data from other counties.

I am rather confused by figures such as discussed in question 1—the Urban Institute estimates costs of illegal immigrants at $1.8 billion, whereas California's own initial estimate was $2.8 billion, which was later revised. How do we know which numbers to believe? Here, the discrepancy is $1 billion, not exactly something to disregard.

Because of the complexity and sheer amounts of data to look at, I'm not sure this article is entirely appropriate to teach from, and I would predict some difficulty in comprehension of the data presented.

7 Probability

We cover probability around the middle of the semester (see Section 12.3), and by that time a few probability examples have already arisen in class. In covering descriptive statistics (see Chapter 3), we discuss the concepts of one- and two-dimensional distributions along with the basic use of the normal distribution. We informally introduce probability calculations in the context of random sampling for data collection (see Section 5.1.1). This chapter has some examples and demonstrations focused specifically on probability. More theoretical probability examples appear in Chapter 15, with some of the easier problems in Section 15.2.

7.1 Constructing probability examples

We like to use examples with some inherent interest. For example, when making simple probability calculations, we prefer to work with examples such as the probability of boy and girl births or of twins (see Section 7.3.1). These sorts of examples are more interesting than poker hands and crap games. Also, using probability models of real outcomes offers good value: first you can do the probability calculations, then you can go back and discuss the potential flaws of the model (see Sections 7.4 and 9.3).

7.2 Random numbers via dice or handouts

7.2.1 Random digits via dice

Some of the demonstrations require random numbers (we have already seen some of these in Chapter 5).

At the beginning of the term, we give each student in the class a 20-sided die on which each of the digits from 0 to 9 are written twice. (These dice can be bought in a game store for about 50 cents each.) Rolling the die once gives a random digit. We have found that creating random numbers in this way is more compelling to students than using a random number table and is a convenient way to simulate a different random number for each student in the class.

7.2.2 Random digits via handouts

If it is too inconvenient or expensive to buy dice, you can prepare handouts of random numbers for your students. It is important that these numbers be easy to use and different for each student. Before the semester begins, use the computer to generate a few hundred random numbers for *each* student and print these out, with one sheet of paper per student, and hand them out on the first day of class. Each student then starts at the beginning of his or her sheet and, whenever a random digit is needed, takes the next digit and then crosses it off so it will not be used again. Sequences of random digits from 0–9 can be generated using any statistics package or on the Web (for example, at `www.random.org/nform.html`).

Another approach is to photocopy the random digits table from a statistics textbook and slice it up, handing two different lines of random numbers to each student on the first day of class, and requiring them to bring these numbers to every lecture.

7.2.3 Normal distribution

When we reach the section of the course on probability, sums of random variables, and the normal distribution, we ask the students how to create a random variable with an approximate normal distribution with pre-specified mean μ and standard deviation σ, using five rolls of the die (or five random digits). After some discussion, they can derive that the sum of five independent random digits has mean 22.5 and standard deviation $\sqrt{41.25} = 6.42$, and we inform them that the distribution is close to normal. (This can be demonstrated by asking each student in the class to roll five dice, and then displaying a histogram of the students' totals on the blackboard.) Thus, the sum of five random digits, minus 22.5, times $\sigma/6.42$, plus μ, has the desired distribution. (The students are required to bring calculators, along with their dice or individual sheet of random digits, to every class.)

7.2.4 Poisson distribution

A more difficult problem that arises in more advanced courses is creating a random sample from a Poisson(λ) distribution. We break this into two tasks: large and small λ. For large λ, we can use the normal approximation, drawing from the distribution with mean λ and standard deviation $\sqrt{\lambda}$ and rounding to the nearest integer. We ask the class: how big must λ be for this to work? Well, at the very least, we do not want to be drawing negative numbers, which suggests that the mean of the distribution should be at least two standard deviations away from zero. Thus, $\lambda > 2\sqrt{\lambda}$, so $\lambda > 4$. For smaller λ, we can compute the distribution function directly, using the formula for the Poisson density function. We can round the density function to two decimal places and then simulate using two random digits obtained by rolling a die twice.

We tell the students that, in practice, more efficient and exact simulation approaches exist; the methods here are useful for developing students' intuitions about distributions, means, and variances.

7.3 Probabilities of compound events

7.3.1 Babies

We enjoy examples involving families and babies. For example, we adapt a standard problem in probability by asking students which of the following sequences of boy and girl births is most likely, given that a family has four children: `bbbb`, `bgbg`, or `gggg`. We give them a minute to consider the question in pairs. Some students mistakenly think that `bgbg` is more likely, but the more sophisticated students realize that all three sequences are equally likely, with probabilities of 1/16.

But, in fact, that is not right either, since births are more likely to be boys than girls. For example, in the United States in 1981, there were 1 769 000 girls born and 1 860 000 boys: 48.7% of births were girls. Using this as an estimate of the probability of a girl birth, and assuming sexes of births are independent with equal probability (a reasonable assumption, in fact), it is easy to compute that $\Pr(\texttt{bbbb}) = (0.513)^4$, $\Pr(\texttt{bgbg}) = (0.513)^2(0.487)^2$, and $\Pr(\texttt{gggg}) = (0.487)^4$.

Another example with the probabilities of boy and girl births appears in Section 8.2.1.

A related topic for probability examples is twins: the historical probabilities that a birth event is identical or fraternal twins are about 1/300 and 1/125, respectively. One can then use a probability tree to calculate the probability of having two boys in a birth event, or two girls, or one boy and one girl.

For a more complicated problem, what is the probability that Elvis Presley was an identical twin? He had a twin brother who died at birth. Given this information, the probability is $\frac{1}{2} \cdot \frac{1}{300} / (\frac{1}{2} \cdot \frac{1}{300} + \frac{1}{4} \cdot \frac{1}{125}) = 5/11$ that this twin was identical.

7.3.2 Real vs. fake coin flips

Students often have difficulty thinking about summary statistics as random variables with probability distributions. This demonstration, which also alerts students to misconceptions about randomness, motivates the concept of the sampling distribution.

People generally believe that a sequence of coin flips should have a haphazard pattern, including frequent (but not regular) alternations between heads and tails. In fact, it is quite common for long runs of heads and tails to appear in sequences of random coin flips.

The demonstration proceeds as follows. We pick two students to be "judges" and one to be the "recorder" and divide the others in the class into two groups. One group is instructed to flip a coin 100 times, or flip 10 coins 10 times each, or follow some similarly defined protocol, and then to write the results, in order, on a sheet of paper, writing heads as "1" and tails as "0" (because "H" and "T" look similar and can be confused when reading them off a sheet of paper). The second group is instructed to create a sequence of 100 "0"s and "1"s that are intended to *look like* the result of coin flips—but they are to do this without flipping any coins or using any randomization device (or consulting with the other group of students)—and to write this sequence on a sheet of paper. The recorder is instructed to copy these sequences onto two blackboards. We announce that the instructor and the judges will leave the room for five minutes while the students create their sequences, and then we will return and try to guess which sequence is from actual coin flips and which was made up.

00111000110010000100	01000101001100010100
00100010001000000001	11101001100011110100
00110010101100001111	01110100011000110111
11001100010101100100	10001001011011011100
10001000000011111001	01100100010010000100

Fig. 7.1 Two binary sequences produced by students in an eighth grade class for the demonstration of Section 7.3.2. Can you figure out which is the actual sequence of 100 coin flips and which is the fake? The answer appears on page 269.

We return to the room, examine the sequences written on the two blackboards, and ask the judges to guess which sequence is real (see Fig. 7.1 for an example). We then identify real and fake sequences; almost always the identification is correct, and the students are impressed. How did we do it? Well, even as the sequences are being written on the blackboard, the students notice a difference: the sequence of fake coin flips looks "random" in an orderly sort of way, with frequent switches between 0's and 1's, whereas the sequence of real coin flips has a "streaky" look to it, with one or more long runs of successive 0's or 1's. (One could distinguish between real and fake sequences using a formal rule based on the longest run length, but we find that we can make the distinction more effectively based on a visual inspection of the sequences, which implicitly takes into account much more information.)

We picked out the real sequence using our experience and knowledge of coin flips. How can this reasoning be formalized? For each of the two sequences on the blackboards, we count the number of runs (sequences of 0's and 1's) and the length of the longest run. We then hand out copies of Fig. 7.2, which shows the *probability distribution* of these two statistics, as simulated from 2000 independent computer simulations of 100 coin flips. The students are instructed to circle on the scatterplot the locations of the values for the sequences on the blackboard. Most of the times we have used this example in class, the sequence of real coin flips is near the center of the scatterplot, and the sequence of fake coin flips has too many runs and too short a longest run, compared to this distribution.

In addition to its "magic trick" aspect, this demonstration is appealing because it dramatically illustrates an important point for the interpretation of data: seemingly surprising patterns (long sequences of heads or tails) can occur entirely at random, with no external cause. Long runs in real coin-flip data surprise students because they expect that any part of a random sequence will itself look "random"—that is, typical of the whole. This can motivate a discussion of the general phenomenon that small samples can be unrepresentative of a population. Familiar examples include biological data (for example, a family can have several boys or girls in a row) and sports (a basketball player can have "hot" and "cold" streaks that are consistent with random fluctuation).

We have also occasionally tried an alternative version of this demonstration in which the students divide into four groups, creating two sequences of real coin flips and two of fake coin flips. Here, it is much harder to tell them apart—it's a

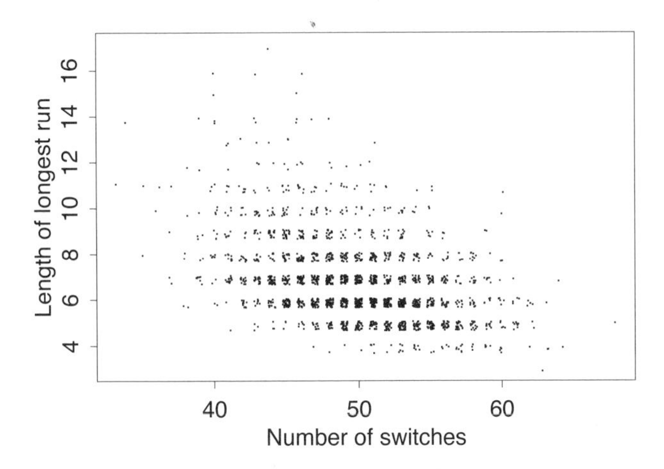

Fig. 7.2 Length of longest run (sequence of successive heads or successive tails) vs. number of runs (sequences of heads or tails) in each of 2000 independent simulations of 100 coin flips for the demonstration described in Section 7.3.2. Each dot on the graph represents a sequence of 100 coin flips; the points are jittered so they do not overlap. When plotted on this graph, the results from an actual sequence of 100 coin flips will most likely fall in an area with a large number of dots. In contrast, a sequence of heads and tails that is artificially created to look "random" will probably have too many switches and no long runs, hence will fall on the lower right of this graph.

matter of picking the one correct answer out of six possibilities, rather than one out of two—and we have occasionally picked it wrong. For maximum dramatic impact, we recommend just two groups so that it is easy to make the correct identification quickly.

After surprising students by identifying the real and fake sequences of coin flips, it is useful to develop their intuition as to why real sequences would be expected to have some long runs of heads or tails, as is indicated by Fig. 7.2. We ask the students what is the probability of having six straight heads? They quickly calculate that it is 1/2 to the sixth power, or 1/64. A sequence of 100 coin flips includes 95 sequences of length six, and thus one would expect to see one or two runs of six heads, as well as one or two runs of six tails. A sequence of seven heads occurs with probability 1/128, which is certainly a possibility given that there are 94 chances.

7.3.3 Lotteries

Lotteries are a popular source of examples for teaching probability. Surely the most important *practical* lesson we can teach our students here is not to play (see Sections 8.2.2 and problem 9 of Section 15.2.1 for illustrations of the futility of gambling). In addition, though, lotteries are a relatively simple and familiar example of extreme probabilities, and class discussion can bring up common points of misunderstanding.

For example, on September 9, 1981, the four-digit lottery number 8092 was drawn in both Massachusetts and New Hampshire. This was described as a 1 in 100 million event. However, it would be more accurate to call this a 1 in 10 000 event, since the coincidence would have been remarked upon had any of the four-digit numbers been picked. (It was not newsworthy that the number was "8092.") This can be illustrated by sketching a $10\,000 \times 10\,000$ grid on the blackboard and labeling the diagonal elements as the cases in which the two states would have the same number.

But, what about 1 in 10 000? Is this still not a surprise? Not if you consider the number of daily lottery drawings (three years is about 1000 days) and the number of possible pairs of states that could be compared. From this perspective, it is no surprise that different states will occasionally have the same lottery number on the same day, and it hardly seems newsworthy.

For a slightly more complicated example, a woman won the New Jersey lottery twice, in 1985 and 1986, and this was particularly surprising given that the chance of winning the lottery was less than 1 in a million each time. (In 1985, the lottery game involved picking six numbers from a set of 39, in which case any individual ticket had a 1 in 3.3 million chance of winning. In 1986, the lottery required picking six numbers from a set of 42, and the probability of winning was 1 in 5.2 million.) The probability of winning both lotteries is then the product of these probabilities, or 1 in 17 trillion.

How could such an unlikely event have occurred, we ask the students? To start with, as in the previous example, the surprising thing is not the first lottery win (after all, someone had to win that lottery) but rather that this woman, after winning the first lottery, won again. Thus the surprising event has a probability of 1 in 5.2 million, not 1 in 17 trillion. Another adjustment can be made because, after winning the lottery the first time, this woman played weekly for years and bought several tickets each week. If she bought ten tickets each week for ten years, that is 5200 tickets, so her probability of winning the lottery again in the next decade is about 1 in 1000. Finally, considering that there are thousands of lottery winners all over (for example, if each of 20 states has a lottery winner every week, that's 1000 winners ever year), if many of them continue to play regularly after winning, it is no surprise that some person, somewhere, will win twice.

7.4 Probability modeling

The next step beyond probability calculation is *modeling*: that is, applying probability distributions to real phenomena. We present some simplified examples here on topics in sports, social science, and engineering.

7.4.1 Lengths of baseball World Series

In the fall semester, we usually cover probability in mid-October, which is a good time to introduce the following example that has often been used as a case for probability modeling. In the baseball World Series, two teams play games

Length	Theoretical probability	Expected number (out of 92 cases)	Actual # of cases
4	1/8	11.5	18
5	1/4	23.0	20
6	5/16	28.8	20
7	5/16	28.8	34
total	1	92	92

Fig. 7.3 Frequency of lengths of best-of-7 baseball World Series, along with expected frequencies under an independent coin-flipping model. In class, we first derive the probabilities, then reveal the actual frequencies, then discuss discrepancies between theory and reality.

until one team has won four games; thus the total length of the series must be between 4 and 7 games. What is the probability of each of these lengths under the assumption that the games are independent events with each team equally likely to win?

We work this out step by step on the blackboard by enumerating all the possibilities for 4-, 5-, and 6-game series. Enumerating 7-game series seems like a lot of effort, so at this point we pause and wait for some student to recognize that Pr(7 games) = 1 − Pr(4 games) − Pr(5 games) − Pr(6 games). Interestingly, 6- and 7-game series turn out to be equally likely; is there a reason for that? We point out that, if the series has gone 5 games and is not over, then the sixth game determines whether a seventh game is played and, under our model, the two outcomes are equally likely.

The probabilities are shown in Fig. 7.3 along with the frequencies of the lengths of the first 92 series (excluding some early series that were not played under the best-of-7 rule), which we show to the students only *after* computing the theoretical probabilities.

When we do reveal the numbers in Fig. 7.3, we are in a position for a lively discussion of two aspects of lack of fit of the probability model:

- There are more four-game series than expected. Why would this be? One possibility is that teams are not always equally matched, so the assumption of 50/50 probabilities might be wrong. To illustrate, we compare the expected proportion of four-game series from the coin-flipping model, $2(0.5)^4 = 0.125$, to the proportion of four-game series we would expect if the better team had a 60% chance of winning each game: $0.4^4 + 0.6^4 = 0.155$. This comes to an expected $0.155 \cdot 92 = 14.3$ four-game series out of 92 cases, which is closer to the observed 18. (It seems unlikely that the probability of the better team winning a world-series game is any higher than 0.6, since the best baseball teams typically win only about 60% of their games against all opponents, and presumably their World Series opponent is better than average.)
- There are many more seven-game than six-game series, despite their having equal theoretical probabilities. Why has this been happening? Student ex-

planations have included home-field advantage, the scheduling of the pitching rotation, and even collusion (a longer series will sell more tickets and advertising).

This is a fun example because it interests many of the students and illustrates how we make approximations that are not quite true, and it is useful to see where the approximations fail. Section 8.5.5 presents a formal test of the model's fit.

7.4.2 Voting and coalitions

What is the probability that your vote will be decisive in a Presidential election? (The answer depends on the closeness of the election and also on which state you vote in.) Or, for a simpler example, suppose a class of 25 students is to vote on some issue (for example, whether the end-of-semester party should have pizza or Chinese food): what is the probability that any given student's vote is decisive in this choice?

How can this probability be calculated? It is worth spending some time discussing it. A start is to assume that each student's vote is equally likely to go in either direction—in this case, the probability that 24 votes are split evenly (so that you, the 25th voter, are decisive) is, from the binomial distribution, $\frac{24!}{12!12!}(0.5)^{24} = 0.16$. What if you think pizza is more popular than Chinese food? If pizza is preferred by 60% of the general population from which this class is considered a sample, then the probability of a 12-12 vote split is $\frac{24!}{12!12!}(0.6)^{12}(0.4)^{12} = 0.10$.

The calculations become more subtle when coalitions are involved. For example, suppose the class of 25 has 15 women and 10 men. Now consider the following voting rule: the voters of each sex will get together and form a pact: they will make a preliminary vote within their group and then, in the general vote of all 25 students, they will all vote as a block in favor of the choice preferred by a majority of their sex. Now what is the probability that your vote is decisive? It is easy to see that the men's votes are irrelevant, since all 15 women have already agreed to vote unanimously. What if you are a woman? Then, your vote is decisive if the other women are divided 7-7, which has probability $\frac{14!}{7!7!}(0.5)^{14} = 0.21$ if all votes are equally likely, or $\frac{14!}{7!7!}(0.6)^{7}(0.4)^{7} = 0.16$ under the "60% prefer pizza" model. Either way, the women benefit from being part of a cohesive coalition.

So, we ask the students, should you bother to vote at all? Certainly there's no point if you're in the position of the men in the block voting system. Otherwise, it depends on how likely it is for the vote to be tied. (Just for a standard of reference, there were 20 000 contested elections for the U.S. Congress between 1900 and 1990, and none of these were tied. But six of the elections were within 10 votes of being tied, and 49 were within 100 votes).

7.4.3 Space shuttle failure and other rare events

The surprising explosion of the Space Shuttle in 1986 presents an interesting example of probability modeling: before the launch, the official estimated probability of the space shuttle failing was 1/1000, but then it blew up. Looking back at the historical data gave a more reasonable probability estimate of about 2%.

The problem was that the 1/1000 probability came from a "fault-tree" model, estimating the probability of each component failing and then combining these to compute the probability that the system as a whole would fail in some way. The inherent problem with this approach is that it did not include the probability of a failure they had not thought of! In fact, any potential problem of which they were aware would have been fixed; hence the probability estimate was bound to be too low.

It is more reasonable to estimate the probability of failure empirically, by considering the failure rate of previous rocket launches. Such a calculation was, in fact, done at the time—yielding a probability estimate of 2%—but it was brushed aside.

For another example of the empirical estimation of the probability of a rare event, the London *Observer* reported,

> In the last World Cup there was a competition offering prizes to anyone who could kick a ball through a hole in a board put up in front of a goal. 'We had to get a group of people kicking balls at the board to work out the odds,' [insurance executive] O'Reilly said.

We discuss other examples of small probabilities in Sections 7.4.2 and 13.1.4.

7.5 Conditional probability

7.5.1 What's the color on the other side of the card?

We have found the following standard probability example to be well-suited to classroom use. Before coming to class, we prepare several sets of three cards: each set has one card that is blue on both sides, one that is pink on both sides, and one that is blue on one side and pink on the other. We keep one set for ourselves, for demonstration purposes, and hand out the other sets to the class (one set for each student). Using our set, we place our three cards in a hat, then select one card and, without looking, tape the selected card to the blackboard.

Suppose the side of the card facing the class is blue. We then ask the class what the probability is that the other side of the card on the blackboard is also blue (if the side facing the class is pink, then we ask for the probability that the other side of the card is also pink). The students make their guesses and give their reasons for them.

The typical guessed probability is 1/2. The actual probability, however, is 2/3, which can be easily seen by considering the card selection as a random choice among the six possible card *sides* that could have been picked. Of these six card sides, three are blue, and two of those three sides have blue on the other side: thus, the conditional probability that the other side is blue is 2/3. This analysis is shown as a probability tree in Fig. 7.4.

We explain the probability reasoning to the students but not immediately. Rather, we start with a simulation, using the sets of cards we have handed to the students, to demonstrate that the probability is greater than 1/2. We ask each student to randomly select one card from his or her set, without looking, and lay it on his or her desk. We ask the students with a blue card face up to

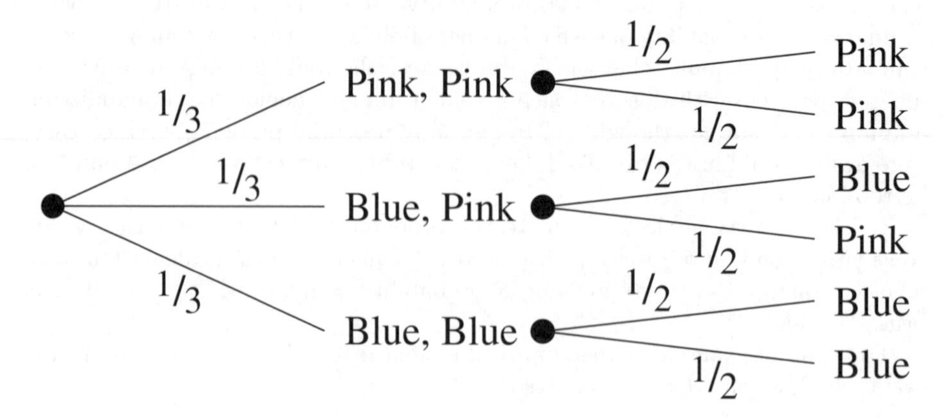

Fig. 7.4 Probability tree for the three-card example of Section 7.5.1. Show this to the students only after performing the demonstration.

raise their hands, and we count the number. For example, in a class of 30, we might see 14 students raising their hands. We now ask these students to look on the other sides of their chosen cards (while keeping their hands in the air) and, if the other side is pink, to put their hands down. We then count the number of hands remaining in the air, which might be 10 in this example. We then estimate the conditional probability of the second card being blue as the fraction of hands that stayed in the air (for example, $10/14 = 0.71$). We repeat this a few more times to convince the students that the outcome they are witnessing is not a fluke.

We then explain the probability reasoning via the tree in Fig. 7.4. This example is fun because the students see, experimentally, that their original guesses were (typically) wrong. At this point, the derivation of the correct answer using probability theory is a useful introduction to sample spaces and conditional probability.

Similar probability trees underly two other popular probability problems.

The "Monty Hall" problem is similar, and students never seem to tire of solving it. A contestant chooses one of three doors. Hidden behind two of the three doors is a donkey, and the third holds a prize. Monty Hall opens one of the two doors that the contestant has not chosen to reveal a donkey, and then proceeds to ask the contestant if she wants to switch doors. (The rules of the game require Monty to open a door that does not have the prize, and then he must give you the option to switch.) It behooves the contestant to switch, as the probability that the door Monty didn't open has the prize behind it is now 2/3.

Another related problem involves three prisoners. Two of the three are about to be released but only the warden knows which two. One prisoner asks the warden to secretly tell him the name of one of the other two prisoners who is to be released. He argues that since he knows that at least one of the others is to be released it won't change anything if he is given this information. But once the warden complies, the prisoner regrets his request, for he now thinks that his chances of being released have dropped from 2/3 to 1/2. When we ask the class what they think of the prisoner's logic, it starts a lively class discussion.

7.5.2 Lie detectors and false positives

In teaching conditional probability, we embed a well-known example in a dramatic setting to get students directly involved with the problem. The scenario is as follows. Through accounting procedures, it is known that about 10% of the employees of a store are stealing. We pick two students to play the role of "managers," and the other students in the class are the "employees." The managers would like to fire the thieves, but their only tool in distinguishing them from the honest employees is a lie detector test that is 80% accurate: if an employee is a thief, he or she will fail the test with probability 0.8, and if an employee is not a thief, he or she will pass the test with probability 0.8.

To simulate these conditions, each employee rolls a die on which are written the digits from 0 to 9 (or random digits can be used; see Sections 7.2.1–7.2.2). If the die roll (or random digit) is in the range 1–9, the employee is honest; if it comes up "0," he or she is a thief. In either case, however, the employee does not reveal this outcome to anyone else. Instead, he or she rolls the die again to determine the outcome of the lie detector test. If the die roll is in the range 2–9, the lie detector gives the correct answer ("pass" for an honest employee, "fail" for a thief); if it comes up "0" or "1," the lie detector gives the wrong answer and records "fail" for an honest employee and "pass" for a thief. The employees who have failed the lie detector test are asked to raise their hands. For example, in a class of 50 students, one would expect about $50 \cdot 0.26 = 13$ students to raise their hands.

The managers are then asked, "How many of those employees do you think are thieves?" A typical response is that about 80% of those who failed the lie detector test are thieves. The students who have raised their hands are now asked to tell their true status—honest or thief—and, in fact, it generally turns out that most are honest! The mistake made by the managers is the well-known fallacy of reversing the conditional probability (that is, confusing $\Pr(A|B)$ with $\Pr(B|A)$), which is also explained in terms of neglecting the "base rate" (in this example, the rate of thieves in the general population).

We then explain the correct reasoning by drawing a probability tree that has branches indicating "honest" or "thief," each of which has branches indicating "pass" or "fail"; see Fig. 7.5. The total probability of "fail" is $(0.1)(0.8) + (0.9)(0.2) = 0.26$, and the conditional probability of "thief," given "fail," is $\Pr(\text{"thief"} \mid \text{"fail"}) = \Pr(\text{"thief" and "fail"})/\Pr(\text{"fail"}) = 0.08/0.26 = 0.31$: if you fail the test, there is only a 31% probability that you are a thief. We also explain by expressing the possible outcomes of the die rolls as a 10×10 table with the row and columns indicating the results of the first and second die rolls, respectively. The top row of the table corresponds to the thieves, and the left two columns correspond to the lie detector giving the wrong answer. It is clear

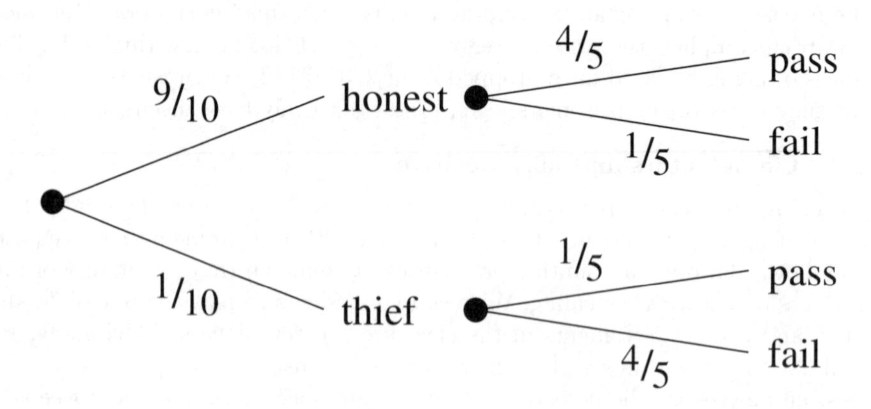

Fig. 7.5 Probability tree for the lie detector example of Section 7.5.2. Show this to students only after performing the demonstration.

that there are 18 ways to be honest and fail the lie detector test, but only eight ways to be a thief and fail.

Another standard example of this phenomenon of a high false-positive rate is in medical testing; whatever the context, we recommend that the student-participation version precede formal presentation of tree diagrams and conditional probability.

7.6 You can load a die but you can't bias a coin flip

"A coin with probability $p > 0$ of turning up heads is tossed ..."
— M. Woodroofe, *Probability With Applications* (1975, p. 108)

"Suppose a coin having probability 0.7 of coming up heads is tossed ..."
— S. Ross, *Introduction to Probability Models* (2000, p. 82)

The biased coin is the unicorn of probability theory—everybody has heard of it, but it has never been spotted in the flesh. As with the unicorn, you probably have some idea of what the biased coin looks like—perhaps it is slightly lumpy, with a highly nonuniform distribution of weight. In fact, the biased coin does not exist, at least as far as flipping goes.

We have designed classroom demonstrations and student activities around the notion of the biased coin. The simple toss of a coin offers opportunities for learning many lessons in statistics and probability. For example, we ask our students: How is a coin toss random? and What makes a coin fair? This starts a discussion of random and deterministic processes. We can design experiments and collect data to test our assumptions about coin tossing, and because flipping coins is such a simple and familiar concept, important issues surrounding experimental design and data collection are easy to spot and address.

Gambling and the art of throwing dice have a colorful history. For example, in the eleventh century, King Olaf of Norway is said to have wagered the Island of Hising in a game of chance with the King of Sweden. King Olaf beat the Swede's pair of sixes by rolling a thirteen! One die landed six, and the other split in half landing with both a six and a one showing. Stories about biased dice have continued to the present day. Ortiz (1984) gives an amusing story of an elaborate confidence game based on a rigged top. What amazes us most about this story is that people are apparently willing to bet with complete strangers in a bar on the outcome of a spinning top.

But for coins, the physical model of coin flipping (discussed in Section 7.6.3 below), which says that the "biased coin," when flipped properly, should land heads about half the time, may explain why we had trouble finding such stories about biased coins. One exception comes from the mathematician John Kerrich, who tossed a coin 10 000 times while interned in Denmark in 1941. He describes his method of tossing, "A small coin, balanced on the writer's forefinger, was given a little flip with the thumb so that it spun through the air for about a foot finally landing on a cloth spread out flat over a table ... if the coin fell heads in one spin it was convenient to balance it head uppermost on the operator's forefinger when preparing for the next, and vice versa." In addition to tossing this coin (which landed heads 5067 times), Kerrich also tossed a wooden disc that had one face coated with lead. Calling this face "tails" and the other "heads," he found the coin landed heads 679 out of 1000 times. As the coin was allowed to bounce on the table, a bias was observed.

7.6.1 Demonstration using plastic checkers and wooden dice

Unfair flipping and experimental protocol

We bring a plastic checker to class and affix putty to the crown side, which we also call the heads side. Then we ask the students whether they think the probability the checker lands heads when tossed is 1/2 or not? Most are positive that the "coin" is biased. We try flipping the checker a few times and varying the way we flip it—high, low, fast spin, no spin, like a frisbee, off kilter, bouncing on the ground, catching it in the air.

The students quickly realize that to make any sense of this probability statement it is important to specify how to flip the checker. We ask them to come up with a list of rules to follow when flipping the coin to make the flips as similar as possible. For example: begin with the crown side up and parallel to the floor; flip the coin straight up, high in the air so it spins rapidly, and with the spinning axis also parallel to the floor; catch the coin midair in the palm of hand.

We proceed from here with designing an experiment to test the hypothesis that p, the probability the coin lands heads, is 1/2. We ask questions such as, how many flips are needed to determine if the coin is not fair? These lead to discussions of hypothesis testing, significance levels, and power.

Checkers and bubble gum

Our in-class demonstration continues with a student activity on biased coins and dice. After showing the students our altered checker, and discussing how to flip it fairly, we give them a chance to make their own biased coins and dice. We divide

3 6 3 1 2 1 4 4 1 2 1 4 3 1 5 5 1 1 1 3
1 1 1 1 1 2 3 1 3 4 1 2 3 4 2 2 5 1 5 1
5 5 5 5 3 2 3 5 1 3 5 1 3 1 1 2 1 3 3 1
3 1 2 5 4 3 2 1 2 3 2 2 3 2 3 5 1 3 1 5
5 3 1 5 1 4 1 2 3 2 1 3 2 1 3 5 1 2 1 4
5 1 3 4 1 5 5 1 1 4 3 5 1 5 3 1 3 5 1 3

Fig. 7.6 The results of 120 rolls of a die that has had the edges on the 1-face rounded. In 120 rolls, the 6, which is opposite the 1, showed only once.

them into pairs and give a plastic checker, wooden die, and piece of sandpaper to each pair. We tell them that they can alter the checkers and dice however they want—for example, they can sand the edges of the die or affix gum to one or both sides of the checker (but then they should let the gum dry before handling the checker). The object is to maximize the probability of tails (or heads) for the checker, and to alter the die so the six sides are not equally probable.

We provide explicit instructions on how to flip and spin the checker and how to roll the die. To roll the die, they must find a smooth level surface and draw a circle on it about one to two feet in diameter. They shake the die in a cup and drop it into the circle. It must remain within the circle when it comes to a rest in order to count as a successful roll. The same circle is used for spinning the checker. The spins are to be contained in the circle, and they must spin quickly before falling. To flip the checker, we follow the rules set up in our earlier discussion. After our in-class demonstration, the students understand the necessity of closely following the protocol.

We also give specific instructions about the order that they are to work on the dice and checkers. First, we have them modify the die. (Although it is easy enough to do, we were surprised at how much sanding was needed to notice a big difference from what is expected in 120 flips.) Next, they alter the coin. The hitch is that we instruct them to modify the checker until they are satisfied that it is biased when spun (say when tails come up 65 or more times in 100 spins). Then they are to flip the altered checker, without making any further modifications to it. The students should find that the alterations have essentially no effect on the flips even though they have a large effect on the spins.

We ask the students to bring their modified coins and dice to the next lecture along with a record of their results. They are to roll the die 120 times, and spin and flip the checker 100 times each. Figures 7.6 and 7.7 contain the results from one student's efforts to modify her die and checker. After rounding one of the corners on her die, she rolled only one 6 in 120 throws of the die. She also found that her altered checker landed tails 77 times in 100 spins (Fig. 7.7), which has a probability of less than one in ten million of occurring with fair spins. But this same altered checker when flipped landed tails 47 times out of 100, a typical result for a "fair coin."

If we have a lot of students in the class then we would expect a few of the pairs to have significant results for the flipping activity. This is an excellent

100 spins of the checker	100 flips of the checker
01101001100000101010	11110101010000001111
10001000001010000100	00111000011000110100
00010100010000010000	01111100111001110111
00010000000001001000	00100010111100110010
00100000000000010010	10111100111101100010
(23 heads, 77 tails)	(54 heads, 46 tails)

Fig. 7.7 The results of 100 spins and 100 flips of a plastic checker which has been altered with putty. Heads are denoted by 1 and tails by 0. (We indicate Heads and Tails by 1 and 0, respectively, because "H" and "T" are hard to distinguish visually.)

opportunity to discuss the notion of multiple comparisons. Some students may insist that their checker has a probability greater than 0.5 of landing tails (or heads) when flipped. Since they have brought their checkers to class, we have them turn over their checkers to us for further investigation.

7.6.2 Sporting events and quantitative literacy

After the students have been "tricked" with the checkers, we discuss the findings of Tomasz Gliszczynski and Waclaw Zawadowski, statistics teachers at the Akademia Podlaska in Siedlce, Poland. These statisticians had their students *spin* the Belgian one-Euro coin 250 times, and they found that it came up heads 140 times. (The one-Euro coins have a common design on the tails side and a national image on the heads side. Belgium portrays King Albert II on the heads side.) As the introduction of the Euro was the largest currency switch in history, this finding received a lot of press coverage. We present some excerpts from news stories for discussion and critique.

> Memo to all teams playing Belgium in the World Cup this year: don't let them use their own coins for the toss ... "It looks very suspicious to me," said Barry Blight, a statistics lecturer at the London School of Economics. "If the coin were unbiased the chance of getting a result as extreme as that would be less than 7%."

> The academics claim Belgium Euro coins have been struck "asymmetrically" and come up tails only 44% of the time. ... With a French Euro and a limited knowledge of physics—is it best to flip a coin in the air or spin it on a table?—we set to work. ... The French flip 56% "tree." The French spin 52% "tree."

> The observation is not to be taken lightly on a sports-mad continent where important decisions can turn on the flip of a coin. ... Gliszynski says spinning is a more sensitive way of revealing if a coin is weighted than the more usual method of tossing in the air. ... But Howard Grubb, an applied statistician at the University of Reading, notes that, "with a sample of only 250, anything between 43.8 per cent and 56.2 per cent on one side or the other cannot be said to be biased." *New Scientist* carried out its own experiments with the Belgian Euro in its Brussels office. Heads came up five per cent less often than tails.

With their experience flipping and spinning their uniquely modified checkers, the students are ready to discuss these news stories. Points that quickly come to surface are: there is confusion between flipping and spinning the coin; two articles report their own experimental results, but they do not supply the number of flips or spins; and the two statisticians quoted do not agree on whether the results are suspicious or not.

7.6.3 Physical explanation

Deterministic physical laws govern what happens in the flip of a coin and the throw of a die, but we consider these events as random. It's hard to separate the random from the deterministic even in something as simple as the coin flip. What makes a coin toss fair?

The uncertainty of the coin's initial state is the key. A coin toss is basically deterministic. The coin obeys Newton's laws of motion, with its final state depending on its angular velocity (rate of spin) and time traveled (which in turn depends on the upward velocity with which it is flipped). For tosses where the coin spins rapidly and goes high in the air, the set of initial velocity values that lead to either heads or tails are of equal size. That is, half of the initial conditions lead to heads and half to tails. So, uncertainty in the initial state (for example, a smooth probability distribution on a range of values for the initial state) leads to the coin landing heads half the time.

Conservation of angular momentum tells us that once the coin is in the air, it spins at a nearly constant rate (slowing down very slightly due to air resistance). At any rate of spin, it spends half the time with heads facing up and half the time with heads facing down, so when it lands, the two sides are equally likely (with minor corrections due to the nonzero thickness of the edge of the coin); see Fig. 7.8. Weighting the coin has no effect here (unless, of course, the coin is so light that it floats like a feather): a lopsided coin spins around an axis that passes through its center of gravity, and although the axis does not go through the geometrical center of the coin, there is no difference in the way the biased and symmetric coins spin about their axes.

Jaynes (1995) describes how to add another kind of spin to the coin like the spin when you toss a frisbee, which enables you (if you are good enough at coin flipping) to have the coin, biased or not, always land heads. To prove his point, he tossed the lid of a pickle jar according to three different methods. First he tossed it with a frisbee-type twist and a very slow spin, the lid landed "heads" 99 out of 100 times. (Heads in this case is the inside of the lid.) Then he tossed the lid so that it landed on its edge and spun rapidly on the floor before falling to one side. This time the lid landed heads 0 out of 100 times, because a lopsided coin tends to fall on the side that makes the center of gravity high, and the center of gravity for the lid was closer to the top. (The lid had a diameter of 2 5/8 inches, height of 3/8 inch, and center of gravity 0.12 inches from the top of the lid.) Finally, when the pickle jar lid was tossed without any bounce or frisbee-spin, it landed heads 54 out of 100 times.

In the first method, the frisbee-style twist on the toss dominates, in the second, the bias takes over, and in the third, we have a "fair coin toss." It does not make sense to say that the coin has a probability p of heads, because the

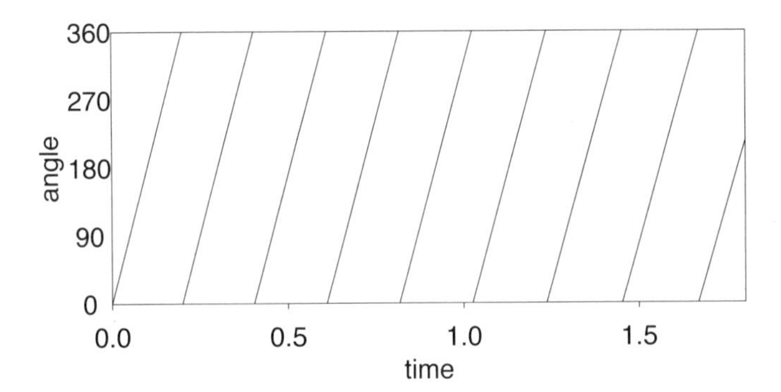

Fig. 7.8 Angular position of a flipped coin as a function of time. Suppose heads was up initially. Then, when the coin is caught, if the angle is between 90 and 270 degrees, it will show "heads" when caught and displayed; otherwise, it will show "tails." The initial condition of the coin is "forgotten" if the uncertainty about when the coin is caught is much greater than the rotation period. In this case, the coin will be in the "heads" region of angular space with probability 1/2, no matter how the coin is weighted.

outcome can be completely determined by the manner in which it is tossed—unless it is tossed high in the air with a rapid spin and caught in the air with no bouncing, in which case $p = 1/2$. If we must assign a probability p to a coin, then that probability must be approximately 1/2 (unless it is double-headed or double-tailed), no matter how it is weighted.

But dice can be "loaded" to make some faces more likely because, among other reasons, dice bounce after being thrown, and weighting and beveling can affect the bounces. As we saw with Jaynes's experiment, if a coin is spun, or if it is thrown and allowed to bounce, it can have a stable probability of heads that is not close to 1/2, and it is easy to alter this probability by shaving the edges of the coin to different angles.

8

Statistical inference

We begin this chapter with a very successful demonstration illustrating many of the general principles of statistical inference, including estimation, bias, and the concept of the sampling distribution. We then present various demonstrations and examples that take the students on the transition from probability to hypothesis testing, confidence intervals, and more advanced concepts such as statistical power and multiple comparisons.

8.1 Weighing a "random" sample

It is well known among statisticians that when you take a "haphazard" sample without using any formal probability sampling, you are likely to oversample the more accessible units. We have found students to respond well to the following demonstration based on estimating the weight of a collection of objects. This demonstration goes beyond the earlier examples of sampling bias in Section 5.1 in that we identify a sampling distribution and its bias and variance.

We pass around the room a small digital kitchen scale along with a bag which, we (truthfully) tell the class, we filled ahead of time with 100 wrapped candies of different shapes and sizes (for example, 20 full-sized candy bars and 80 assorted small candies). We divide the class into pairs and tell each pair of students to estimate the total weight of the candies in the bag by first selecting a "representative or random sample" of 5 candies out of the bag, then weighing the sample and multiplying by 20 to estimate the total weight. We ask each pair of students to write their measurement and estimate silently (so as not to influence the other students), then put their sample back in the bag, shake up the bag, and pass to the next pair. At the end, we tell the students, we will weigh the bag, and whoever has the closest estimate will get to keep all the candy.

This demonstration takes about two minutes to explain, and then it proceeds while the lecture takes place, thus giving all the students the opportunity to participate without taking away lecture time. As usual, we have the students work in pairs so they will focus more seriously on the task.

Once all the pairs of students have weighed their candies, we ask each pair to state their estimated total weight; we write all these weights on the blackboard and then display them as a histogram, as shown in Fig. 8.1. This histogram illustrates the *sampling distribution* of the estimated weights. We then get another student to weigh the entire bag (a digital kitchen scale with enough accuracy

Estimated weights		
2040	2600	2760
3240	2600	4200
3500	2827.7	3500
4200	2840	3060
2640	3600	2640
2880	2520	

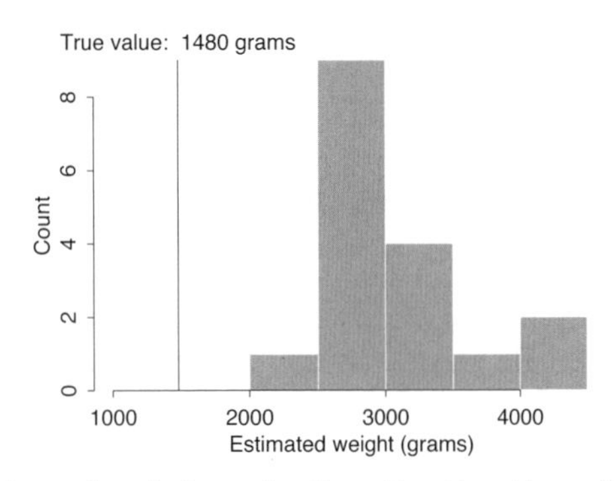

Fig. 8.1 Results of 17 pairs of eighth graders independently estimating the weight of a bag of 100 candies of varying size by selecting a sample of 5, weighing them (and returning them to the bag), and multiplying the weight by 20. (The estimate of 2827.7 was from a student trying something clever.) The histogram represents the *sampling distribution* of the estimates, and the spread in the histogram shows the *variance* of the estimate. The true weight was 1480 grams; thus the sampling distribution also includes a large *bias*. Students tend to overestimate the weight, even when they are motivated to guess accurately, because it is easier to grab the larger candies out of the bag.

to weigh 5 candies and enough range to weigh all 100 can be bought for about $50) and state the total weight. It is invariably lower than most or even all of the sample-based estimates, and this shocks the students.

Why did this happen? The students realize that the larger candies are more accessible (and also are more likely to remain on the top of the bag after it has been shaken). Even though they tried to get a representative or random sample, they could not help oversampling the large candies.

This example leads to the topic of random sampling. We ask the students how they would take a random sample of size 5 from the 100 candies. In addition, this is an excellent way to introduce the concepts of bias and variance of a sampling distribution, as shown in Fig. 8.1. We can then discuss how the bias and variance would change if (a) we switched to a random sampling approach, or (b) we increased the sample size from 5 to 10 or 20.

8.2 From probability to inference: distributions of totals and averages

A crucial and always-confusing topic in statistics is the sampling distribution of sums and means of random variables. It is helpful to explore these ideas with examples and student participation as well as algebra.

8.2.1 Where are the missing girls?

The natural probability that a baby is a girl is about 0.487 (see Section 7.3.1), and this probability has historically been very stable. The outcomes are also independent across births (except for the rare occasions of twins and triplets). Hence, students can easily calculate summaries such as the mean and standard

Bet	Example of winning numbers	Payoff
Even	2, 4, 6, 8, ..., 36	\$1 for each \$1 bet
1st 12	1, 2, 3, 4, ..., 12	\$2 for each \$1 bet
Column	3, 6, 9, ..., 36	\$2 for each \$1 bet
Line	4, 5, 6, 7, 8, 9	\$5 for each \$1 bet
Quarter	14, 15, 17, 18	\$8 for each \$1 bet
Three numbers	16, 17, 18	\$11 for each \$1 bet
Split play	30, 33	\$17 for each \$1 bet
Single play	25	\$35 for each \$1 bet

Fig. 8.2 A sample of bets available at a Nevada roulette table. The table has 38 numbers (1–36, 0, and 00) that are equally likely to occur. Each of these bets has expected values of $-2/38$. To place a "line" bet on 4–9, the wager is put at the intersection of the line between 4 and 7 and the horizontal line below the bottom row of the numbers. To place a "quarter" bet on 14–18, the wager is put at the common corner of the boxes surrounding the four numbers.

deviation of the proportion of girls in 100, or 1000, or 10 000 births. For example, in 10 000 births, there is a 95% probability that the proportion of girls will be in the range $[0.482, 0.492]$. In a million births, there is a 95% probability that the proportion of girls will be in the range $[0.486, 0.488]$.

It has been noticed, however, that in some countries the proportion of girl births is much lower, for example, 47.6% in South Korea, 46.1% in China, and below 45% in parts of India. Demographers think the discrepancy in girl births is due to infanticide or to selectively aborting female fetuses. This is an interesting example of the use of statistical regularity to learn about social processes.

We can also use this example to motivate a discussion of sample size: how many births would be needed to detect a given difference in probabilities?

8.2.2 Real-time gambler's ruin

Games of chance, especially roulette, are common examples for showing the distribution of sums. We turn these typical problems into a lively class demonstration where we wipe out the students at roulette.

We begin by passing out a diagram of the roulette table and a chart of the payoffs for various bets (Fig. 8.2). Then we ask for strategies on how to place bets, if you had \$20 to bet. Students volunteer their strategies, and we write them up on the board with their names next to their strategies. We also ask them to explain their strategies. Some want to play for as long as possible, and others want to make money fast or try out a friend's "system." Here are some of their strategies: put all \$20 down on Red; bet \$1 at a time on #17; cover the board with multiple bets on different sets of six numbers; bet \$1 on red, double the bet if you lose, and continue to double the bet until you win, then start over with a \$1 bet. If this last system worked then you would be guaranteed to make money, but there's a hitch and that is that you don't have the resources to keep doubling your bet indefinitely.

We are ready to start spinning the wheel once we have about 6 strategies on the board. Each person to volunteer a strategy is given $20 in chips. In addition to the roulette wheel, we bring to class three boxes of 100 chips: 50 white ($1), 25 red ($5), and 25 blue ($10). We enlist students to help make change, spin the roulette wheel, and make the payoffs. The gamblers come to the front of the class and start placing bets. As soon as they are wiped out they must sit down. After about a dozen spins of the wheel, one or two students are left standing. These are usually the ones with a strategy of placing only one bet. We tally the net gain for each gambler, and then we add all of their gains together and switch the sign to find the casino's net gain. It is surprising to see how easy it is for the group to lose money.

To drive the point home we finish with the following calculation. Suppose there are 100 roulette wheels, and each is spun 50 times an hour, 20 hours a day for ten days. On each spin of each wheel, $1 is bet on Red. We work out the probability that the net gain for the casino is less than $50 000. It is less than 1 in 15 million.

8.3 Confidence intervals: examples

Confidence intervals are complicated, and we introduce them in two parts, beginning in this section with some simple examples illustrating their connection with probability distributions, means, and standard deviations. We follow with demonstrations (described in Section 8.4) that use class participation to illustrate the general concept of sampling distributions of confidence intervals.

Simple one-sample and two-sample intervals can be constructed from just about any of the data collection activities previously done in the class. For example, the class can construct confidence intervals for the average handedness score (see Section 3.3.2) in the population, using the mean and standard deviation of their responses. We begin by constructing a confidence interval using the normal distribution (treating the standard deviation as known) and then go back and redo using the t distribution.

Here we discuss two particularly interesting examples derived from in-class demonstrations. For each example, we set up the problem and then allow students to work in pairs for two minutes to construct the confidence interval. Then we discuss the result and consider various perturbations (for example, how would the confidence interval change if the sample size were changed, or if we were to change the confidence level up or down from 95%).

8.3.1 Biases in age guessing

On the first day of class, the students learned some of the difficulties in guessing ages of people from photographs (see Section 2.1). For example, the first column of guesses in Fig. 2.2 shows that all ten groups of students overestimated the age of the 29-year-old person on card 1. The errors of their guesses varied from 4 to 14 years, with a mean of 9.7 and a standard deviation of 3.1. In contrast, the guessing errors for card 2 ranged from -7 to 2, with an average of -2.8 and a standard deviation of 2.9.

We ask the students: given that these are only samples of size 10, what can we say about overall biases in guessing the ages of the people on these cards? Considering the 10 groups of students as a random sample from a population yields 95% confidence intervals for the bias as $[9.7 \pm 2.26 \cdot 3.1/\sqrt{10}] = [7.5, 11.9]$ for card 1 and $[-2.8 \pm 2.26 \cdot 2.9/\sqrt{10}] = [-4.8, -0.8]$ for card 2. (The coefficients of 2.26 come from the t distribution with 9 degrees of freedom.) The confidence intervals give us some sense of the size of the biases and our uncertainty about them.

8.3.2 Comparing two groups

In the classroom experiment described in Section 5.3.1, we looked at students' estimates of the percentage of countries in the United Nations that are in Africa. Students who are given the "anchoring" value of 10 percent tend to guess lower than those given the anchor of 65 percent (see Fig. 5.16). As we discussed with the class at the time, this is an experiment, and the difference in average responses between the two groups is an estimate of the "treatment effect."

We discuss with the class: how certain are we of this treatment effect? The question can be answered with a confidence interval. For example, with the data displayed in Fig. 5.16, the estimated difference is $21 - 18 = 8$, the standard deviation of this difference is $\sqrt{13^2/20 + 15^2/23} = 4.3$, and so a 95% interval from the normal distribution is $[8 \pm 2 \cdot 4.3] = [-1, 17]$. Or, the t distribution with 19 degrees of freedom can be used (a conservative choice since the standard deviations are estimated from groups of size 20 and 23), in which case the 95% confidence interval is $[8 \pm 2.1 \cdot 4.3]$, which, once again, is $[-1, 17]$.

Usually when we do this experiment in class, the result is statistically significantly different from 0—for example, for another class, 15 students received the "$X = 10$" treatment, and they had an average response of 17 with a standard deviation of 10. The other 16 students were told "$X = 65$," and their responses averaged to 32 with a standard deviation of 19. The 95% normal-approximation confidence interval for the treatment effect was then $[4, 26]$. Because of sampling variability and small sample sizes, we never know ahead of time if a demonstration will "work," in the sense of producing statistically significant results. This is a point that can be made with the students, possibly leading to a discussion of how the demonstration could be constructed to be more likely to give statistically significant results. This leads the students to think about experimental design in the context of a specific example.

8.3.3 Land or water?

We ask the class how they might estimate the proportion of the earth covered by water. After several responses, we bring out an inflatable globe. If we were to take a random sample of points on the globe, then the proportion that touched water would be a reasonable estimate of the overall proportion of water covering the earth. Better yet, we can use a confidence interval to provide an interval estimate of the overall proportion. We explain that the globe will be tossed around the class, and we instruct students to hit the globe with the tip of their index finger when it comes to them. When they do, they are to shout "water!" if their finger touches water, or "land!" if their finger touches land. After the class starts to tire of volleyball, we can use the results to construct a confidence interval for the proportion. A discussion can follow this demonstration about the interpretation of the interval, and about possible biases in the sampling procedure (what are we assuming about consecutive hits of the globe, the weighting of the globe, and so forth), leading to a formal connection with probability, independence, and the applicability of the binomial distribution in practical problems.

For a different twist on the problem, students can be asked how one could use random numbers from the computer to pick a random spot on the globe. Any point is characterized by a latitude (which must be between 90° North and 90° South) and a longitude (between 180° West and 180° East). So can you pick a random spot on the globe by picking a latitude between -90 to 90 at random, and a longitude between -180 and 180 at random? No, because then points near the poles will be oversampled ... One must think seriously about the geometry of the problem to get the right answer here.

8.3.4 Poll differentials: a discrete distribution

In introductory statistics courses, a key use of probability theory is to derive the distribution of the mean of a set of independent random variables. This is usually illustrated in the general case with a continuous measure (for example, estimating the average height of a population of students) and also with the binomial distribution (for example, estimating the proportion of people in a population who would respond "yes" to a particular survey question).

A problem with these standard examples is that they obscure the probabilistic derivation: in the continuous example, means and variances are generally specified or estimated from data, rather than being computed using probability theory. The binomial mean and variances are generally derived using probability, but this derivation is quickly forgotten and simply becomes another formula.

An applied example that is slightly more complicated than the binomial is the "lead in the polls." For example, on August 18–19, 2000, the Gallup poll surveyed 1043 adults and asked them their preferences in the upcoming Presidential election: 47% supported Al Gore, 46% supported George Bush, and the remaining 7% supported other candidates or had no opinion. So the differential between Gore and Bush was 1%.

What would be an appropriate confidence interval for the differential? (As usual in these introductory statistics examples, we assume simple random sampling and ignore design effects and nonresponse corrections.) The problem can be solved with a little trick: for each respondent i, let the response be

$$y_i = \begin{cases} 1 & \text{for Gore supporters} \\ 0 & \text{for other or no opinion} \\ -1 & \text{for Bush supporters.} \end{cases}$$

Then Gore's lead is $\bar{y} = 0.01$, with estimated standard error $s/\sqrt{1043}$, where $s = \sqrt{0.47(1-0.01)^2 + 0.07(0-0.01)^2 + 0.46(-1-0.01)^2} = 0.96$.

8.3.5 Golf: can you putt like the pros?

As a homework assignment or an in-class activity, students can compare their abilities on some task to the professionals. For example, if you have an indoor golf kit, you can have students try five-foot putts. Before they try the putts, ask them how likely they are to get them in, and how well they think professional golfers can do.

A study from golf tournaments found that the pros made $208/353 = 59\%$ of their five-foot putts. Under the binomial model, if the success probability at a given distance is θ, then the number of successes y has a binomial distribution with parameters n and θ, and y/n has a mean of θ and a standard deviation of $\theta(1-\theta)/n$. An estimate for θ is then y/n, with an approximate standard deviation of $\sqrt{(y/n)(1-y/n)/n}$. So the success rate of professional five-foot putts is estimated at $208/353 = 0.59$ with a standard error of $\sqrt{0.59(1-0.59)/353} = 0.026$. And thus a 95% interval for the success rate is $[0.59 \pm 2(0.026)] = [0.54, 0.64]$.

A similar computation can be done from students' shots, and then the students and the pros can be compared, with an estimate and confidence interval for the difference in their abilities. Is it reasonable that the students are as good as the pros? Probably not, but the students are shooting on a flat surface indoors, which should help them.

8.4 Confidence intervals: theory

8.4.1 Coverage of confidence intervals

At the beginning of the lecture, we pass out blank slips of paper to the students and tell them to write their weights (in pounds) on their slips and put them in a hat. When the hat is filled, we ask a student volunteer to copy all the weights onto a sheet of paper. While the rest of the demonstration is going on, we ask this student to compute the mean and standard deviation of the weights. (Or, if the volunteer does not have a calculator that computes means and standard deviations, he or she can approximate these by the median and half the width of an interval that contains the central 68% of the data.)

Meanwhile, the hat (filled with the slips of paper) is passed around the room. Each pair of students is told to mix up the slips in the hat, pick out 5 slips at random, write the numbers, put the slips back in the hat, and pass the hat to the next pair. By the time this is done, the student volunteer will have finished computing the mean and standard deviation; we ask him or her to report the standard deviation (but not yet the mean) aloud.

Intervals from the normal distribution

Each pair of students should then use the mean of their sample, along with the known standard deviation, to create a 68% confidence interval for the average weight of all the students in the class (with samples of size 5, the normal distribution may not be perfect, but at least the students can do their computations fairly quickly). This can be all done while the lecture is proceeding. (Students are required to bring calculators to every class.) When all the pairs of students

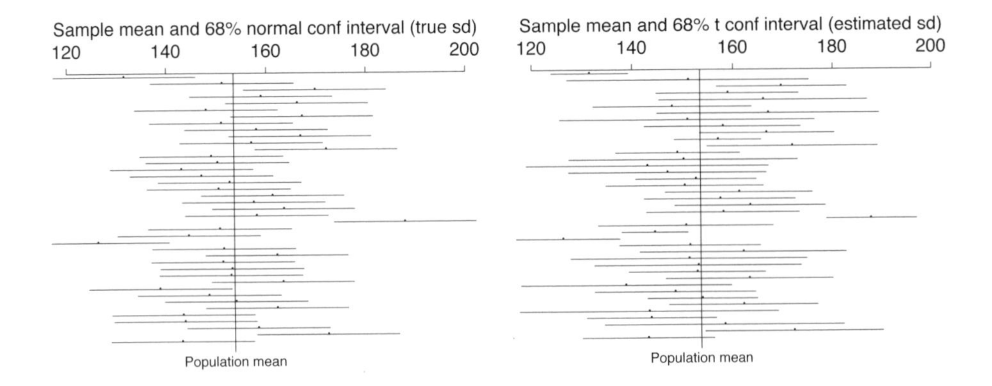

Fig. 8.3 (a) Example of confidence intervals (mean ± 1 standard error) for the average weight of students in a classroom (see Section 8.4.1). Each segment represents the 68% interval created by a different pair of students based a sample of size 5, using the normal distribution and with the population standard deviation set to its true value. The vertical line shows the true population mean (not known by the students when they are constructing their intervals). (b) Similar plot where intervals are constructed from the t distribution using estimated standard deviations. The intervals now vary in width, but approximately 68% still include the true value.

are done, they go up to the blackboard and plot their confidence intervals on an axis (as horizontal segments, stacked vertically). Figure 8.3a shows an example. (Alternatively, we can pass around the room a sheet of transparency paper with a horizontal axis labeled, along with an indelible marker, and each pair of students can draw their 68% interval on the transparency, which, when completed, can be displayed from a projector.)

Approximately 68% of the intervals should contain the true value, which is ... (we ask the volunteer for the mean of all the numbers in the hat), which we can draw as a vertical line on the blackboard display. The students should realize that the exact number of intervals that contain the true value should follow the binomial distribution. This demonstration dramatizes that the confidence interval is itself a random quantity, subject to sampling variability. The display is particularly compelling when the sampling distribution is created by the students themselves.

The instructor can of course replace "weight" by the response to some more controversial question about which the students might be particularly curious. Or the instructor can use one of the body measurements (for example, arm span) described in Section 4.2.1.

Intervals from the t distribution

We return to this example when we cover the t distribution. We then ask each student to calculate the standard deviation of his or her sample of size 5 and to then use this to get a t-interval for the unknown mean (for the t with 4 degrees

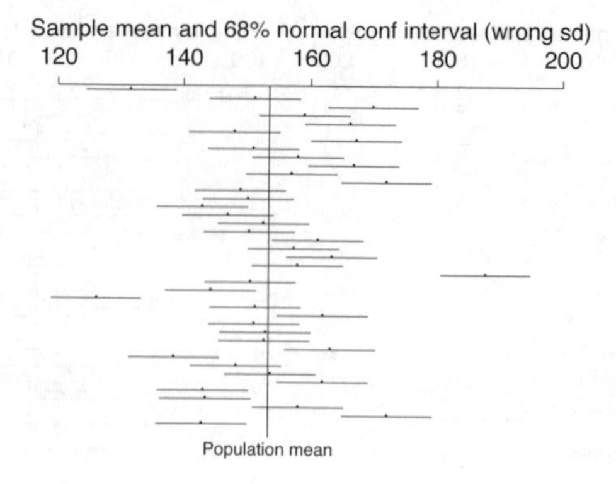

Fig. 8.4 Example of *wrong* confidence intervals (mistakenly using 1/2 of the true standard error) for the average weight of students in a classroom (see Section 8.4.2). Each horizontal line represents the interval from the sample of a different pair of students. Construct this plot with students' intervals but *without* drawing the vertical line indicating the true value. The students should be suspicious since there is *no* value for which 68% of the intervals could be correct. Compare to Fig. 8.3a.

of freedom, the 68% interval is the mean ± 1.13 standard deviations). Different students' intervals will now have different widths, but 68% of them should still contain the true population mean (see Fig. 8.3b).

8.4.2 Noncoverage of confidence intervals

We repeat the demonstration of the previous section, but this time with a different question. It is good for students to have yet another example of this basic confidence-interval calculation. This time, however, we give it a twist by giving the students a false value for the standard deviation—after gathering the responses, we compute the standard deviation but then secretly divide it by 2 before telling it to the class. The students' confidence intervals will then *not* have the correct coverage probabilities, and displaying all the intervals on the blackboard gives a picture as in Fig. 8.4, but without the vertical line since we have not yet revealed the true population mean. We ask the students if anything appears wrong and, after some discussion, point out that there is *no* true value that is consistent with these intervals—that is, no vertical line goes through approximately 68% of the students' nominal 68% intervals. We reveal to them that we have lied, and also use this to motivate the t-interval (discussed at the end of Section 8.4.1), in which the students use their own data to determine the widths of their intervals.

8.5 Hypothesis testing: z, t, and χ^2 tests

Rather than introduce new examples here, we simply return to examples we have already covered in class and replace informal comparisons with formal hypothesis testing. We usually start by referring to the demonstration in which we identified the real and fake sequences of coin flips (Section 7.3.2). This was certainly a hypothesis test, and the question arises of how to quantify it. We do not go further with that example but rather use it as a general motivation for introducing the theory of hypothesis testing, which we introduce in the context of data examples we have already discussed.

8.5.1 Hypothesis tests from examples of confidence intervals

Our next step is to construct hypothesis tests from statistical statements about confidence intervals. We already have several of these from data collected from students:

- Section 4.2.1 describes how we use data on students' heights and hand spans to illustrate a principle of correlation. The reported heights from these data can also be compared to the known average values in the United States, as given in Section 3.6.1 on page 29. For example, consider the data shown in Fig. 4.5 on page 44. The women in this sample are consistent with the U.S. average, but the men are over an inch taller on average, and the difference is statistically significant. (The 95% confidence interval for the mean is $[70.5 \pm 2(3.2)/\sqrt{29}] = [69.3, 71.7]$, which excludes the population mean of 69.1 inches.) This leads to a discussion of the reason for the discrepancy: perhaps the men in the class were overstating their heights?
- The confidence interval for the percent of the earth covered by water (Section 8.3.3) can be compared to the true value of 71%. If the confidence interval excludes the true value, then there is evidence that the sampling procedure is biased. For example, suppose the students tap the ball 45 times, and 35 of the taps touch water. The estimate is then $35/45 = 0.78$, and the approximate 95% confidence interval is $[0.78 \pm 2\sqrt{0.78(1-0.78)/45} = [0.65, 0.90]$. This interval contains 0.71, so the data are consistent with unbiased sampling.
- Conversely, a hypothesis test can be constructed from the binomial distribution given the true value of 71%, and then the class can construct a p-value. For example, with 45 trials, the number of successes has a distribution with mean $0.71 \cdot 45 = 31.9$ and standard deviation $\sqrt{0.71(1-0.71) \cdot 45} = 3.0$. We ask the students to draw this normal distribution in their notes, then insert the observed value from the classroom demonstration. For example, if 35 out of 45 hits were "water," then the p-value is the probability of a z-score being greater (in absolute value) than $(34.5 - 31.9)/3.0 = 0.87$. (The value 34.5 is used rather than 35 for the continuity correction since the data are discrete.) From the normal distribution, this p-value is $2(0.19) = 0.38$, which is not statistically significant. This should be clear to the students since the z-score of 0.87 is much less than the 95% cutoff value of 2.
- Similarly, the confidence interval for the bias in age guessing, from Section 8.3.1, becomes a hypothesis test when this confidence interval is compared to zero. Or, more directly, the confidence interval for the estimated age can be compared to the true age. In either case, the students can use a t-interval with standard error estimated from the data.

- The experiment in which two groups of students guess the proportion of countries in the United Nations that are in Africa (see Section 5.3.1) is ideal for testing the hypothesis of no difference between groups. As described in Section 8.3.2, the data collected from students (as in Fig. 5.16) can be used to construct a confidence interval for the treatment effect (as estimated by the difference in means between the two groups). The hypothesis test is simply a check to see if this confidence interval includes zero, and a p-value can be constructed based on the z-score of the observed difference compared to 0, divided by the standard error.

8.5.2 Binomial model: sampling from the phone book

The simplest scenario in which to introduce formal hypothesis testing is the binomial distribution: checking if some observed data set, consisting of y successes out of n trials, is consistent with a model in which the trials are independent, each having some specified probability p of success. We introduce this problem with data that students have generated earlier in the course.

As discussed in Section 5.1.1, when covering data collection, we coordinate a demonstration in which each pair of students in the class has to sample 10 random numbers with replacement between 1 and 126 (as part of the task of sampling addresses and telephone numbers from the phone book). Figure 5.3 shows a filled-out sampling form that made us suspect a mistake, since 7 out of 10 of the numbers were 100 or more, and under random sampling we would expect only $27/126 = 21\%$ of the numbers to be at least 100.

Could this observed difference have occurred by chance? We lead the students in a discussion of how this can be checked, and then we introduce the formal hypothesis test, which can be performed using either the exact binomial distribution or the normal approximation. From the binomial distribution, the p-value can be written as,

$$\begin{aligned}\Pr(y \geq 7) &= \binom{10}{7}0.21^7 0.79^3 + \binom{10}{8}0.21^8 0.79^2 + \binom{10}{9}0.21^9 0.79 + \binom{10}{10}0.21^{10}\\ &= 0.0013.\end{aligned}$$

Another way to test the hypothesis is using the normal approximation. Here, the expected number of successes in 10 trials (assuming the sampling had been done correctly) is 2.1, the standard deviation is $\sqrt{10(0.21)(0.79)} = 1.3$, so an approximate 95% interval for the number of successes is $[2.1 \pm 2(1.3)] = [-0.5, 4.7]$. This is clearly an approximation (since the number of successes cannot be negative) but is still useful.

The hypothesis test shows that the observed proportion of 7 out of 10 is much too large to be reasonably explained by chance, so we can confidently conclude that the 126 numbers were *not* equally likely to be sampled.

An alternative hypothesis, as described at the end of Section 5.1.1, is that the numbers were selected so that it is equally likely for any number to be in the range 1–99 or 100–126. In this case, we would expect 5 numbers in the sample to be at least 100, with a 95% interval of $[5 \pm 2\sqrt{10(0.5)(0.5)}] = [1.8, 8.2]$. Thus, the observed value of $y = 7$ is consistent with this other model.

8.5.3 Hypergeometric model: taste testing

The experimental protocol in the taste testing experiment (Section 5.3.3) implies a probability model—the hypergeometric distribution—for the number of successes, under the null hypothesis that the taste-tester is simply guessing at random. That is, if the expert has no special skills, and the eight cups of soda are identical in appearance and temperature, then the expert would just be guessing which cups contained regular or diet soda. The expert does know there are four of each type, however, and so will select four cups as regular. There are $\binom{8}{4} = 70$ possible ways to choose four cups as regular, and they are all equally likely under the null hypothesis. This means the probability of getting them all right is $1/70 = 0.014$. But it's pretty easy to get all but one correct. This probability is $16/70 = 0.23$. Our expert must correctly identify all eight cups to convince us that he or she is not just guessing.

We discuss with the class the problems that would arise if the diet and regular soda were different temperatures, different colors, or if the cups were not identical. Another point we discuss is why we use eight cups, why not 20, or two? Aside from the problem of the expert wearing out from all of the tasting, we compute various p-values for these versions of the test.

8.5.4 Benford's law of first digits

Section 5.1.2 describes the distribution of student-collected data on first digits of telephone numbers and addresses. Reasonable models for these data are a uniform distribution on digits 0–9 for the telephone numbers and a logarithmic distribution on digits 1–9 for the addresses; the latter is known as Benford's law, as is discussed in Section 5.1.2.

You can check whether the actual data collected fit the models using χ^2 tests, with 9 degrees of freedom for the telephone numbers and 8 for the addresses (since they have 10 and 9 categories, respectively). For example, Fig. 5.5 on page 53 displays data from 90 telephone numbers and 85 addresses (some of the sampled telephone book entries had no addresses listed). For the telephone numbers, the χ^2 statistic is 6.9, which is close to the expected value of 9 for 9 degrees of freedom. For the addresses, the χ^2 statistic is 14.1, which is higher than expected but not statistically significant. (The 95th percentile point for the χ^2_8 distribution is 15.5.)

8.5.5 Length of baseball World Series

In Section 7.4.1, we describe how we lead our classes through a probability model based on coin-flipping in order to model the length of the baseball World Series. As discussed in Section 7.4.1, the model does not fit the data perfectly; in particular, there have been more 4-game and 7-game series (and correspondingly fewer 5- and 6-game series) than would be predicted by the model in which the game outcomes are independent with probability 0.5.

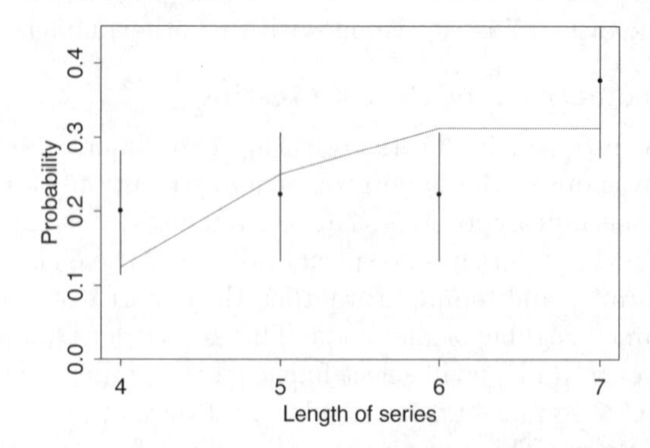

Fig. 8.5 Estimated probabilities of the baseball World Series lasting 4, 5, 6, or 7 games (see Section 7.4.1). The points are empirical estimates based on the sample data, the vertical bars are 95% intervals based on the binomial distribution at each point, and the dotted line shows the theoretical probabilities based on the coin-flipping model.

When we reach the point in the course where we cover statistical inference, we can check to see if these discrepancies are statistically significant. Figure 8.5 displays the empirical proportions of 4, 5, 6, and 7-game series, along with 95% binomial-theory confidence intervals. More formally, we can perform a χ^2 test: the test statistic is $\sum_{i=1}^{4}(\text{Observed}_i - \text{Expected}_i)^2/\text{Expected}_i = (18 - 11.5)^2/11.5 + \cdots + (34 - 28.8)^2/28.8 = 7.8$.

By comparison, the χ^2 distribution with 3 degrees of freedom has an expected value of 3 and a 95th percentile of 7.8. Thus, the discrepancy between model and data is just at the borderline of statistical significance at the 5% level.

8.6 Simple examples of applied inference

Once the basic ideas of inference have been introduced, they can be applied in many ways in the context of small projects or assignments, as we illustrate here.

8.6.1 How good is your memory?

An easy way to gather data in class is simple memory tests, as we have already illustrated in Section 4.3.1. It is well known that accuracy of short-term memory decreases rapidly when people are asked to remember more than seven items. We investigate this phenomenon using the students in the class.

We start by giving some background to students about short-term memory, anchoring the discussion with familiar points such as the relative ease of remembering 7-digit telephone numbers and the difficulty of memorizing 10 digits. We also point out individual variation in abilities such as remembering names and faces. We then tell the students that we will read aloud a sequence of seven random digits and, after 15 seconds, ask them to write the sequences on paper. We do this and count the number of students who got all seven digits correct.

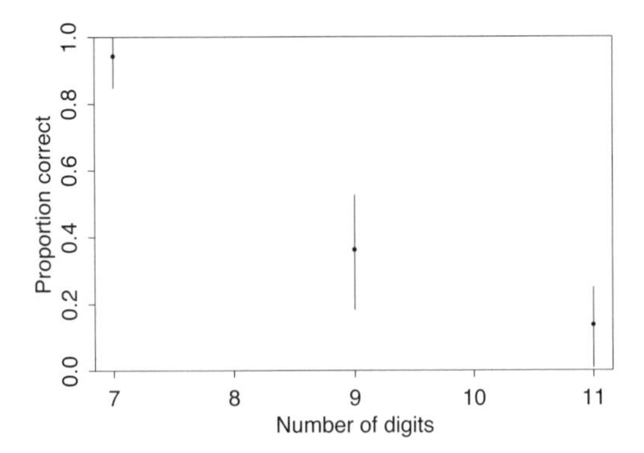

Fig. 8.6 Estimates and 95% confidence intervals, $[y/n \pm 2\sqrt{(y/n)(1-y/n)/n}]$, for the proportion of students who can accurately remember, after a 15-second delay, all of a set of random numbers read to them. Data came from a class of 31 students. The probabilities are plotted vs. the length of the set of numbers. This example illustrates the kind of applied statistical inference in which standard errors are used but not to create a single confidence interval or hypothesis test.

We then repeat with random sequences of 11 digits and of 9 digits and display the results on the blackboard. For each estimate, we display 95% error bars of $\pm 2\sqrt{p(1-p)/n}$. Figure 8.6 presents sample results: there is a decline in average performance as the number of items increases, and the size of the changes, compared to the standard errors, indicate that this pattern is clearly not explainable by chance.

Figure 8.6 is an example of inference that is not a simple confidence interval or hypothesis test but rather has aspects of both. The estimated curve and error bars act as confidence intervals in giving us a sense of the accuracy of the point estimates; at the same time, the picture as a whole implicitly rules out any models of short-term memory that are not consistent with the error bars.

8.6.2 How common is your name?

A fun mini-project or extended homework assignment is for students to estimate how many people in the country have their name (this could be done with first or last names). First, the class should break into pairs to discuss how this could be done, for example using telephone books. The estimate would only be approximate since it would necessarily involve extrapolation to the entire country, including people not listed in the book. Rules must also be set up for what constitutes an equivalent name (for example, Cathy, Kathy, and Katherine). The students would then individually gather the data and estimate the frequency of their names as homework. Finally, once the assignments have been collected and graded, the class can compare and discuss the estimated relative frequencies of the different names. They can also comments on each others' estimation methods; for example, if a student simply counts the occurrences of his or her name in the Manhattan telephone book and extrapolates, this might not be a good representation of the frequency in the entire United States.

8.7 Advanced concepts of inference

8.7.1 Shooting baskets and statistical power

We introduce the comparison of proportions with the following demonstration. We ask for two volunteers: one student who is considered to be good at basketball shooting and one who is considered to be a poor shot. They will take twenty shots each, throwing a tennis ball into a trash can. We also pick two students from the class to be "judges"; they will decide who is the better shooter. We pull out a bag of ten tennis balls, set up two trash cans at opposite sides of the room, and stand each shooter 13 feet away from a can. One shooter takes ten shots, then the other takes ten, then they repeat. We keep score on the blackboard while the judges gather up the missed shots. The results for the two shooters are compared; a typical outcome is 7/20 successes for one and 9/20 for the other.

Do the judges conclude that the second shooter is better? Suppose the shots are independent with probabilities of success p_1 and p_2. The students are led to constructing a confidence interval for $(p_1 - p_2)$. For example, if the data are 7/20 and 9/20, then the 95% confidence interval for the difference becomes $[(0.45 - 0.35) \pm 2\sqrt{0.45(1-0.45)/20 + 0.35(1-0.35)/20}] = [0.1 \pm 0.3] = [-0.2, 0.4]$.

What do the students conclude from their confidence interval? Who would they bet would make more shots in the next 20 tries? The next 200? If Bayesian methods are being covered, the students can discuss how to use the information that the students' initial self-evaluations differed.

If, as typically happens, the difference between the success rates is not statistically significant, this is a good time to introduce the idea of the statistical *power* of an experiment. We ask the class how many tries would be necessary to be likely to find a statistically significant result. 50? 100? 200? The class is led through a power calculation, beginning with guesses of the true probabilities. It becomes clear that, even if the true difference is quite large, 20 is most likely too small a sample to distinguish between the abilities of the two shooters. For example, if the true probabilities of success of the two students are 0.4 and 0.5, and each student shoots 100 baskets, then the standard deviation of the observed difference in proportions is $\sqrt{(0.4)(0.6)/100 + (0.5)(0.5)/100} = 0.071$, so that the true difference is still less than two standard errors away from zero. This has obvious consequences for experiments in other contexts (such as medical treatments) as well as real-life conclusions that we draw from small samples. Conversely, when discussing the possible results from very large samples, the students can discover the distinction between statistical and practical significance: with a huge sample size, even tiny differences can become statistically significant.

8.7.2 Do-it-yourself data dredging

In the general population, IQ is normally distributed with a mean of 100 and a standard deviation of 15. (Why is IQ normally distributed? Because the test scores are transformed so that they have a normal distribution. But that's another story.) We tell the students that we will determine their IQs. But instead of giving each student a test—that would take a lot of time—we'll have each student roll dice to simulate a random draw from the distribution. They know how to use 5 die rolls or random digits to simulate a draw from the normal distribution with mean 22.5 and standard deviation 6.42 (see Section 7.2); what is the transformation required to get a mean of 100 and standard deviation of 15? After some discussions, the students recognize that subtracting 22.5, multiplying by 15/6.42, and adding 100 will do the trick. The students roll the dice and compute their IQs. It might not be the correct IQ for each student, but the distribution is roughly normal. Now we do some comparisons. If we were to compare the average IQ of men vs. women in this group, would we find a difference? Yes—the difference would almost certainly not be exactly zero. Would it be statistically significant at the 5% level? After some discussion, the students realize that if the experiment were performed many times, with the same number of students, only 5% of the samples would have differences extreme enough to be "statistically significant." What about comparing freshmen to upperclassmen? front row vs. back row? Same answer. We now ask the students with IQs above 110 to hold up two hands, those between 90 and 110 to hold up one hand, and those below 90 to hold up no hands. Given this information, we together construct a division of the class that has almost all the high-IQ students on one side and almost all the low-IQ students on the other. It is important that the divisions of the class be based on some external criteria such as position in class, whether students wear glasses, hair color, etc. (for example, comparing men in the front row to women in the back two rows). We get the IQs for the two groups and compare and, sure enough, the difference is statistically significant! We can construct an amusing story to explain the difference (for example, the smarter women sit in the back rows because they do not need to follow the lectures carefully). But of course it is *not* real; the IQs were created by rolling dice.

We discuss the well-known implications of "data dredging" for scientific studies. For example, consider a drug company that is testing 1000 new treatments. Even if they all have no effect, 50 of them will appear to be statistically significant at the 5% level. Another example is given in Section 10.5.2.

8.7.3 Praying for your health

We were reading the Web-based magazine *Salon* one day and saw an article reporting on a study of the effectiveness of prayer on the health outcomes of 990 patients at a critical-care unit. The article continues:

> But does it do any good? Everybody's got an opinion but nobody knows for sure, because the faith–health dichotomy has never received much in the way of serious scientific scrutiny.
>
> Until now. A massive study published in the Oct. 25 issue of the Archives of Internal Medicine (a journal of the American Medical Association) showed that heart patients who had someone praying for them suffered fewer complications than other patients.
>
> ...

Dr. Harold G. Koenig, director of Duke University's Center for the Study of Religion/Spirituality and Health, has spent his entire professional life looking at how spirituality affects a person's physical well-being. This particular study is significant, he says, "because it's published in an AMA journal, it has a huge sample, and it shows significant results."

The prayed-for patients had better results, on average, than the control patients, and the difference was significant at the 4% level. On the other hand, three tests were performed on the data (see Section 8.7.2), and one might be skeptical about the findings, given that neither the patients in the study nor their doctors knew about the prayers being conducted on their behalf.

We give copies of the news report and the journal article to the students at the end of class one day with the instructions to read and prepare discussion points for the next lecture. We divide the students into two groups: the "spiritual" group and the "skeptic" group. Within each group, the students break into pairs, and then each pair is required to come up with an argument on their side of the issue (either supporting or opposing the claim of the article that intercessory prayer is good for your health).

In performing this demonstration, we have had some resistance; for example, one student said that because she is religious, she believes in prayer, and so she did not want to argue on the "skeptic" side. We discussed this with the class and pointed out that, even if you think an effect is real, this does not mean that any given study is flawless.

9 Multiple regression and nonlinear models

Near the end of the semester, we cover multiple regression, linking statistical inference to general topics such as lurking variables that arose early in the course. Many examples can be used to illustrate multiple regression, but we have found it useful to come to class prepared with a specific example, with computer output (since our students learn to run the regressions on the computer). Section 9.1 steps through a continuation of the study of height and income from Section 4.1.2. In Section 9.2 we discuss how to involve the class in models of exam scores. It is also a good strategy to simply find a regression analysis from some published source (for example, a social science journal) and go through the model and its interpretation with the class, asking the students how the regression results would have to differ in order for the study's conclusions to change.

We conclude the chapter with two examples of nonlinear models that can be used in more advanced classes: a nonlinear regression for golf putting in Section 9.3 and an illustration of the limitations of regression modeling in Section 9.4.

9.1 Regression of income on height and sex

We continue the example of predicting income from height in the national survey of adults, previously considered in Section 4.1.2. The linear regression of earnings on height yielded a positive slope for height (see Fig. 4.3), implying that taller adults have higher earnings. It seemed possible that this positive slope was explained by the lurking variable of sex: men tend to be taller and have higher incomes than women.

9.1.1 Inference for regression coefficients

The first thing to check in the original regression of earnings on height is whether the slope is statistically significant. The regression table in Fig. 4.4 shows an estimated slope of 1560 (in units of dollars per year per inch), a standard error of 130, and a 95% interval of $[1300, 1820]$. This is clearly statistically significant.

9.1.2 Multiple regression

We control for sex as a lurking variable by including it in the regression analysis; the result appears in Fig. 9.1. We walk the students through the regression output.

```
. summ
Variable |       Obs        Mean   Std. Dev.         Min         Max
---------+-----------------------------------------------------
    earn |      1379    20014.86    19763.75           0      200000
     sex |      2029    1.631345    .4825589           1           2
  height |      2021    66.56111     3.81942          57          82
. regress earn height sex
  Source |       SS       df       MS               Number of obs =     1379
---------+------------------------------            F(  2,   1376) =   101.73
   Model |  6.9335e+10     2  3.4667e+10            Prob > F      =   0.0000
Residual |  4.6892e+11  1376   340785180            R-squared     =   0.1288
---------+------------------------------            Adj R-squared =   0.1275
   Total |  5.3826e+11  1378   390606004            Root MSE      =    18460
------------------------------------------------------------------------------
    earn |      Coef.   Std. Err.       t     P>|t|     [95% Conf. Interval]
---------+--------------------------------------------------------------------
  height |   550.5448   184.5701     2.983   0.003      188.4756     912.614
     sex |  -11254.57   1448.892    -7.768   0.000     -14096.85   -8412.295
   _cons |   1617.938    14059.5     0.115   0.908     -25962.44    29198.32
------------------------------------------------------------------------------
```

Fig. 9.1 Stata output from the multiple regression of earnings on height and sex. Students should focus on the coefficient estimates, their standard errors, the R-squared, and the root mean squared error. Compare to the simple regression of earnings on height (Fig. 4.4 on page 42).

1. What are the codings for men and women? From the summary at the top of Fig. 9.1, we can see that `sex` ranges from 1 to 2. The average value is 1.63, and we know that there are more women than men in the population, so a reasonable guess is that men are coded as 1 and women as 2.
2. The summaries also show some problems with the data. Women are 63% of the sample, compared to about 52% of the adult population, so they are overrepresented in this sample. Also, there are missing responses: height is recorded for only 2021 of the 2029 respondents, and earnings are known for only 1379 people. The regression is fit to the 1379 people for whom all three variables are recorded. We ignore the missing data for the rest of the analysis, but it is important to be aware of the issue.
3. The coefficients for sex is significant: comparing two people with the same height, on average the woman's earnings are \$11 000 less than the man's. (This is the effect of increasing `sex` by 1 with the other predictor unchanged.)
4. Even after controlling for sex, height is predictive, with an increase of \$550, on average, for every additional inch.
5. The coefficients for sex and height are both statistically significant (their confidence intervals exclude zero).
6. The residual standard deviation indicates that, given height and sex, you can predict earnings to an accuracy of about \$18 000, and the R-squared tells us that height and sex explain 13% of the variance in earnings. (That is, $1 - (18\,460/19\,763)^2 = 0.13$.)

Next, students work in pairs and draw plots of earnings vs. height, with separate regression lines for men and women. The estimated regression equation is,

$$y = 1600 + 550 \cdot \text{height} - 11\,300 \cdot \text{sex}.$$

We explain how to interpret this by considering men and women separately. For men and women, the `sex` variable is 1 and 2, and so the regression lines are,

$$\begin{aligned}
\text{for men:}\quad y &= 1600 + 550 \cdot \text{height} - 11\,300 \cdot 1 \\
&= -9700 + 550 \cdot \text{height} \\
\text{for women:}\quad y &= 1600 + 550 \cdot \text{height} - 11\,300 \cdot 2 \\
&= -21\,000 + 550 \cdot \text{height}.
\end{aligned}$$

These lines are parallel, both with a slope of 550, with intercepts of -9700 (for men) and $-21\,000$ (for women). The intercepts themselves are irrelevant, though, since heights are never zero. At a height of 60 inches, the lines go through the points $-9700 + 550 \cdot 60 = 23\,300$ (for men) and $-21\,000 + 550 \cdot 60 = 12\,000$ (for women). This is a bit of reasoning that the students will appreciate more if they have to figure it out in pairs rather than simply seeing it presented by the instructor on the blackboard.

Compared to the previous regression that did not include sex as a predictor (Fig. 4.4 on page 42), the coefficient for height has decreased (this makes sense, since some of the apparent effects of height are explained by sex) but is still positive, and the residual standard deviation has decreased, but only slightly (from 18854 to 18460). The nonzero coefficient for height (after controlling for sex) is interesting and worth a minute's discussion in class.

Finally, the low R-squared implies that most of the variation in income cannot be explained by height or sex. We ask students if this is a good or bad thing. It is actually good, because factors such as qualifications, skill, and experience should be much more important than height and sex in determining income. If the R-squared were higher, this would mean that incomes were highly predictable given only height and sex.

9.1.3 Regression with interactions

The next step in the regression is to consider different slopes for men and women: for example, perhaps height is beneficial to men but not for women. Figure 9.2 shows the results for a regression that includes the interaction.

The estimated regression equation is,

$$y = -41\,000 + 1200 \cdot \text{height} + 16\,000 \cdot \text{sex} - 400 \cdot \text{height} \cdot \text{sex}.$$

Again, we interpret this equation by considering the sexes separately:

$$\begin{aligned}
\text{for men:}\quad y &= -41\,000 + 1200 \cdot \text{height} + 16\,000 \cdot 1 - 400 \cdot \text{height} \cdot 1 \\
&= -25\,000 + 800 \cdot \text{height} \\
\text{for women:}\quad y &= -41\,000 + 1200 \cdot \text{height} + 16\,000 \cdot 2 - 400 \cdot \text{height} \cdot 2 \\
&= -9000 + 400 \cdot \text{height}.
\end{aligned}$$

```
. gen hght.sex = height*sex
. regress earn height sex hght.sex
  Source |       SS        df       MS              Number of obs =     1379
---------+------------------------------           F(  3,   1375) =    68.22
   Model |  6.9738e+10      3  2.3246e+10           Prob > F      =   0.0000
Residual |  4.6852e+11   1375   340739573           R-squared     =   0.1296
---------+------------------------------           Adj R-squared =   0.1277
   Total |  5.3826e+11   1378   390606004           Root MSE      =    18459
------------------------------------------------------------------------------
    earn |      Coef.   Std. Err.       t     P>|t|     [95% Conf. Interval]
---------+--------------------------------------------------------------------
  height |   1176.099   603.7528     1.948   0.052     -8.277402    2360.475
     sex |   16012.95   25099.33     0.638   0.524     -33224.19    65250.08
hght.sex |   -403.668   370.9507    -1.088   0.277     -1131.359    324.0226
   _cons |  -41191.84   41776.58    -0.986   0.324     -123144.6    40760.89
------------------------------------------------------------------------------
```

Fig. 9.2 Stata output from the multiple regression of earnings on height, sex, and the interaction of height and sex. Compare to the simpler regressions in Figs. 4.4 and 9.1.

The estimated slope is thus 800 (that is, \$800 per inch) for men and 400 for women. But the lines cannot yet be interpreted because the intercepts apply to the meaningless condition of zero height. At a height of 60 inches, the expected earnings are $-25\,000 + 800 \cdot 60 = 23\,000$ for men and $-9000 + 400 \cdot 60 = 15\,000$ for women.

We graph these two non-parallel lines on the blackboard, a picture that shows the regression predictions more directly.

9.1.4 Transformations

The regression of earnings on height (or on sex and height) is also a good topic for a discussion of transformations. The regression model assumes linear effects, for example, each inch of height is worth \$550. Does this make sense? Perhaps the effect should be proportional, for example, each inch adding $x\%$ to your earnings. If so, what would x be? The average income in the sample is \$20 000, so an increase of \$550 is a proportional change of $550/20\,000 = 2.75\%$.

A fixed percentage (rather than a fixed dollar) effect of height on income can be modeled by regressing the *logarithm* of income on height. One would expect the coefficient to be approximately $\log(1.0275) = 0.012$ (assuming the base 10 scale of logarithms).

So why did we not fit the model on the log scale, we ask the students. Various suggestions are given, but most important is the practical problem that some people have zero earnings, and you can't take the logarithm of zero. Thus, more complicated methods would be needed to transform earnings. Such models exist (for example, mixture and tobit models), but we do not discuss them in our introductory statistics courses.

9.2 Exam scores

9.2.1 Studying the fairness of random exams

Section 5.3.2 describes an experiment in which we randomly assigned two versions, A and B, of the midterm exam to the students (without their knowledge) and then, after the grades were recorded, compared the average scores on the two exam forms. This motivated a discussion in class on the fairness of adjusting the grade. We can usefully formulate this a regression problem.

Simply comparing the two means is equivalent to regressing exam score on an indicator variable that equals 0 for students who took exam A and 1 for those who took exam B. Because of the randomization, this is a reasonable estimate of the effect of the exam. However, it is also possible that some of the observed difference can be explained by differences in abilities of the students who took the two exams. Randomization balances this factor on average, but only on average.

We ask the class, How could we tell if better students were taking exam B? What other information do we have about their abilities? One thing is their scores on their homeworks. A difference in difficulty between the two exams should appear as a nonzero coefficient on the variable "exam type," in a regression of exam score, after controlling for homework score. Including the control variable should improve our estimate of the relative difficulties of the exams.

9.2.2 Measuring the reliability of exam questions

Psychometric tools can be used for more elaborate analyses of exam grades. The goals of these psychometric methods vary, but they include assessment of individual exam questions; development of fairer and more effective measures of student abilities; and assessment of the effectiveness of the course.

For an introductory course, these ideas can be developed using simple plots and summaries. For example, for each exam question, one can compute two histograms—one for the students who got the question correct, and one for the students who got it wrong—of total scores on the other questions of the exam. If the students who got the question correct did worse, on the rest of the exam, than those who missed the question, then it is worth a closer look. This comparison can be formalized by comparing means and checking for statistical significance.

The psychometric properties of continuous exam scores can be shown using graphs, as in Fig. 9.3. Here, each of the six questions on an exam seems consistent with the other five. A scatterplot with a strong negative correlation would indicate a question where the better students did worse, which would be troubling. These sorts of methods are used to evaluate questions on standardized tests.

In a more advanced course, more formal psychometric tools can be used, such as fitting muliple linear regressions (for continuously graded questions) or logistic regressions (for items scored as correct/incorrect) predicting each student's performance on each question using his or her total score of the other questions as a predictor. Here it is desirable for the slope of the regression to be positive, implying that better students do well on this particular question (or, more precisely, that high scores on this question are associated with high scores on the rest of the exam).

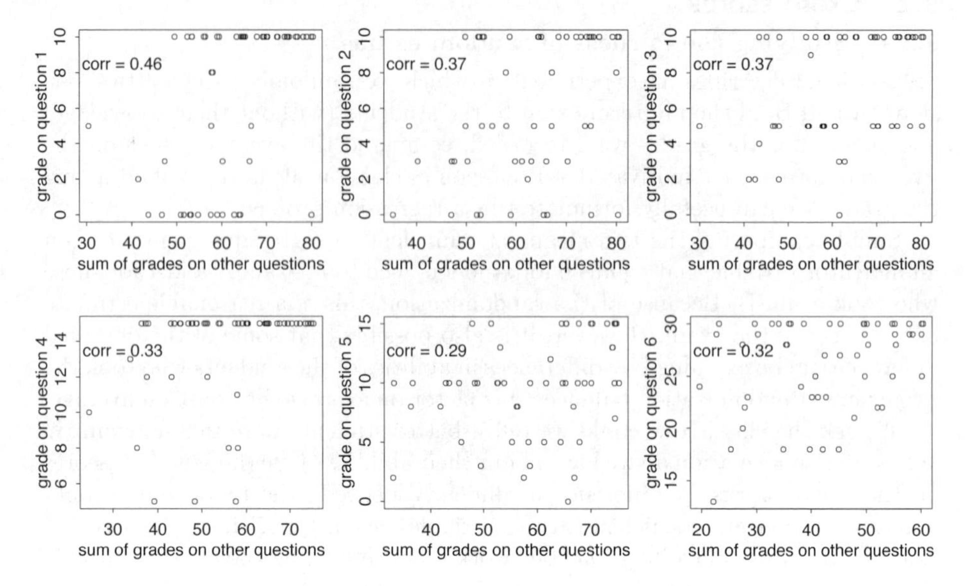

Fig. 9.3 A psychometric study of the grades on a statistics midterm exam. For each question, a scatterplot shows each student's grade on that question and his or her grade on the rest of the exam. The correlations are also given. If any question had a negative correlation, we would be worried.

These comparisons are similar to the correlations and regressions of Section 3.6.3 but more specifically focused on finding problems and patterns in exam questions, rather than simply illustrating the statistical methods. The goal is to give some feeling of applied statistics in an area—exams—with which students are familiar. This can also lead to discussion of how to make exams better, the different goals of an exam, and so forth.

Psychometric theory also gives suggestions on how to construct an exam. For example, given an exam of fixed total length, which is better: a few long questions or many short ones? For the purpose of accurately estimating student abilities, it is probably better to use many short questions, because averaging over more items reduces the standard error of the mean. Similar measurement issues arise in other demonstrations such as age guessing (Section 2.1).

9.3 A nonlinear model for golf putting

Golf is a harder game than it looks, or else five feet is further than we think. A study of professional golf players found that they made fewer than 60% of their five-foot putts. Figure 9.4a shows the success rate of golf putts as a function of distance from the hole. We use these data to motivate a derivation of a probability model. It is a fun example because it involves some trigonometry, and it gives the students a sense of the interplay between mathematics, probability, and statistics.

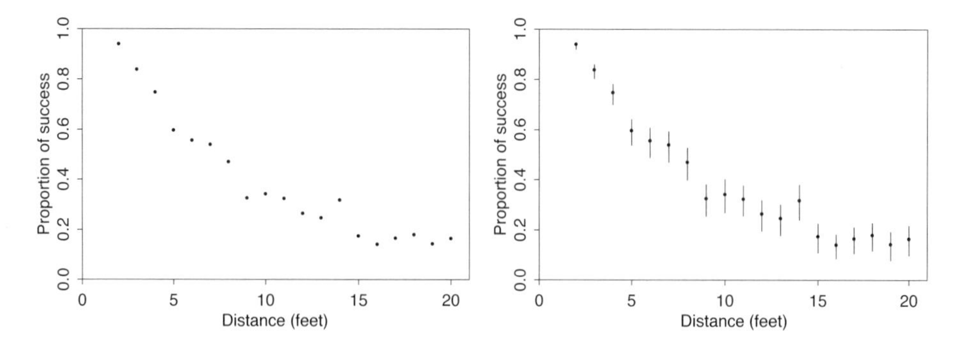

Fig. 9.4 (a) Success rate of golf putts, as a function of distance from the hole; (b) the same figure, with vertical lines showing 95% error bars based on the normal approximation to the binomial distribution.

9.3.1 Looking at data

We start by estimating the standard error of the estimated probability of success at each distance, so that we have a sense of how closely our model should be expected to fit the data. Each point in Fig. 9.4a is an estimate of the form y/n (for example, $208/253 = 59\%$ of five-foot putts were made) with an estimated standard error of $\sqrt{(y/n)(1-y/n)/n}$ (see Section 8.3.5). Figure 9.4b repeats the graph with ± 2 standard errors, which correspond to approximate 95% intervals.

9.3.2 Constructing a probability model

We now ask what sort of model could fit the data in Fig. 9.4. With the class, we discuss the possibilities. Clearly a linear regression is inappropriate, given the evident curve in the data. What about a quadratic? This runs into problems because the probabilities are bounded between 0 and 1. What should happen at the extremes of the distance x from the hole? The probability of success must have an asymptote and approach 0 as $x \to \infty$. Also, shots from zero distance must go in, and so the success probability at distance 0 must be 1.

We then sketch an idealized golf shot (see Fig. 9.5) on the blackboard. Simple trigonometry shows that the shot goes in the hole if its angle is less, in absolute value, than the threshold angle $\theta_0 = \sin^{-1}((R-r)/x)$. The students can work in pairs or small groups here to discover this relation.

How does this translate into the probability of a successful shot? The only random variable here is θ, and so we need to assign a distribution to it. A normal distribution seems reasonable (why?), presumably centered at $\theta = 0$ (assuming that shots do not systematically list to the left or the right), in which case the only parameter is the standard deviation, σ (see Fig. 9.6).

From this distribution, the probability the ball goes into the hole is,

$$\Pr(\text{success of a shot from distance } x) = 2\Phi\left[\frac{1}{\sigma}\sin^{-1}\left(\frac{R-r}{x}\right)\right] - 1,$$

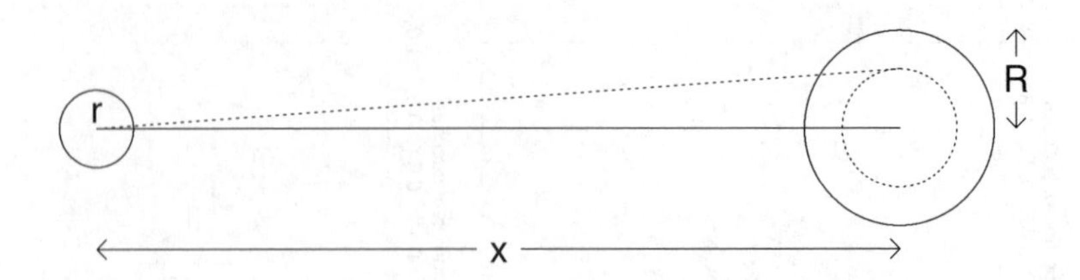

Fig. 9.5 A simple geometric model for golf putting. The ball has diameter $2r = 1.68''$ and the hole has diameter $2R = 4.25''$. The shot goes in the hole if the error in its angle is less than $\theta_0 = \sin^{-1}((R-r)/x)$; this is the angle between the solid and dotted lines in the picture.

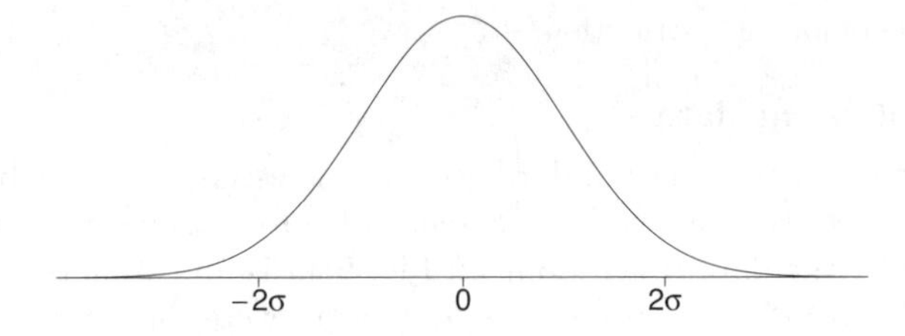

Fig. 9.6 Assumed normal distribution for the angle of error of the golf shot: $\theta = 0$ is a perfect shot, and the shot goes in the hole if $|\theta| < \theta_0$, where the threshold error, θ_0, depends on the distance x from the hole.

where Φ is the normal cumulative distribution function. (If $x < R - r$, then $\sin^{-1}((R-r)/x)$ is not defined, but in this case the model is not needed since the ball is already in the hole!)

The unknown parameter σ can be estimated by fitting to the data in Fig. 9.4. (We fit this using nonlinear least-squares, but we do not go into this in class—we just say that we fit the curve to the data.) The resulting estimate is $\hat{\sigma} = 0.026$ (which, when multiplied by $180/\pi$, comes out to $1.5°$), and the fitted curve is shown overlain on the data in Fig. 9.7. The model fits pretty well.

9.3.3 Checking the fit of the model to the data

The fit of the curve is not exact, however. As the vertical bars in Fig. 9.7 indicate, several of the 95% confidence intervals do not intersect the curve, and we can formally check this with a χ^2 test. We introduce the notation $i = 1, \ldots, 19$ for the data points in Fig. 9.4, and x_i, n_i, y_i for the distance to the hole, the number of shots attempted at this distance, and the number of successes, respectively. The Pearson χ^2 test statistic is then,

$$\chi^2 = \sum_{i=1}^{19} \frac{(y_i - \mathrm{E}(y_i))^2}{\mathrm{var}(y_i)},$$

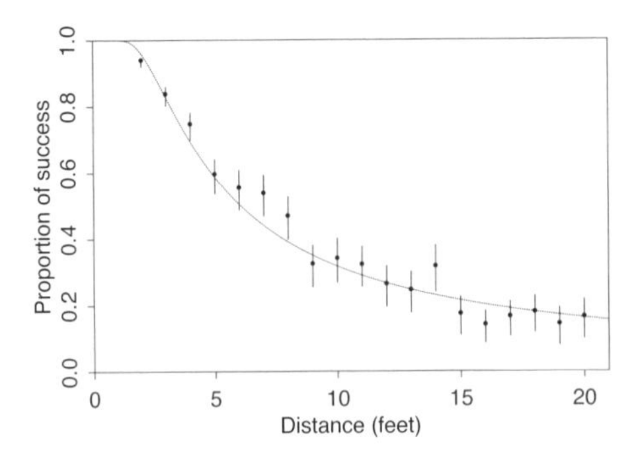

Fig. 9.7 Success rate of golf putts, as a function of distance from the hole, along with a fitted curve, Pr (success of a shot from distance x) $= 2\Phi(\sin^{-1}[(R-r)/x]\sigma) - 1$, with $\sigma = 0.026$ (that is, 1.5°) estimated from the data.

where $\mathrm{E}(y_i) = n_i \Pr(\text{success from distance } x_i)$ and $\mathrm{var}(y_i) = n_i \mathrm{E}(y_i)(1 - \mathrm{E}(y_i))$. The expected values $\mathrm{E}(y_i)$ must be calculated given the estimated parameter σ.

The value of the test statistic is 58 for the data and fitted curve in Fig. 9.7. The nonlinear least-squares method of estimating σ that we have used ensures that the approximate distribution of this statistic if the model is correct is χ^2_{18} (19 data points minus one degree of freedom for the parameter σ estimated from the data). Our result is clearly statistically significant (the 95th percentile of the χ^2_{18} distribution is 29).

So the model does not fit perfectly. Nonetheless, it seems pretty reasonable. At this point, we ask the students if they can see any reasons why the model might not be correct. One serious flaw is that the model does not allow for shots that miss because they are too short. The model also ignores that chance that a ball can fall in if it goes partly over the hole. In addition, the binomial error model assumes that the probability of success depends only on distance, which ignores variation among golf greens, playing conditions, and abilities of pro golfers. Including these complexities into the model would be difficult, but we explain that the current model, for all its flaws, yields some insight into putting and into why the curve of success probabilities looks the way it does.

This demonstration can be elaborated upon by actually bringing a putter and a golf ball into class, having the students take shots, and marking off the distribution of their error angles.

9.4 Pythagoras goes linear

A fundamental problem in statistics is that it can be easy to fit the wrong model to data but difficult to notice the problem. A memorable way to convey this to students is to give them a specially prepared dataset of several observations with an outcome variable y and two predictors x_1 and x_2 with the instructions to fit y as a function of x_1 and x_2. They will, of course, fit a linear regression model,

x1	x2	y
0.4	19.7	19.7
2.8	19.1	19.3
4.0	18.2	18.6
6.0	5.2	7.9
1.1	4.3	4.4
2.6	9.3	9.6
7.1	3.6	8.0
5.3	14.8	15.7
9.7	11.9	15.4
3.1	9.3	9.8
9.9	2.8	10.3
5.3	9.9	11.2
6.7	15.4	16.8
4.3	2.7	5.1
6.1	10.6	12.2
9.0	16.6	18.9
4.2	11.4	12.2
4.5	18.8	19.3
5.2	15.6	16.5
4.3	17.9	18.4

```
Call: lm(formula = y ~ x1 + x2)
Coefficients:
                 Value Std. Error t value Pr(>|t|)
(Intercept)  0.7145  0.6531     1.0941  0.2892
         x1  0.4939  0.0779     6.3381  0.0000
         x2  0.8641  0.0341    25.3647  0.0000

Residual std err: 0.849 on 17 degrees of freedom
Multiple R-Squared: 0.9743
```

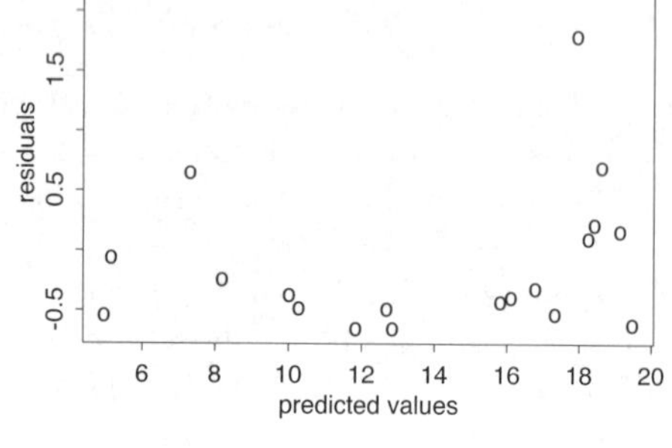

Fig. 9.8 A small dataset we give to the students and the linear regression output from S-Plus. The residual plot looks odd, but the overall fit is excellent ($R^2 = 97\%$). However, the model is not appropriate for the data, as discussed in Section 9.4. Do *not* simply give students this figure; rather, give them the data and allow them to model y as a function of x_1 and x_2. After coming up with this regression model on their own, they will be better able to appreciate its strengths and failings when they are later told the true model that generated the data.

and in this case, it will fit quite well, with an R^2 of 97%. The data, regression fit, and residual plot are shown in Fig. 9.8.

Unbeknownst to the students, however, the data were simulated from the "Pythagorean" model, $y^2 = x_1^2 + x_2^2$. We used the following code in S-Plus, using the `runif` command, which draws a random sample from a uniform distribution:

```
x1 <- runif (n=20, min=0, max=10)
x2 <- runif (n=20, min=0, max=20)
y  <- sqrt (x1^2 + x2^2)
```

It is striking that the linear model, $y = 0.71 + 0.49x_1 + 0.86x_2$ (see Fig. 9.8) fits these data so well.

What is the point of this example? At one level, it shows the power of multiple regression—even when the data come from an entirely different model, the regression can fit well. There is also a cautionary message, showing the limitations of any purely data-analytic method for finding true underlying relations. As we tell the students, if Pythagoras knew about multiple regression, he might never have discovered his famous theorem!

10
Lying with statistics

We prefer the term "statistical communication," but the phrase "how to lie with statistics" is a good hook to get students thinking about the issues involved. We try throughout to dampen the natural cynicism that comes with this topic and emphasize that, to most effectively tell the truth, you must avoid lying by accident as well as on purpose. We assign readings and also discuss several ways of lying with statistics that are not covered in usual treatments of the topic.

10.1 Examples of misleading presentations of numbers

We illustrate the difficulties of statistical communication by clipping newspaper articles and discussing them in class. (The misleading is not always done by the newspaper; in many cases, newspapers report on lying done by others.) When presenting an example, we break the class into groups of four, give them two minute to discuss, and then move to a general class discussion. In this section we give several examples of articles we have used in class.

10.1.1 Fabricated or meaningless numbers

The simplest form of lying with statistics is to simply to make up a number, such as Senator McCarthy's proclaimed (but nonexistent) list of 205 Communists, or, for a more recent example, "Patriot Missile Hits Revised from 41 to 4" (see Fig. 10.1). More subtly, numerical measurements can be used dubiously, as in the article, "Survey: U.S. Kids Reading Well" displayed in Fig. 10.1: it is not at all clear if there is a reasonable way to compare reading ability in Finnish, Hungarian, English, Chinese, and so forth. Comparison may be possible, but without more detail, it is not clear how to interpret these rankings. Amusingly, the two articles in Fig. 10.1 appeared on the very same page of the newspaper (which of course reveals the possibilities of using newspaper articles as source material for statistics classes). These examples are useful in class because the students are told so much about subtle ways of misleading with statistics that it is refreshing to remind them that simple fabrication or conceptual errors are possible too.

10.1.2 Misinformation

Perhaps the most common error involving statistics is to make a claim that is contradicted by available statistical information. It is not difficult to find such ex-

A-8 Wednesday, September 30, 1992 ★ ★ SAN FRANCISCO EXAMINER

Survey: U.S. kids reading well

No relation between skills, economics

By Dianne Henk
ASSOCIATED PRESS

ALBANY, N.Y. — Schoolchildren in the United States rank near the top of the class in a 31-nation study of basic reading skills.

Finnish students finished first in both age groups tested, ages 9 and 14, said the International Association for the Evaluation of Educational Achievement.

The United States was second among 9-year-olds and ninth among 14-year-olds. The United States trailed by small margins among 14-year-olds, said Alan Purves, a member of the research team and an education professor at the State University of New York at Albany.

The study found reading levels in a country are closely related to economic development, health and adult literacy.

"It's not necessarily whether you're richer; it's whether the kids' families have more books or not," Purves said.

"Parents clearly can help, particularly by providing books in the home," he said. "And it looks as if they can help by encouraging kids to read and reading to kids."

The study found American schools do a good job of teaching basic reading skills, although students were weaker on reading such things as maps and charts, Purves said.

The test did not address critical reading skills, such as grasp of poetry and philosophical argument.

Girls scored consistently higher in all countries and for both ages.

The report supported findings that extensive television viewing hurts reading scores.

Viewing up to two hours of television daily did not show much of an effect, Purves said. But "there's a downward slope when you get beyond about 2½ hours," he said.

READING SKILLS

Rankings from the International Association for the Evaluation of Educational Achievement for basic reading skills of 9-year-olds and 14-year-olds. Not all countries participated in both age group studies.

9-year-olds

1. Finland
2. **United States**
3. Sweden
4. France
5. Italy
6. New Zealand
7. Norway
8. Iceland
9. Hong Kong
10. Singapore
11. Switzerland
12. Ireland
13. Belgium (French)
14. Greece
15. Spain
16. West Germany
17. Canada (British Columbia)
18. East Germany
19. Hungary
20. Slovenia
21. Netherlands
22. Cyprus
23. Portugal
24. Denmark
25. Trinidad-Tobago
26. Indonesia
27. Venezuela

14-year-olds

1. Finland
2. France
3. Sweden
4. New Zealand
5. Hungary
6. Iceland
7. Switzerland
8. Hong Kong
9. **United States**
10. Singapore
11. Slovenia
12. East Germany
13. Denmark
14. Portugal
15. Canada (British Columbia)
16. West Germany
17. Norway
18. Italy
19. Netherlands
20. Ireland
21. Greece
22. Cyprus
23. Spain
24. Belgium (French)
25. Trinidad-Tobago
26. Thailand
27. Philippines
28. Venezuela
29. Nigeria
30. Zimbabwe
31. Botswana

Patriot missile hits revised from 41 to 4

GAO downgrades weapons' success against Iraqi Scuds; congressman says U.S. was misled

By David Evans
CHICAGO TRIBUNE

WASHINGTON — Patriot missiles shot down four Iraqi Scud missiles during the Persian Gulf war, far fewer than the 41 successes in 42 engagements claimed at war's end by the Defense Department, according to the latest analysis of the missile's combat performance.

In a shot-by-shot review of each missile launch and ground damage reports, General Accounting Office investigators found that Iraqi missiles were intercepted in only 9 percent of the Patriot engagements. The GAO is the investigative watchdog for Congress, and the audit of Patriot performance was done on behalf of the House Government Operations Committee.

In releasing the report Tuesday, committee Chairman John Conyers Jr., D-Mich., said: "We have watched the claims for this missile drop from 100 percent during the war to 96 percent in official statements to Congress, to 80, 70, 52, 25, and now we're under 10 percent and dropping.

"The Patriot may have hit only a few Scud warheads, and there are doubts about these. . . . The public and Congress were misled," Conyers declared.

Maj. Peter Keating, an Army spokesman, said: "The GAO report does repudiate the critics who asserted there wasn't a single-warhead kill during the war."

The Pentagon benefited enormously from the initial impression of a near-perfect missile defense. Congress increased the Patriot budget by hundreds of millions of dollars last year and pumped an additional $1 billion into the "Star Wars" global missile-defense program.

The GAO report caps a months-long debate in which independent experts have criticized the Army for inflating the Patriot's wartime performance.

The GAO report said the Army now expresses "high confidence" that 25 percent of the Patriot engagements hit Scuds, but even this claim cannot be supported by the evidence. The Army relied heavily on ground damage reports and "probable kill" messages flashed by the Patriot missiles just before they detonated.

In both cases, the data are unreliable.

The GAO investigators took a more conservative approach, giving credit for a successful intercept only if a recovered Scud contained Patriot fragments or if radar tapes showed a rapid slowdown of a Scud, indicating falling debris after an intercept.

Several of the million-dollar Patriots were fired in each engagement when it appeared that a Scud had been launched. Of the total of 158 fired during the gulf war, it now appears that half were launched at non-existent targets or at falling debris from Scuds that broke up on re-entry.

Fig. 10.1 Two articles, appearing on a single page of the *San Francisco Examiner*, illustrating potentially misleading numbers. In the top article, it is not explained how one can accurately compare reading ability for students using different languages. The bottom article discusses the possibility that the Army fabricated statistics during the Gulf War.

amples once you looking out for them. For example, we came across the following statement in the *Economist* magazine:

Back in Vietnam days, the anti-war movement spread from the intelligentsia into the rest of the population, eventually paralyzing the country's will to fight.

This common belief is in fact false: as described in Section 3.5.2, the highly educated people in the United States were in fact *more* likely to support the Vietnam War (see Fig. 3.8 and the discussion on page 266). If the students have worked this example when covering descriptive statistics, it is enlightening for them to see a reputable magazine make the same error they made themselves earlier in the semester.

In this case, the magazine's statistical error was to make a false claim without checking against readily available information on public opinion. This sort of mistake illustrates why statistical data are gathered in the first place.

10.1.3 Ignoring the baseline

A common error, whether accidental or intentional, is to compare raw numbers without adjusting for expected baseline differences. For an obvious example, it is no surprise that California has more teachers than Arizona, given that California has many more residents. A more reasonable comparison would be teachers per person in each state, or perhaps the number of teachers divided by the number of children between 5 and 18 years of age. As this simple example illustrates, it is not clear what should be the baseline, but for most purposes it is important to make some sort of adjustment.

Not adjusting for baseline occurs all the time. For example, it is commonplace for dollar trends over time to be reported without adjusting for inflation. There is some debate over the most appropriate price adjustment (see Section 3.7.2) but it can't be right to use raw dollars. A more subtle adjustment problem appears in Fig. 10.5 on page 157.

Figure 10.2 illustrates how baselines can be ignored in a map. In this map of Berkeley, California, areas were shaded that had more than 200 thefts and 75 burglaries in the previous year. We show this map to the students and ask what is wrong here; eventually they realize that it would be more appropriate to compare crime rates per population. For example, the large shaded region on the left of the map contains relatively few people; in fact, much of the shaded area is in the San Francisco Bay.

The class can continue by discussing appropriate measures of population—that is, the denominator in the crime rate—for example, perhaps burglaries should be measured per household and robberies measured with respect to the number of pedestrians who frequent the area.

10.1.4 Arbitrary comparisons or data dredging

A more subtle error is selection of data, which is also illustrated by the map in Fig. 10.2. (Once again, there is so much educational potential from a single newspaper clipping.) Setting aside the problems with population variation, the decision about where to set the threshold for shading appears arbitrary. What if

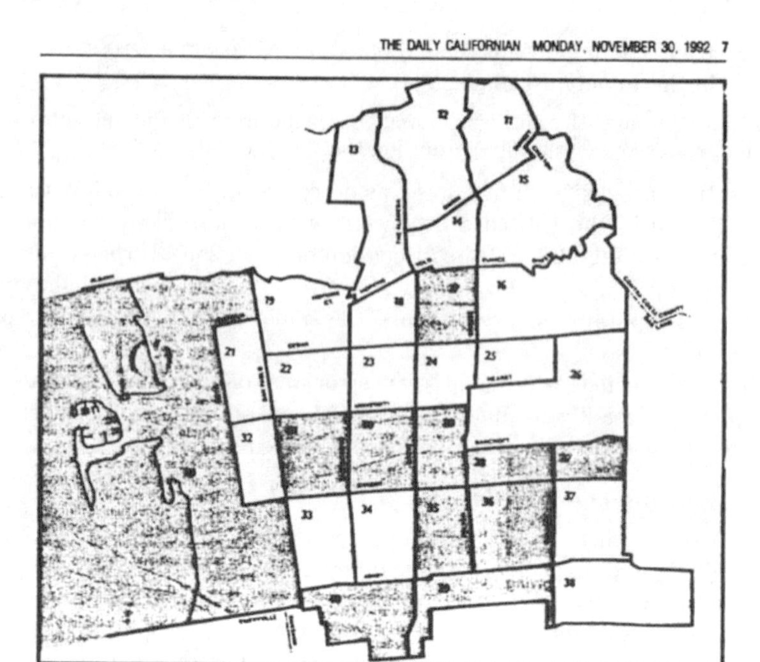

Fig. 10.2 A map, from the student newspaper of the University of California at Berkeley, illustrating several potential statistical fallacies. Areas are shaded or not shaded according to total crimes, without adjusting for possible differences in population in the different areas. In addition, the threshold—200 personal thefts and 75 burglaries—appears somewhat arbitrary, suggesting that the shading on the map might be open to manipulation.

the rule were changed to "more than 300 thefts and 50 burglaries" or "more than 100 thefts and 100 burglaries" or "more than 250 thefts or burglaries"? Unless this was some sort of preset standard, this thresholding rule seems subject to manipulation in a way similar to multiple comparisons (see Section 8.7.2).

We ask the students how they could improve the map to allay these suspicions. One suggestion has been to construct two smaller maps—one for thefts and one for burglaries—and to use four levels of shading to indicate ranges of theft or burglary rates.

For a more lighthearted example of selection, we tell the students of a comment we saw in the newspaper several years ago: "The team whose city has the tallest free-standing structure has won six of the past ten [baseball] World Series." This statement was obviously intended to be humorous, but it is interesting to debunk it. First, from the structure of the statement, we can suppose that the team with the tallest free-standing structure *lost* the World Series eleven years earlier (otherwise, the statement would presumably have been "seven of the past eleven" or "eight of the past twelve" or whatever). We thus have "six out of the

The Boston Herald Friday June 9 1989 21

Hub schools' test scores improve

By ZACHARY R. DOWDY

THE SCORES are in and Boston public school students in grades one through nine are performing at or above the national median in reading and math, but grades 10 through 12 are lagging.

School Superintendent Laval Wilson said the "Boston school system is making good progress," adding that "for the third year in a row, the Boston Public Schools has improved in reading and math."

Metropolitan Achievement Test scores released yesterday by the School Department show that the school system has jumped from having two grades performing above the national percentile to having nine grades above that percentile in reading over a four-year period.

The report said that math scores improved from having five grades above the national percentile to nine above that percentile.

Wilson said new reading programs like Focus One, an intensive reading program targeted at first graders and a new "coordinated reading program" have helped improve student performance.

Wilson said that, since 1986, first-grade performance has skyrocketed 20 points, from having students reading 11 points below the national percentile to having students reading 9 points above it.

The report also showed that students in grades 10 through 12 are not performing on the level of their peers on the national level.

The 11th grade indicated a sore spot in the school's higher levels, scoring 10 percentile points below the national percentile in reading, a 5-point increase, and eight below it in math, an 8-point increase.

Fig. 10.3 Article from the *Boston Herald* reporting improved school test scores. There is more to the story, however; see Section 10.1.5.

past eleven," which is as close to 50/50 as can be, given that you can't win half a World Series.

10.1.5 Misleading comparisons

We show our students several ways in which numbers can be juxtaposed so as to imply misleading conclusions. A favorite example, as always, involves test scores (a topic that always fascinates students). Figure 10.3 shows an example: a newspaper article we encountered several years ago in the *Boston Herald* reporting improved scores in local schools.

But there is more to the story. It is typical for school systems that test students every year to use the same standardized test form for several years before switching to a new edition. Possibly from familiarity with the test forms, average scores tend to rise during the years that a form is retained and then drop when the new test is introduced. This was described in an article the same day

DOG BITES

Go Figure At the Hall of Justice, the police recently released stats showing an 87 percent increase in homicides committed by youth. But the rise may be the result of phony number-crunching, charges the Center on Juvenile and Criminal Justice. It says the police compared 1992 stats of arrests for crimes to 1993 stats of arrests for crimes *and* attempted crimes. The cops deny the mistake but say they will redo the numbers to make sure they're right.

Honest Error Over in the Mayor's Office, spokesman Noah Griffin has been boasting that 2,100 units of affordable housing have gone up during Frank Jordan's tenure. But the figure, which has appeared in the press to show Jordan's commitment to reducing homelessness, is way off, says the mayor's Housing Director Ted Dienstfrey. "I'm embarrassed," he said, explaining the mistake was the result of a confusing memo he sent to Griffin. Documents show the real number is closer to 720 units — down from 900 during Art Agnos' last two years in office.

Cooked Club At Candlestick Park, the much-ballyhooed Giants' report showing the franchise contributes $93 million to the S.F. economy has been cooked like a hot dog, says San Franciscans for Planning Priorities. Among the flaws used to inflate the figure, the report applies employee incomes to the total figure, but many of the employees live and spend their money out of town. The report, however, could help the Giants in their quest for a new ball park.

More children than cops are shot in U.S.

Report paints shocking picture of guns' effect on kids — urges 'cease-fire'

By Katherine Seligman
OF THE EXAMINER STAFF

A child is shot to death in America every two hours, according to a new report that says youngsters are far more likely than police officers to be gunned down.

In its annual report on the state of the nation's children, the Children's Defense Fund says gun violence is taking a record toll of children. The report released Thursday by the influential advocacy group called for a "cease-fire" in the gun war on children.

"Never before has our country seen or permitted the epidemic of gun death and violence that is turning our communities into fearful armed camps and sapping the lives and hopes of our children," says the report. It pulls together data and research from state, local and federal sources.

The problem has prompted the local group Coleman Advocates to start a letter-writing campaign to urge Mayor Jordan to begin comprehensive youth programs instead of just expanding detention centers.

"We care about the victims of crime but we can't pretend this isn't a product of years of neglect," said Coleman's Carol Callen. "We need to reinvest in children."

Nationwide, 13 children are killed and at least 30 wounded by guns every day, according to the report. That's the equivalent of a classroom filled with children every day.

A police officer dies of gun violence about every 5½ days, the report says.

Juveniles account for an "appallingly high" rate of both those who kill and those who are victims. Juvenile arrests for murder and manslaughter rose by 93 percent between 1982 and 1991. And 79 percent more 10- to 17-year-olds used firearms to kill between 1980 and 1990.

"Our worst nightmares are coming true," said Children's Defense Fund President Marian Wright Edelman. "After years of epidemic poverty, joblessness, racial intolerance, family disintegration, domestic violence and drug and alcohol abuse, the crisis of children having children has been eclipsed by the greater crisis of children killing children."

Among the report's findings:

- ▶ In 1991, 5,356 children and teens died from gunshot injuries.
- ▶ A child in America is 15 times more likely to be killed by gunfire than a child in Northern Ireland.
- ▶ The number of children killed by guns from 1979 to 1991 — 50,000 — equals the number of U.S. battle casualties in the Vietnam War.

The report calls for stronger gun-control laws, safety plans to protect children and programs that keep kids off the streets.

The report finds that children have made little progress in recent years. Child poverty is still edging upward, with 21.9 percent living in poverty in 1991.

Three times as many children were reported abused or neglected in 1992 than in 1980. And 68 percent more lived in foster care in 1992 than a decade earlier.

California continues to have double the national rate of children in foster care and children who are neglected or abused.

Fig. 10.4 Newspaper clippings illustrating arbitrary and perhaps meaningless comparisons. The article on the left (from the *San Francisco Weekly*) shows several examples of official reports with questionable numbers. The article on the right (from the *San Francisco Examiner*) makes us wonder whether we should be happy or sad that more children than cops are shot.

by the rival newspaper, the *Boston Globe*:

Hub students improve in national test; Nine grades at or above average

Boston public school students who took the Metropolitan Achievement Test, one of the main yardsticks to measure academic performance, scored at or above the national average this year in nine out of 12 grades.

Some school officials touted the results, saying that the scores were the best overall showing since 1980, when officials first kept systemwide testing statistics. Last year, eight out of 12 grades met or exceeded the national average on the Metropolitan test.

"I'm very pleased that our focus on reading and math has paid off," Superintendent Laval S. Wilson said.

This sounded pretty good, but then there was a disclaimer:

But other school officials and some parents, skeptical about test scores being used as propaganda to boost the image of the school system, were cautious in interpreting the results.

The Metropolitan test scores, they said, have been climbing for the past three years because the same version has been recycled, and teachers and students are increasingly familiar with that version.

Paula Georges, director of the Citywide Educational Coalition, said that whenever a new version of the Metropolitan is introduced, scores drop nationwide. Until then, she said, "the trend is for them to go up."

But the school officials still want to take credit:

[Wilson] said the Metropolitan results, taken together with a leveling off of the drop out rate and increased promotion rates, suggest "student learning is improving."

...

Joyce Grant, deputy superintendent for curriculum, and Mary Russo, director of reading, said a major reason for the boost in scores is the effort of individual teachers.

Some other strange numerical comparisons appear in the articles shown in Fig. 10.4. For example, it is sad that 13 children were being killed per day, but it is not at all clear why this should be compared to the rate at which police officers are shot. The comparison later in the article to Northern Ireland is more reasonable (although it might be even more relevant to compare all violent deaths rather than restrict to gunshots).

10.2 Selection bias

A favorite source of statistical errors is selection bias, which can generally be categorized as a sample being unrepresentative of the population because some units are much more likely than others to be represented, with the more likely units differing from the unlikely units in some important way. Before getting to this topic, we like to discuss simpler methods of lying with statistics (see above) to make it clear that no great sophistication is required to mislead.

10.2.1 Distinguishing from other sorts of bias

Covering selection bias in class has two benefits: in addition to reminding the students of an important source of error, it gets them thinking systematically about sampling and probability as applied to real-world settings. Whenever we introduce a selection bias example, we like to stop and ask the students to calculate or guess the probability of selection for different units in the population. We can illustrate with many examples that we have already covered in class or homework:

- Surveys of World Wide Web users that overrepresent frequent users (Section 5.1.3)
- Counts of number of siblings in families, in which larger families are more likely to be selected because they have more children that could end up as students in the class and thus be counted (Section 5.1.6)
- Sampling from a bag of candy; larger candies are more likely to be selected (Section 8.1).

We also explain that there are all sorts of biases in statistics that are *not* selection bias. We have already considered in class: measurement error (as in the age-guessing demonstration on the first day of class (Section 2.1) or the United Nations experiment described in Section 5.3.1); lurking variables (as in the regression of earnings on height, ignoring sex, in Section 4.1.2), biased survey questions (see Section 5.1.3), inappropriate comparisons in observational studies (such as the biased estimates of the effect of coaching on SAT scores, described in Section 5.4.3), and multiple comparisons (in the study of the effects of prayer, described in Section 8.7.3). These all illustrate biases, but they are not *selection* bias because, in all these cases (with the partial exception of the SAT coaching example), the problem is not with sample selection but with the measurements or analysis performed on the units selected.

10.2.2 Some examples presented as puzzles

Having given students examples of selection bias and clarified the concept, we are ready with several more examples, which we have drawn from the statistical and scientific literature. We present these as puzzles: for each, we briefly describe the phenomenon; the class then tries to figure out the explanation. (Answers are given on page 272.)

1. *The most dangerous profession.* In a study in 1685 of the ages and professions of deceased men, it was found that the profession with the lowest average age of death was "student." Why does being a student appear to be so dangerous?
2. *Age and palm lines.* A study of 100 recently deceased people found a strong positive correlation between the age of death and the length of the longest line on the palm. Does this provide support for the claim that a long line on the palm predicts a long life?
3. *The clinician's illusion.* When asked to judge the severity of a syndrome among their patients, clinical psychiatrists tend to characterize the syndromes as much more serious and long-term, on average, than are estimated by surveys of patients who have the syndrome.
4. *Your friends are (probably) more popular than you are.* Sociologists have conducted surveys in which they select random people and ask for a list of

the people they know, and then they contact a sample of the friends and repeat the survey. The people sampled at the second stage have, on average, many more friends than do the people in the original sample. This suggests that, on average, your friends are more popular than you are.

5. *Barroom brawls.* A study of fights in bars in which someone was killed found that, in 90% of the cases, the person who started the fight was the one who died.

10.2.3 Avoiding over-skepticism

When concluding the discussion, we find it useful to remind the students that the existence of statistical biases does not necessarily invalidate a finding. For example, several years ago a widely publicized study found that the average age of death of left-handers was about 8 years less than that of right-handers. (The researchers started with a sample of death certificates and then interviewed close relatives or friends to find out the handedness of the deceased.) The authors of the study went on to speculate about reasons why left-handers may be more likely to die at a younger age.

This research was criticized as biased because the frequency of left-handedness may have changed appreciably over time: if many people born 60 or more years ago were forced to use their right hands, then the probability of being left-handed would be higher for younger people, and this would result in a lower average age of death for left-handers, even in the absence of any greater risk of death for the individuals.

This claim of bias is potentially reasonable—however, the researchers on the original study did other work that suggests this bias is small, and that left-handers really do die younger, on average, than right-handers. The difference of 8 years in the raw data is probably an overestimate, but we should be wary about simply discarding the research—it would be better to estimate the magnitude of the bias and try to correct for it. We do not claim to have the answer here; it is better to keep an open mind and consider different ways of studying the problem.

As always, students must learn to be both skeptical and constructive.

10.3 Reviewing the semester's material

We cover statistical communication in the last week of class, and it provides a useful framework for reviewing all that came before.

10.3.1 Classroom discussion

It is possible to lie (or to make mistakes) by ignoring some key statistical principles. These are typically covered in detail during the semester, and when we cover "lying with statistics," we find it useful to mention these in class, asking students to quickly make up examples of each.

- Correlation does not imply causation (see Section 4.2.2).

- "Statistically significant" does not necessarily mean "important" (we can illustrate this with a hypothetical example of a very small effect that can be discovered with a very large sample size).
- Not "statistically significant" is not the same as zero (recall the example of Section 8.7.1, where 20 shots each are not enough to identify large differences in abilities between basketball shooters).
- Response bias in sample surveys (see Section 5.1.3).
- Misleading extrapolation (for example, the world record times in the mile run (Fig. 3.1 on page 20), or the regression of earnings on height in Section 4.1.2).
- Problems with observational studies (for example, the comparison of SAT scores before and after coaching, described in Section 5.4.3).
- Regression fallacy (as illustrated with several examples in Section 4.3).
- Aggregation (for example, the way the regression of earnings on height changes when sex is included in the model; see Section 9.1.2).

We illustrate many of these examples with newspaper clippings—it is fun and not difficult to get your own, possibly with the help of your students. The point here, however, is not to slam the press. In fact, as we discuss in Chapter 6 in the context of the statistical literacy assignments, we are generally happy about how the newspapers report on scientific and technical issues. But they are not perfect, and they can often let their guards down when their news is not coming from a scientific source.

10.3.2 Assignments: find the lie or create the lie

As a homework problem, we assign the problem of finding the two most important statistical errors in the article shown in Fig. 10.5. This is a challenging problem—in fact, many of our colleagues cannot readily find the errors.

In addition, we give the students the following homework assignment:

Do one of the following:

- Find an article in a recent newspaper or magazine that lies with statistics in some way. Explain what the "lie" is and how you would correct it.
- Find an article in a recent newspaper or magazine with numerical information that does *not* lie or mislead. Using the data in the article, create your own misleading "lie."

Include a photocopy of the article with your homework solutions. You get double credit for this problem if your article is *not* used by any other student in the class.

Other possibilities include students working in groups to find the most outrageous statistical "lie" or to produce the most outrageous "lie" themselves.

10.4 1 in 2 marriages end in divorce?

Class discussions can be structured around commonly-quoted, but not necessarily well-understood, numbers. For example, we hear statistics like "1 in 2 marriages end in divorce," but how can you really estimate that? It's tricky. If you look at the number of marriages in 1995 per 1000 adults, that number is roughly twice

26 CHICAGO SUN-TIMES, THURSDAY, MARCH 31, 1994

RICH CHAPMAN/SUN-TIMES

Deputy Fire Commisioner John Ormond (from left), Walgreens district manager Kermit Crawford, Fire Commissioner Raymond Orozco and Deputy District Chief John Schneidwind announce Wednesday that Walgreens is donating 1,000 smoke detectors for areas plagued by fires.

Walgreens Donates Smoke Alarms

By Phillip J. O'Connor
Staff Writer

Eighteen people have been killed in Chicago fires so far in March and there were either no smoke detectors or non-working detectors in all of them, Fire Cmdr. Raymond E. Orozco said Wednesday.

So far in 1994, there have been 38 fire deaths—one more than during the comparable period last year—and in 29 of them there were either no smoke detectors or non-working detectors, Orozco said.

The March toll included three recent fires that killed a total of 13 people, he said.

Orozco cited the statistics at a news conference at which he announced Walgreens Drug Stores has donated 1,000 smoke detectors to the Fire Department, which will distribute them in areas where fire deaths have occurred.

Orozco plans to have members of fire companies that actually fought the deadly blazes go door-to-door and distribute them to citizens whose homes are not protected by detectors.

"Smoke detectors save lives," Orozco said. He urged residents to check batteries in their home smoke detectors over Easter weekend when they change clocks to daylight savings time, which begins early Sunday.

He also urged citizens to help prevent injury and death by developing a home escape plan in the event of a fire and to practice good fire safety. "A fire, prevented, cannot hurt anybody," he said.

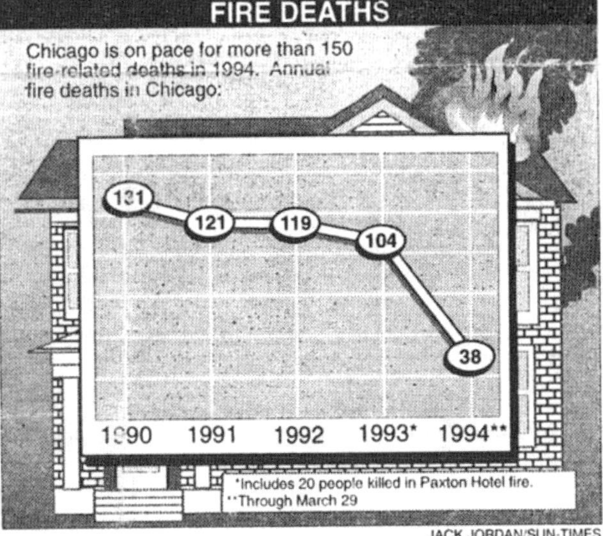

JACK JORDAN/SUN-TIMES

Fig. 10.5 Find the two most important statistical errors in this newspaper article. Answers appear on pages 272–273.

the number of divorces in that year. But, those getting divorces in 1995 are by and large not the same people as those getting married that year. Since so many marriages are ongoing, do you have to wait until one member of a married couple dies before you can count that as a nondivorced marriage? For example, only about 1 in 7 women who married in the early 1940s eventually divorced. But it is not very satisfying to make statements about people who married sixty years ago. Often assumptions are made that the divorce rate in a current year continues indefinitely into the future. A careful study of marriage longevity would need to consider life expectancy and control for age at marriage and length of marriage.

10.5 Ethics and statistics

Ethics is a topic that we find interesting, but students in statistics classes seem to be wary of it. We recommend preparing for a class discussion about ethics by gathering some relevant newspaper clippings; here we give examples of several areas in which ethical issues arise in statistical data collection and analysis. As always, we set up each problem and then ask the students to discuss in pairs as a prelude to general class discussion.

10.5.1 Cutting corners in a medical study

On March 15, 1994, the *New York Times* reported on a Federal investigation of a Canadian researcher at St. Luc's Hospital in Montreal. The investigation found violations of the scientific guidelines that govern the way the study was carried out. According to the *Times* article:

> A Federal investigation found that Dr. Poisson had falsified data in his part of the study that helped change the way breast cancer is treated. That influential study concluded that full mastectomies were not necessary to prevent the spread of early forms of the disease in many women.
>
> ...While insisting that he did little more than tell "white lies" that he believed would not change the conclusions of the study, Dr. Poisson signed an agreement with the Federal Food and Drug Administration acknowledging that he had falsified results of other studies.
>
> In a written reply to the United States Public Health Service's Office of Research Integrity, which spearheaded the Federal investigation, he stated, "I always feel sorry for a nice case to be denied the right to enter a good protocol just on account of trivial details: a difference of a few days in the date of surgery because the patient took a long time to decide."
>
> "When I lost a patient who did not wish to participate," he said, "I always took it as a personal defeat, knowing that the best protocol with the best biostatisticians is useless unless enough patients are registered" ...
>
> Dr. Poisson said he should have read the fine print of the study design more carefully to avoid the irregularities.

We provide students with a copy of this excerpt from the newspaper article, and because the discussion can get heated, we ask them to write their answer to the following question: On the basis of Dr. Poisson's remarks, why do you think the federal investigators were concerned about the experimental justification for the claim that "full mastectomies were not necessary to prevent the spread of early forms of the disease in many women"?

We give them five minutes to work in pairs to write a short answer and then we collect their written responses and lead a discussion on the topic. We use these written answers to express opinions of students who might be too shy to enter the discussion. In our discussion, we quote from Dr. Poisson's letter to the *New England Journal of Medicine*, where he stated, "My sole concern at all times was with the health of my patients ... For me, it was difficult to tell a woman with breast cancer that she was ineligible to receive the best available treatments because she did not meet 1 criterion of 22, when I knew this criterion had little or no intrinsic oncologic importance." We also provide the opinion of Dr. Broder, Director of the National Cancer Institute, which funded the study. In his testimony before the House Subcommittee on Oversight and Investigations, he said, "we consider the entire data-set from St. Luc to be a total loss to the American taxpayer." These are delicate issues.

10.5.2 Searching for statistical significance

As another example, a colleague who works at a university statistical consulting service reported the following story. A company wanted to get a drug approved, but their study appeared to have no statistically significant results. (See Section 8.7.2 for a classroom demonstration of multiple comparisons.) The researchers at the company broke up the data into subgroups in about 15 or 20 ways, and then they found something significant. Is this data manipulation? What should the statistician do? In this case, the company reported the results and their stock went up 50%.

10.5.3 Controversies about randomized experiments

A fundamental ethical problem in statistics arises in experimentation, for example in the context of studies of experimental drugs for treating AIDS. On one side, organizations such as the National Institutes of Health insist on randomly assigning treatments (for example, by flipping a coin for each patient to decide which treatment to assign). The advantage of randomized experiments is that they allow reliable conclusions without the need to worry about lurking variables. However, some groups of AIDS patients have opposed randomization, instead making the argument that each patient should be assigned the best available treatment (or, to be more precise, whatever treatment is currently believed to be the best). The ethical dilemma is to balance the benefits to the patients in the study (who would like the opportunity to choose among available treatments) with future patients (who would be served by learning as soon as possible about the effectiveness of the competing treatments).

The issue is complicated. On one hand, the randomized study is most trustworthy if all the patients in the study participate; if they are not treated respectfully, the patients might go outside the study and try other drugs, which could bias the estimates of treatment effects. On the other hand, the patients might be benefiting from being in an experimental study: even if the treatment is randomized, the patients are getting close medical attention from the researchers. Current best practice is to design studies so that all subjects will be expected to benefit in some way, but still keeping the randomized element. For example, a study can compare two potentially beneficial experimental treatments, rather than comparing a treatment to an inert "control." But there will always be conflicts of interest between the patients in the study, the scientists conducting it, and the public at large.

10.5.4 How important is blindness?

Other ethical issues arise in the blindness or double-blindness of experiments. In order to achieve blindness, studies in psychology often use deception. For example, in the United Nations experiment described in Section 5.3.1, the students were not told that the anchoring values of "10" and "65" were an experimental manipulation.

For another example (among many), we bring up in class the topic of stereotyping in job interviews. Are some applicants discriminated against because of their race, ethnicity, or gender? Once students have given some opinions on the topic, we steer the discussion toward the question of how this sort of bias could be measured statistically. One approach is to compare the success of people of different groups in job interviews. This would be an observational study, and we ask the students what sort of lurking variables would need to be controlled for, and how this could be done (see Section 5.4 for examples in which we discuss these issues).

Another way that job discrimination has been studied is to run a randomized experiment in which the experimental subjects are people who might be in a position to evaluate job applicants. We ask students what might be the treatments: for example, the subjects might be given hypothetical resumes, identical in all respects except that the ethnicity of the job applicant is selected at random (similarly to how the anchoring values were assigned at random in the United Nations experiment). This experiment must be done blindly, which necessarily involves deceiving the experimental subjects. (If they are told ahead of time that this is a study of ethnic stereotyping, their heightened awareness may affect their judgments of the job applicants.) Once the experiment is over, the subjects can be told of the true purpose of the experiment, but the practice of misleading them is still controversial and, some would say, unethical. On the other hand, it is important for society to learn the extent of problems such as racial bias, and certain information can be gathered using deception that would be difficult to gather any other way.

In other settings, blindness can pose medical risks. For example, in early studies of heart bypass operations, the treatment was compared to a control regimen of medical (non-surgical) intervention. This posed difficulties both for blindness and double-blindness (a patient knows if he or she has had heart surgery, and so does a doctor making subsequent evaluations). The solution chosen was to perform a "sham operation" on the patients receiving the control treatment—that is, to open up their chests, do nothing, and sew them back up—so they would be externally indistinguishable from the patients who received actual surgery. This certainly seems to be an ethically questionable way to achieve blindness. On the other hand, if it allows doctors to learn more about the effectiveness of heart bypass surgery, maybe it is worth it? This is a topic for student discussion.

10.5.5 Use of information in statistical inferences

Criminal justice

Another sort of ethical issue concerns the acceptable uses of information in making inferences and decisions. Members of ethnic minorities are more likely to be stopped and questioned by police. In what sense is this a reasonable policy given crime statistics? This has been a controversial area, both in stopping people on the street and in their cars on highways. Using the information about ethnicity is generally considered unethical, partly because of its obvious unfairness. There is general agreement that upholding the principle of equal treatment under the law is more important than potential short-term efficiency of police procedures.

In a criminal trial, a jury is not supposed to use knowledge of prior arrests or convictions to draw conclusions about the guilt of a suspect. From a statistical standpoint, if there is evidence that a person has committed previous crimes, he or she is probably more likely to have committed the crime in question—but legally and ethically, it is not considered acceptable to use this information. Similar issues exist in many settings.

Background information and course grades

For another example, consider a course in which a "pre-test" is given at the beginning of the semester to assess general background knowledge. The students are told that the pre-test will not count in their grade, but they are requested to try their best. Then, during the semester, there are some scheduling difficulties, and the midterm exam for the course has to be canceled. At the end of the semester, the instructor considers using the pre-test score in place of the missing midterm grades. Is this fair? Students generally think not, and in this setting, most instructors would not use the pre-test. However, ignoring this information probably means that the students' final grades will be less accurate indicators of student abilities: if a student did better on the pre-test, he or she is probably a better student, even after controlling for the grades in other aspects of the course. The following two goals conflict: (a) keeping the promise to the students (or, even more generally, in using information before the course began to determine grades) and (b) producing final grades that best reflect students' abilities.

The issue becomes even more complicated when the background information being used is not under the control of the student; examples that can be predictive of grades include ethnicity and parents' education levels.

Models for guessing on multiple-choice exams

Consider a test with several true/false questions. If all the students answer a question correctly, then presumably they all know the answer. Now suppose that half the students get a certain question correct. Then, how many students do you think knew the correct answer (we ask the class)? 50%? One possibility is that *none* of the students knew the correct answer, and they were all guessing.

Now consider a question that is answered correctly by 80% of the students. If a student chosen at random knows the correct answer with probability p, or guesses with probability $1-p$, then we can write, approximately, $p+0.5(1-p) = 0.8$, which yields the estimate $p = 0.6$. Thus, a reasonable guess is that 60% of the students actually know the correct answer and 40% were guessing. The conditional probability of a student knowing the correct answer, given that he or she answered the question correctly, is $60\%/80\% = 0.75$.

Where is the ethical dilemma here? Consider now the task of giving each student a total grade for the exams. The reasoning above suggests that they should get *no* credit for correctly answering the question that 50% of the students answered correctly (since the evidence is that they were all guessing), and they should get 0.75 credit for answering the question that 80% answered correctly, and so forth. Thus, the amount of points that a question is worth should depend on the probability that a student's correct answer was not due to guessing. The ethical question is of the fairness of deciding the grading system after the exam has been taken by the students. Is it fair for two students to get the same number of questions correct on the exam but different total scores, because the "circumstantial evidence" suggests that one of the students was more likely than the other to be guessing?

Misleading information

In other settings, it might not be wise to gather information if there is a suspicion that it will used inappropriately. For example the class activity of Section 7.5.2 demonstrated a scenario in which a person can fail a lie detector test and still have a greater than 50% chance of being honest. (This happens whenever the probability of error for the lie detector is greater than the probability that a person is a liar.) A similar problem arises in imperfect medical tests of a rare disease: it is possible that most people who test positive actually do not have the disease.

This information can still be useful if used correctly: for example, a failure on a lie detector test can be grounds for further investigation, and a positive test for a disease can motivate further testing. However, if the tests are (inappropriately) assumed to be perfect, then they can result in innocent persons being accused, or healthy persons being unduly alarmed (or even given inappropriate medical procedures). The ethical dilemma is whether to gather information that can be useful in the right hands but misleading if interpreted crudely.

Privacy and confidentiality

Tradeoffs between individual and social benefits occur in many settings where data are gathered on a population. For example, medical researchers can use data on the personal histories of disease sufferers to study factors associated with disease incidence, potentially saving future lives if they discover important relationships. But such a study might require the use of private medical records, which is open to abuse. For example, if the research were funded by an insurance company, it would be important to make sure that the confidential records could not be used to deny people health insurance. Similar issues have arisen when studying HIV and AIDS: it is important for public health authorities to be aware of trends in prevalence of these conditions, but rules requiring doctors to report new cases to the authorities can backfire, by causing some people at risk to avoid testing.

More generally, any study of the general population can place burdens on the persons being studied. Participation rates in public opinion polls have been declining for decades, partly because market researchers and others have saturated people with unsolicited telephone calls. When we talk about sample surveys in a statistics class, we should remember that the respondents are not simply objects of study, but participants (who are often giving their time for free) in a research project.

Part II

Putting it all together

11
How to do it

In this chapter we give some tips on how to prepare your own activities and how to adapt an activity you have read or heard about to your particular class. We also provide examples of mistakes we have made in carrying out activities, ideas for managing group work in class and in longer projects, and resources for finding and developing new activities.

11.1 Getting started

There is often resistance to doing class-participation activities if they are new to an instructor. We can get a bit scared when trying out new demonstrations, but they always seem to go well, especially by the second or third try, once the details have been tuned. For example, in the coin-flipping demonstration in Section 7.3.2, we recently added the bit about having two students try to guess which sequence is fake and give their reasons.

11.1.1 Multitasking

A natural fear when considering in-class demonstrations and other student-participation activities is that they will take valuable class time away from serious learning. We address this problem by, where possible, running these activities simultaneously with the other classroom material. For example, students can pass a scale and bag of candy around the room and weigh objects (Section 8.1) while a lecture is proceeding. Data can be gathered from students by passing out a form during the lecture, and the times when students are working in groups can be used by the instructor to put material up on the blackboard or to go around the room answering questions. When students are providing data from a simple calculation, such as finding a confidence interval for a proportion (Section 14.3), we prepare in advance a list of all possible outcomes, which makes it easy to display the results quickly.

11.1.2 Advance planning

Successful demonstrations require planning and organization, both for the students and the instructor. We prepare our students by giving them handouts and preceding any complicated activity with simpler exercises illustrating the same point, as described in Section 1.3. For our part, after a few times of showing up in class to find that we left an essential prop back at the office, we no longer leave the preparation until the last minute.

To help us get organized, we include in our class notes a list of materials needed for the activity, an outline of the five parts of the activity as described in Section 11.2.1, and any notes on how to carry it off, just as we might write notes in the margins of a cookbook to improve a recipe.

Before the start of the term, we prepare a packet of materials for the course, which we photocopy for students to purchase. It typically includes news stories, templates, instruction sheets for demonstrations, survey forms, and so forth (see Section 12.1).

It can also be good to practice a new demonstration before trying it in full format in front of a class. Otherwise, unforeseen problems can occur. Once we found an interesting demonstration of a sampling technique in a teaching journal. It came complete with a list of materials required (coins and dice) and a game board and instruction sheet ready to be photocopied for handouts. We read it over and it looked simple enough so we decided to try it out on our students with out making a dry run. In class, we were able to field questions as they arose, but 15 minutes into the activity the students still had not collected enough data to understand the sampling procedure, and they were beginning to get exasperated. We called it off. Fortunately, we had run plenty of demonstrations before then and the students kept their faith in us and our activities.

11.1.3 Fitting an activity to your class

There is no single format for a successful activity. Considerations that enter into choosing an activity for our class are the class size, classroom layout (for example, can chairs be rearranged?), board space, and the presence of a teaching assistant. The first time we try a demonstration, we are often worried about how it will turn out, but after adjusting over two or three semesters, we can generally adapt it to our teaching style.

Some activities are better suited for small classes. The largest class we have worked with is 80, but others have reported success using some of these activities in classes of over 200. An activity that requires students to share a piece of equipment can be unwieldy to carry off in a large class. Also, in larger classes, we limit our activities to those where the students remain seated working in pairs, except for when a few volunteers write answers on the board. In all cases, no matter what the class size or seating arrangement, we strongly advocate group work.

If you have a teaching assistant attending your lecture, then he or she can assist you, for example by doing lengthy computations on the side. For classrooms with plenty of blackboard space, you may want to bring extra chalk and partition the board so different groups can work on problems simultaneously.

11.1.4 Common mistakes

We have learned a few lessons from our mistakes and the mistakes of others. We relay a few of the more obvious ones to you here in hopes that they will help you avoid some.

Fig. 11.1 The class can be divided into groups for a discussion or class activity.

Missing the details

It is important that in-class examples are relevant to the material being covered at that point in the course, and changing seemingly minor details can lose that focus. For example, when covering the binomial distribution, we planned to ask students to work on the problem, "If 1000 babies are born in a hospital, how many girls would we expect?" This is a simple way to introduce the binomial distribution (with $p = 0.487$). But in class we got confused and stated the problem as, "1000 babies were born in a hospital, and 487 were girls. What is the estimated probability of a girl birth?" We had absent-mindedly switched it from a probability problem to a statistics problem, totally missing the point! Even the experienced cook can be better off writing a recipe and following it step by step.

Another time, one of us tried the demonstration of Section 8.2.2, in which a roulette board is set up and several students play until they are wiped out. We made several mistakes in our implementation, including forgetting the steps of (a) having students identify gambling strategies ahead of time, and (b) enlisting other students in the class to make change and keep track of the bets. As a result, the demonstration lacked focus, went slowly, and did not effectively teach the key statistical point, which is the role of a negative expected value in a long sequence of trials. In contrast, when we get the details right, we can focus on the instructional material, not the mechanics of the demonstration itself.

Fig. 11.2 When students work in groups, they can bounce ideas off each other, as in the random sampling demonstration of Section 8.1.

Losing the punchline

None of these activities are so compelling that a student will wait until the next class meeting to hear the end of the story. In fact, they become quite indignant over the possibility of not completing the activity before the bell rings. Not to mention, it makes us look disorganized.

Sensitive subjects

Although students enjoy collecting data on themselves, some subjects are taboo. Take body weight for instance, even students who are not outliers might be embarrassed to reveal their true weight in a class histogram.

When performing experiments on students, be careful in maintaining safety and anonymity of the students. For example, when we let our students design their own taste testing experiments (Section 5.3.3), one group of students made a comparison of rum and cokes with different brands of rum. After reading their report, visions of drunk students came to mind. Since then, we have required all experiments to receive instructor approval before being carried out.

Rewards

We read the evaluations of an instructor who routinely tossed candy to students who had correctly answered a question. Several evaluations were scathing. One likened the event to being "treated like monkeys." Nonetheless, we have had success using prizes in games of chance. Also, in large classes, we occasionally have dollar days, where a volunteer comes to the front of the class and partakes in an experiment, game, or other activity. We ask the student to introduce him or herself to the rest of the class, they participate in the activity, and if they win then we reward them with a dollar. The class applauds them for their efforts, and we never seem to run out of volunteers.

11.2 In-class activities

In this book, we have tried to describe the classroom demonstrations and examples in enough detail that it is clear how the students will be directly involved in learning the material. Here we gather some of our thoughts on what makes an activity work well.

11.2.1 Setting up effective demonstrations

It is important to be clear on why any activity is being done, in several senses:

1. *Concept*: what statistical topic is being covered?
2. *Mechanics*: exactly what are the students expected to do?
3. *Punchline*: what is the key point of the activity?
4. *Context*: how does this message connect to other topics in the course and relevant areas outside statistics?
5. *Follow-up*. Students take activities more seriously if they are connected to the formal structure of the course. We encourage class participation by including questions pertaining to the surveys discussed in class on exams and homeworks.

Surprises

Surprise is often at the heart of a good demonstration or example. For example, when students generate 100 real and 100 fake flips of a coin (Section 7.3.2), we surprise them by not being fooled by the fake flips. The surprise comes from common misconceptions about randomness: people generally believe that a sequence of coin flips should have a haphazard pattern, including frequent (but not regular) alternations between heads and tails, and we use this misconception to distinguish between the two sets of coin flips. Other demonstrations and activities that have surprise twists include activities on family size (Section 5.1.6), weighing a random sample (Section 8.1), the mile run (Section 3.2.1), regression of height and income (Section 4.1.2), and heights of men and women (Section 3.6.1).

In addition to being fun, the shock of confusion, followed by recognition, is a good way for students to learn and remember the material. These double takes are also effective at promoting classroom discussion. In each activity, the students work on the problem before we tell the whole story, and when the surprise twist is revealed, the students are now intrigued with the problem. They want to find out what went wrong with their seemingly correct procedure, why their data do not match some standard, or what was the flaw in their logic.

Props

For many of our demonstrations, we bring props to class. We collect them and keep them on hand for our demonstrations. For example, to demonstrate probability problems, we have a set of lockable boxes, jumbo colored dice, and a toy roulette wheel. The props are fun to use; they make the problem more concrete, and using them takes little additional class time over simply describing the problem in words.

11.2.2 Promoting discussion

One natural idea for finding discussion topics would be to solicit them from our students. To do this, we hand out index cards and each student to write a question he or she wants answered or a topic of special interest. We use these as sources of examples and discussion topics. Unfortunately, students do not provide enough suggestions to use on a regular basis, so it is important for us to come up with our own. Section 11.5 lists several sources of ideas for examples and discussion topics.

The newspaper is always a good starting point for finding topical examples. We often bring newspaper clippings and excerpts of journal articles to class (see Chapter 6 and Sections 5.1.3, 14.1.1, and 14.5) and have students work in pairs, where they read the handout and write responses to questions (these assignments may be as simple as writing one criticism and one positive comment). We use their answers to begin a class discussion. For example, sometimes we have the students hand in their responses (without names on them), and without sorting or looking through the papers, we take the top sheet from the stack, read it aloud to the class, and discuss the student's comments. We continue reading papers, taking them one at a time from the stack until we have heard and discussed a variety of ideas. Alternatively, we have groups focus on different questions, write their answers on the board, and act as discussion leaders for their questions. More elaborate examples and problems can become discussion projects if we assign them to the students ahead of time so they will be prepared, as in the prayer example in Section 8.7.3.

Fig. 11.3 Students are working in pairs on the telephone-book sampling demonstration described in Section 5.1.1.

11.2.3 Getting to know the students

Activities work best if you know your students well enough to call on them by name to answer questions, to supply data, or to write answers on the board. Although we call on students by name to answer questions we raise in class, we do not pressure them into providing an answer. If they signal an unwillingness to answer then we move on to another student. They will get other chances later.

We have a few techniques for learning names. In the first few weeks of class when students volunteer to answer a question, we ask them their names. Then to reinforce, we try to call on them by name again in the same class period. We also have students fill out index cards with their names, nicknames, and other information, such as their major and possible discussion topics. After each class, we go through the cards reviewing the names of students who answered questions during class that day. As a third trick, we return homeworks by calling out the students' names and personally handing each student his or her homework. Some instructors take photos of students and use these to learn their students' names.

11.2.4 Fostering group work

When we develop a new activity, we try to turn it into group work wherever possible. We schedule a lot of group work at the beginning of the term. Through these group activities, students get to know each other, we set the expectation for class participation for the rest of the term, and students figure out who they will work with on projects outside the classroom. As the semester progresses, we

Fig. 11.4 While a demonstration is going on, the instructor can circulate around the room answering questions. Here a class of eighth graders is doing the age-guessing demonstration described in Section 2.1.

often have students work together in class on short problems.

Typically we set a time limit for the groups to work on an activity, say 5 to 10 minutes. To keep students focused, we walk around the classroom, asking and answering questions. When many groups encounter the same point of confusion, we write hints or points of clarification on the board. Other times, we break up a problem into parts and give students 2 to 3 minutes to work on each part, getting the class together at each point to discuss the answer so far and keep all the groups of students on track.

If the activity involves answering questions or drawing graphs, we invite particular groups to write solutions on the board. Sometimes we ask more than one group to write up their responses in order to generate class discussion or to show that there is more than one way to do the activity. We specifically invite groups to the board, to make sure everyone gets a chance to participate at some point in the course.

We have found that groups often work at different speeds; some will not have completed the prescribed task in the allotted time, and others will finish quickly. If everyone's response is required (say, for body measurements on students, as in Section 4.2.1) then we multitask: for example, we start recording data, resume the lecture, or lead a class discussion while the last ones finish up. If we want to record the data on the board (for example, when we draw the coverage of a set of confidence intervals, as described in Section 8.1) we begin going around the classroom soliciting each group's interval. We mark the interval with their names, and in later discussion we refer to the interval by name. If students are filling in a worksheet or drawing graphs (for example, the histogram of soda consumption in Section 3.3.3) then we sometimes pose additional questions to those who have finished already, or have students start writing up solutions before everyone has finished.

We always leave time to go over the activity with the class as a whole. We lead a class discussion that brings out the key point of the activity and ties it to the topic we are studying.

11.3 Using exams to teach statistical concepts

Exams are an important teaching tool partly because students directly learn from working out the problems during the exam and partly from the learning that occurs while studying and preparing beforehand. Here we review some ways in which we have used exams to teach statistical concepts in a third way, as direct experience, by harnessing students' interest in their grades. We have had success using these techniques to involve the students in class discussions, and we believe there is the potential for much more work in this area.

Our courses for undergraduates typically include one or two midterm exams and a final. Students are of course very interested in their exam scores, and so we have developed some tricks to channel this interest into thinking about statistics. The activities involve both experimentation using exams and statistical analysis and adjustment of exam scores. These demonstrations must be conducted with care: students take their grades seriously, and it is important to make it clear that their exam grades are not manipulated in an arbitrary or unfair manner.

Sections 3.5.1 and 3.6.3 illustrate the use of guessed and actual exam scores in scatterplots and regressions. Section 4.2.2 uses exam lengths and scores to illustrate the difference between correlation and causation. Further examples appear in Section 10.5.5 in the context of the ethical uses of statistical information. Section 5.3.2 describes a more elaborate demonstration in which students are randomly given different midterm exam forms, with the results of this experiment motivating a class discussion when the graded midterms are returned to the students.

We also use exams for demonstrations of more advanced topics. Section 9.2 uses the search for unfair exam questions as a way to introduce psychometric modeling, and Section 13.1.5 discusses a probabilistic scoring rule that teaches the concept of calibration of probability estimates.

11.4 Projects

At different times we have used various formats for group projects:

- *Directed projects* are based on well-defined problems. For example, in Section 5.3.3, students design and conduct an experiment modeled after Fisher's experiment of a lady tasting tea. Another example (Section 16.1) has students analyze data from a case study in which they are provided suggestions on how to address a question of interest.

When we first started to develop these projects, our assignments were too prescribed. For example, our data analysis projects were one-page instruction sheets with sample code. We found the students followed the instructions to the letter without exploring on their own. As a result, we replaced the instruction sheet with more general suggestions where the student figures out which statistical techniques are appropriate for analyzing the data.

- *Active homeworks.* Occasionally in class discussions, there will be some number or set of numbers that is unknown but relevant to the discussion topic (for example, the distribution of soda consumption in Section 3.3.3). The task of looking up these numbers can be assigned to pairs of students. By the end of the class, all the students will have done some of this library research.
- *Research projects.* With the right kind of assistance, students can read and report on original research papers. The statistical literacy project described in Chapter 6 is one example where students choose a study reported in the press to investigate. They then track down and read the report or journal article on which the news story is based, and write a report describing the study. Other examples are described in Sections 14.6.1 and 15.8.
  The biggest problem we found with this project was that the student papers were too much like abridged versions of the documents they read. Their papers did not demonstrate that they had applied what they learned in the course to their reading. The papers improved dramatically when we provided more guidance. For example, we now give the students a set of questions or a template to aid them in reading their papers. We also monitor their progress more closely, requiring them to show us their work at an intermediate stage in the project.
- *Class projects* have the entire class working together on the same project (see Section 5.2, for example). We use the project as a theme that runs throughout the course. We divide the project up into subtasks, and students form small groups to work on these. Although the project takes most of a semester to complete, the effort required from each student is limited to a few hours over one or two weeks. The students learn both by doing and through the advice of the instructor. The instructor's involvement at each step helps secure a successful outcome, and it sets a good example of how to carry out independent research projects.
  We oversee the project by scheduling groups to report on their work to the rest of the class, and we use this reporting time to solicit input from all of the students on each aspect of the project. We organize class discussions on each topic and give handouts to clarify each group's job and to focus our in-class discussions. Each group writes a summary of its contribution, which appears in the final report.
- *Independent projects.* Here, students work in small groups on data-collection and data-analysis projects of their own creation, with the work spread over several weeks. This sort of project is described in detail in Section 11.4.2.

11.4.1 Monitoring progress

For all of these projects, it is essential for the instructor to keep track of how the students are doing and intervene when appropriate. Typically, we introduce the projects early in the semester and set deadlines for completion of the main pieces. To get them started, we provide students with a list of past projects, and dedicate class time to reviewing these projects and discussing what we expect.

Students usually work on the projects in groups of three to five. When they are choosing a project from a list, we help them form groups by writing the list on the board and having students stand near the item that interests them. They form their groups, exchange contact information, and sign up for appointments to plan the project with us.

We have found it essential to discuss the project in its early phases. At this meeting, the students are expected to bring a completed research plan, template, or outline, and, where appropriate, a copy of a research paper. We make sure their plans are adequately formulated, we discuss strategies for completing the project, and we identify a few important aspects of their work to which they should pay special attention. We sometimes schedule a second meeting with the groups to review the near complete version of the report, and make plans for oral presentations.

Encouraging student writing

The writing process is an effective way for students to piece their ideas into a coherent story and develop arguments in support of their thesis. The importance that we place on writing skills is reflected in the grading of the projects; typically 25% of the grade for an assignment is based on the report's organization, clarity of argument, and correctness of the statistical statements.

It is especially effective to require students to write for a real-life setting. Overall, the students enjoy such concrete assignments, and it helps them better focus on the problem when they have an audience, albeit imagined, other than the instructor.

We dedicate class time to a discussion of how to prepare the written report. We emphasize the importance of making clear, careful arguments, and we encourage students to revise and proofread their manuscripts. Communicating ideas through charts and graphs is a particularly important part of a statistician's communication skills, and we dedicate one to two lectures to this topic, where we carefully review examples of how to miscommunicate with statistics (see Chapter 10).

Some of our favorite writing resources are Gopen and Swan's (1990) article on how to write clear scientific arguments, Tollefson's (1988) booklet containing examples of common grammatical errors, and Wainer's (1984) examples of poor data displays. Appendix A in Nolan and Speed (2000) provides a synopsis of these writing tips, which we distribute to our students before writing their reports. In addition, Davis (1993) provides many good ideas on how to help students write better.

Presentations

We sometimes ask the groups to present their projects to the class. We give the students a lot of leeway in designing their presentations; some lead discussions modeled after our in-class activities, others use role-playing to present their findings, or they hold a debate where they present conflicting statistical evidence to make their arguments. To help students prepare, we schedule an appointment with each group to review their handouts and plan the class presentation. Sometimes the presentations are done on a voluntary basis, and other times we ask a few groups who have done especially good jobs to present their work.

Grading projects

We give students a clear picture of our expectations for their written report by handing out samples from past reports. We provide multiple examples of good introductions, analyses, arguments, and conclusions. We also place restrictions on the length of a report (number of words), the number of charts and graphs, and a list of appendixes to be included in the report.

We provide feedback to the students through our written comments on the papers. Our grading is done holistically, based on four aspects: composition, basic analyses, graphs and tables, and advanced analyses. For composition, we look for an organized presentation, persuasive argument, descriptions made in the student's own words (not in the words of a reference), and statistically sound statements.

11.4.2 Organizing independent projects

Independent projects can be fun, but they have to be well organized. The biggest problem we have encountered is that students' ideas tend to be too vague and sometimes overly ambitious, and then when the time comes for data collection, the students still don't have a clear plan and end up being sloppy. We try to fix this by focusing more clearly on the data-collection aspect of the project, requiring a statement of purpose, pilot data collection, a writeup, and a formal protocol before they go ahead and collect the main data. Also, we emphasize that the focus of the project is on gathering clean data, not on a fancy analysis. If you want a project in which students learn data analysis, we suspect you're better off with more structured projects such as described in Chapter 16.

Finally, some students can get overwhelmed by the idea of a final project. We emphasize that it is the equivalent in effort of several homework assignments, but with the work divided among four students. All students should participate in all aspects of the projects, but the writing tasks can be divided: one student writes the project proposal, another writes the summary of the pilot data, one writes the formal data collection protocol, and so forth.

For our introductory statistics class, we divide our project over several weeks, with the following steps.

- Brainstorming session in groups during class time
- Forming a project proposal: a goal or research hypothesis and a plan of data collection and analysis

For each part of the project, hand in a single assignment for the entire group, giving the names of all the group members. Your group will all share the grade for all parts of the project. Make a copy of each part of the assignment before you hand it in so you can refer to it while doing the next parts.

1. *Form groups of 4 (due lecture 4):* Form a group and decide on a group name.

2. *Research topic (due lecture 6):* Decide on a question that interests you and how you plan to collect data to answer it. Why do you care about this problem and why do you think it is worth studying? (For the project, you must collect your own data; you can't download data from the internet or copy from a published source.) Write no more than 1 page.

3. *Meet with instructor (between lectures 6 and 8):* Set up a time with your instructor for a meeting of approximately 15–30 minutes to brainstorm about your project. At least 3 of the group members (but preferably all) should attend. At the end of the meeting, you should have a specific research question and a plan of what sort of data to gather. For example, if you are gathering observational data, you will decide where to go and what specific information you will record. If you will interview people, you will decide what population you are studying and what questions you will ask. If you are performing an experiment, you will decide the experimental conditions and what will be measured.

4. *Project description and preliminary data collection protocol (due lecture 10):* Write about 2 pages describing the research question you have decided to study, why it is worth studying, and what data you will collect. You need a detailed data-collection protocol that you will use in collecting your pilot data (see below).

5. *Pilot data (due lecture 12):* Spend about 15 minutes each collecting *pilot data*—that is, a small set of preliminary data that you will use to learn about potential difficulties in your data collection process. For this part of the assignment, you must hand in: a copy of your pilot data, along with a list of the problems you encountered while collecting the data.

Fig. 11.5 First part of instructions for student-organized group projects (schedule based on a 26-lecture course). Continued in Fig. 11.6.

- Collecting pilot data and setting up a formal data collection protocol, an idea the students should be somewhat familiar with after doing one of the statistical literacy assignments described in Chapter 6.

- Collecting data

- Analyzing data with the computer package that we have been using in the course (in our case, Stata). Many of the students want to use Excel, but that has not worked well—it usually leads to unnecessarily simple data analyses (typically, the computation of means and their presentation as bar graphs).

Figures 11.5–11.6 display the handouts we give to students describing the project. The steps include group work, written assignments, and face-to-face discussions with the instructor.

6. *Formal data collection plan (due lecture 16):*

   Write a 2–4 page data collection plan that includes the following:

   - A formal *protocol*—that is, exact instructions for the data collection, including rules for how to classify difficult observations, deal with missing data, etc.
   - A *design*—where and when the data will be collected. If randomization is used in sampling, clearly define the population and how the sample will be selected. If randomization is used in assigning experimental treatments, clearly define the possible treatments and how the randomization will be applied. (Even in observational data collection, random sampling should be used. For example, if you are observing people at a certain cafe in the afternoon, your population is the set of all possible dates and hours when you might take observations, and you should randomly select from that set.)

   The instructor will read the data collection plan, make comments, and return it by lecture 18.

7. *Collected data (due lecture 22):*

   Collect the data according to your plan. Each member of your group should spend at least 2 hours collecting the data (this is in addition to any time spent in preparation for the data collection). Once you have collected your data, tabulate it in a computer file or spreadsheet that can be read into Stata. Do not include the pilot data here.

   For this part of the assignment, you must hand in a sample of your raw data and the table with all the data collected.

8. *Meet with instructor (between lectures 22 and 24):* Set up a time with your instructor for a meeting of approximately 15–30 minutes to consider ideas for how to analyze your data.

9. *Statistical analysis (due lecture 26):*

   Analyze your data (you *must* use Stata for your graphical and numerical analyses). You can try as many things as you want, but your final output should be:

   - A single graph (or set of small graphs on a single page) that displays all your data as informatively as possible.
   - A single statistical analysis (for example, a set of confidence intervals or a linear regression) that addresses your research question as directly as possible.

   Explain how your statistical results relate to your research question. What problems did you have in collecting your data? How might your findings be biased (this could include measurement bias, sampling bias, nonresponse, nonblindness of experimental treatments, etc.)?

   Your final results might be inconclusive. Calculate an estimate of how large your sample size would have to be in order to get statistical significance. If you could do this study again, how would you do things differently?

Fig. 11.6 Second part of instructions for student-organized group projects (schedule based on a 26-lecture course). Continued from Fig. 11.5.

11.4.3 Topics for projects

It is important for students to form groups and start thinking about project topics right away and to get timely advice from the instructor. One semester the projects did not go well because almost all the groups of students did surveys. It would be fine if some groups did this sort of project, but a survey is a limited research tool if it does not involve any experimentation. For example, in one typical project, a pair of students interviewed about 100 students and collected data on their age, family background, and marriage plans (whether they expected to marry and, if so, at what age). The analysis included various scatterplots, correlations, and two-sample comparisons. Even if this sort of project is done very well, it is unlikely to yield interesting results. In addition, there was a sameness to seeing survey after survey. Even in the context of observational studies, the rest of the class would find it more interesting to see a variety of study topics, for example animals, the news media, raindrops, or just about anything different.

How can we inspire students to work on more interesting projects? The best projects came from psychology students who were working on experiments for courses in their department. One strength of these projects were that they were motivated by specific research hypotheses of interest and were not simply data-collection exercises. We hope that by requiring students to think of a research goal or hypothesis *before* designing their study, they will be less likely to gather unanalyzable data. For example, in the study mentioned above, if the students were asked to come up with an interesting hypothesis, they would have to go beyond simple ideas such as "people with different religions have different marriage plans" (yes, that is a research hypothesis but not a particularly interesting one).

It often takes some discussion to isolate the fundamental question underlying a research idea. For example, students often have the idea of gathering from students data on grade point average and study habits, or alcohol consumption and grade point average, or number of hours studied and amount of time spent each week exercising. Students have a natural interest in questions like the effectiveness of studying, differences between athletes and nonathletes, and campus drinking. However, these sorts of questions rarely yield interesting results. Correlations between the measurements are usually weak and, in any case, it is not clear how to interpret the associations. With this in mind, we intercept these projects at the beginning stage and, in our office meetings with each group, ask "Why" questions that push toward more focused ideas.

It is also probably a good idea to hand out ahead of time a list of some potential project topics. After students have done projects in your course for a while, you can augment the list with titles of successful student projects in past years. The intention is not for students to pick from this list (which includes some less-than-exciting topics) but to give a sense of the breadth of possibilities so they can design a data collection project on a topic that interests them. Here are some ideas:

- Observing conversational styles.

- Timing how long it takes various objects to fall different heights (not too exciting, but it causes them to think hard about measurement protocols).
- Simple experiments on self and others (for example, systematically varying eating, sleeping, or studying plans)—this is hard to do in a short time, though.
- Studying friendships and relationships: what time of year do romantic relationships start, how many close friends do people have, how long do friendships last, and so forth.
- Evaluating the accuracy of weather forecasts, sports betting odds, or point spreads (see Section 13.1.6).
- Watching a bunch of violent movies and counting how many times the person with the gun is on the left side or the right side of the screen. (This was suggested by one of our graduate student instructors. No student has tried it yet, but we would be curious to see the results.)
- Trying out one of the classroom demonstrations (for example, the age-guessing example of Section 2.1) on students not in the statistics class. In this case, the students should vary the experimental conditions (for example, trying it on individuals, groups of two, and groups of three, to see to what extent larger groups guess more accurately).
- Analyzing quiz and exam grades for a different course (assuming the instructor of the other course is willing to supply the data). This idea (along with the evaluation of weather forecasts and sports odds) is not a data-collection project, but maybe it is close enough if the analysis is done well.
- Measurements—have people guess their measurements and then compare to actual values. Students doing this as a project can compare and see if different groups of people have different sorts of biases.
- Measuring waiting times for buses/trains and queue lengths: students gather some data and try to see the mathematical relations here.
- How many books are on the library shelves? Similar sampling problems: frequencies of letters in different languages, ... see Section 5.1.
- Taste-testing (see Section 5.3.3): for this to be a good project for a group of students, they would have to do it on many experimental subjects.
- Survey questions on some interesting topic—for example, ask students where they met their friends, how many close friends they have, whatever. It can be hard to get random samples here. One possibility is an email survey, or an in-person survey in dorms. Students should avoid the convenience sample of, for example, students walking out of the student union.
- A survey of some subpopulation of particular interest to the group of students: for example, math majors, or athletes, or Catholic students.
- Psychological experiments: for example, one semester, a group of students had the idea of putting up a "No Eating or Drinking" sign in the library study room, where eating and drinking were indeed prohibited but often done anyway. They counted the number of students eating and drinking

under the "no sign posted" and "sign posted" conditions, with several replications (15-minute periods) of each over a few weeks.

Because they see the project in the context of a statistics class, students are often too focused on numerical measurements. When students conduct surveys, often the most interesting data are qualitative free-form responses. For example, a student survey about textbook-buying habits included the following responses:

> I use the Columbia bookstore even though it is more expensive. There I can use my Columbia Card Flex points. This way my loans, rather than I, pay for the books.

> Labyrinth would be even better if they got rid of the pseudo-intellectual staff who know that Lacan is in a separate "Lit Crit" section, but still get paid the same as Mickey D employees.

Students do not generally realize that it is perfectly acceptable to gather qualitative data and then code it for a statistical analysis (as is described in research-methods texts such as Babbie, 1999).

Direct observations are another source of the interesting and unexpected. For example, Stilgoe (1998) describes some project ideas from his classes in landscape history:

> One [student] has just noticed escape hatches in the floors of inner-city buses and inquired about their relation to escape hatches in the roofs of new school buses. Another has reported a clutch of Virginia–Kentucky barns in an Idaho valley and wonders if the structures suggest a migration pattern. A third has found New York City limestone facades eroding and is trying to see if limestone erodes faster on the shady sides of streets. A fourth has noticed that playground equipment has changed rapidly in the past decade and wonders if children miss galvanized-steel jungle gyms. Another has been trying to learn why some restaurants attract men and women in certain professions and repel others, and another (from the same class years ago) has found a pattern in coffee shop location. Yet another reports that he can separate eastbound and westbound passengers at O'Hare Airport by the colors of their raincoats.

All these topics can involve statistical data collection.

11.4.4 Statistical design and analysis

In projects for introductory statistics classes, students typically learn most from the design and data collection stages. It is crucial for them to learn the difficulties of actual data collection, and we do not let the students download data from the Internet. Data can be collected through surveys, experiments, or observationally. Experiences in data collection are often discouraging, but sometimes the students have unexpected success; for example,

> We had originally intended on having 300 samples, but there were a number of people who were unavailable when we went to their rooms and a very small number of people refused to take part in the survey. It was surprising how eager other people were to take part, as floormates would often brag about their head circumferences to each other after being measured.

If the data collection involves sampling, we have found it necessary to enforce the requirement that the students perform random sampling from a defined population. It is helpful to hand out a successful past example, such as Figure

	Sunday	Monday	Tuesday	Wednesday	Thursday
11:00 pm	Sabrina	Therese	Sandra	85	113
11:15 pm	Sabrina	30	58	86	114
11:30 pm	3	Therese	Sandra	Sandra	115
11:45 am	4	32	60	88	116
12:00 am	Sandra	33	61	89	117
12:15 am	6	34	62	90	Sandra
12:30 am	Sandra	Therese	63	91	119
12:45 am	Sandra	36	64	92	Sandra
1:00 am	9	37	65	93	121
1:15 am	10	38	66	94	122
1:30 am	11	Therese	67	95	123
1:45 am	12	40	68	96	124
2:00 am	13	41	69	97	125
2:15 am	14	42	70	98	126
2:30 am	15	Therese	71	99	127
2:45 am	Yves	Therese	72	100	128
3:00 am	17	45	Yves	101	129
3:15 am	Yves	46	74	Sabrina	130
3:30 am	19	47	75	Sabrina	Yves
3:45 am	20	Therese	76	104	132
4:00 am	Yves	49	Yves	105	Yves
4:15 am	22	50	78	106	Yves
4:30 am	23	51	79	107	135
4:45 am	24	Therese	80	108	136
5:00 am	25	53	81	109	Sabrina
5:15 am	26	54	82	110	Sabrina
5:30 am	27	55	83	111	139
5:45 am	28	56	84	Sabrina	Sabrina

Fig. 11.7 Sampling plan from a group of four students who were studying the use of the school library during school nights. The students divided the time into 140 15-minute slots and then took a simple random sample of 32 of these slots.

11.7, which is an example of simple random sampling of school library hours. We can stimulate a class discussion of design issues by asking the following sorts of questions:

- Why did the students sample 15-minute blocks instead of full hours?
- Why is the sample unbalanced with respect to times of night and days of week? Will the imbalance cause problems?
- Is it a problem that the names are not assigned to time slots randomly (for example, Therese is only assigned to Mondays; Yves is only assigned times between 2:30 and 5:00 am)?

In the absence of direct guidance from the instructor, students typically take convenience samples. For example, one group, misunderstanding the idea of random sampling, wrote,

To ensure randomization, we handed out surveys at many different places, and at different times. Moreover, by choosing to sample a relatively large population, we were able to ensure that the average results of many individual results would produce a stable result (law of large numbers—reduce bias, increase randomization).

A different group used a more careful design:

Our population is defined as Columbia College sophomores and juniors. This population is listed in the facebooks ... In selecting our sample, we will first divide the population into two strata: males and females. Next, each student will be assigned an integer value. These numbers will be assigned separately for each stratum ... We will use two sets of random numbers to select 200 people from each stratum.

After the subjects are randomly selected, each of the four members of the group will survey 100 students at their dormitories. The locations for individual students will be obtained using the online Columbia directory. We will personally hand each subject the questionnaire ... If anyone refuses to complete a survey or if they cannot be reached, we will replace that subject with another of the same sex, using the randomization procedure outlined above.

When it comes to analysis, remember that most introductory statistics courses (including ours) do not cover multiple regression or analysis of variance until the very end. Thus, the design and analysis of the project will necessarily be focused on graphical displays and simple comparisons. Students can make reasonable graphs (in Stata, they typically create histograms and scatterplots). Their most common errors are to simply present raw computer output and to report unrounded results (for example, reporting a confidence interval of -58.5962 ± 191.1593). We also assure the students that it is acceptable to report whatever they have found; they do not need to try to search for or construct statistically significant results.

11.5 Resources

After you have tried your hand at some of the activities found in this book, you might want to develop some of your own demonstrations, examples, and projects. As discussed in Chapters 6 and 10, the popular press can be an important motivator for learning statistics and also a source of interesting examples. We are continually clipping articles from newspapers and magazines to hand out to our introductory classes. At the beginning of the semester, we give out a course packet with many of these handouts, which we discuss at the appropriate points in class.

But you can cook up your own demonstrations from just about anything—like a box of spaghetti. There are a great many resources available to get your creative juices flowing. These are found in books, periodicals, Web sites, and from conversations with other people. Even if you're only teaching introductory statistics, we recommend you read over the chapters in this book on demonstrations and projects for more advanced courses to spark ideas of how to develop similar class-participation activities for your own course.

11.5.1 What's in a spaghetti box? Cooking up activities from scratch

As an example of how demonstrations can be constructed from any raw materials, we came up with some ideas using a box of spaghetti:

1. How strong is a piece of fettuccini? Break pieces of fettuccini into different lengths. Lay a piece of fettuccini crossways between two supports. Hang a plastic bag from the middle of the piece of pasta, and add pennies to the bag one at a time until the noodle breaks. Record the length of the noodle and the breaking weight.
2. How much do noodles weigh after being cooked? (We have always thought that capellini doesn't fill you up as much as spaghetti.) Boil two ounces of various types of noodles (for example, capellini, spaghetti, linguini, and fettuccini) for the recommended time. Record the thickness (or volume) of a piece of uncooked noodle, and the weight of the cooked pasta.
3. How long is a randomly broken piece of spaghetti? Color one end of several pieces of spaghetti. Stand each piece on end (with the colored tip up) and slowly push down until it breaks. Record the length of the broken piece with the colored end.
4. How many pieces of spaghetti are in a box? Spill a box of spaghetti on a tiled floor, and use cluster sampling techniques (each tile is a cluster) to estimate the number of pieces in a box.
5. Can you taste the difference between different brands of pasta? See Section 5.3.3 for details on taste tests.
6. How long is a cooked noodle? See Section 15.3 on Buffon's noodle for estimating the length of a noodle by throwing it on a lined floor and counting the number of lines it crosses.

Other questions that have crossed our minds: How many pounds of pasta does the average American eat in a year? How has pasta consumption changed over the past 50 years? Does spaghetti stick to the wall when it's done? Does adding oil to the water keep it from sticking? How far can you toss a javelin-noodle? How far can you stretch a limp noodle?

11.5.2 Books

In recent years, many textbooks have been written that cover introductory statistics in a way that integrates theory, application, homework, and computer assignments. We have used Moore and McCabe (1998) as a text for some of our introductory classes. Utts and Heckard (2001) is a useful supplementary text, since it has a large collection of interesting examples. Books are also available on a variety of special topics. For example, Gastwirth (2000) is a collection of articles on statistics and the law, and Finkelstein and Levin (2001) is an introductory statistics textbook based on legal examples.

In addition, some books have been written specifically focusing on student activities for statistics, including Charlton and Williamson (1996), Scheaffer et al. (1996), and Rossman and Van Oehsen (1997). Some of the ideas in these books are reviewed in Moore (2000). You may find it useful to read through these books for ideas on how to keep students participating in class. For more sources of interesting examples and topics for class discussion and student projects, take a look at the books of Hollander and Proschan (1984), Anderson and Loynes

(1987), Chatterjee et al. (1995), and Pearl and Stasny (1992), along with the classic collection edited by Tanur et al. (1972, 1989). In addition, many popular books on uncertainty and statistics (for example, Sprent, 1988) contain interesting discussions that can be adapted for classroom use. For probability, Mosteller (1965) provides a collection of standard problems and examples, which includes all the favorites and some more obscure examples, with a readable presentation that is rigorous without being mathematically pedantic.

The books by Davis (1993) and Bligh (2000a, b) describe many strategies for organizing class discussion, group work, and projects for general college courses. We have found Bligh's discussions of "buzz groups" to be particularly useful. Case (1989) is a collection of advice for teaching assistants and instructors of mathematics classes. Other references appear in the Notes at the end of this book.

11.5.3 Periodicals

Technical journals in statistics and other fields are full of interesting examples; the key input required by the instructor is to structure these so that they bring realism to the lecture without overwhelming the students with details. At this point, the available example material may seem overwhelming. We mention all these sources to show how, once we feel confident in integrating examples and projects into our lectures and assignments, we find ourselves surrounded by relevant topics. For example, Section 2.4 discusses how we use newspaper articles to motivate the study of statistics, and Chapter 10 contains many examples of news articles illustrating points of statistical communication.

For more ready-made material, *Chance* is a nontechnical magazine with statistical applications aimed at statisticians, students, and general audiences. Every issue typically has examples that are good for class discussion. In addition, several journals regularly publish literature on statistics teaching ideas and methods. *Teaching Statistics*, published in England, is for teachers of students aged 9–19, and the online *Journal of Statistics Education* is for "postsecondary teaching of statistics." The *Mathematics Teacher* is for secondary mathematics but includes some probability and statistics examples as well. The *American Statistician* and the *Journal of Educational and Behavioral Statistics* are more general journals of research and exposition with "Teacher's Corner" sections that include ideas for college statistics teaching.

In addition, there are many publications in the field of education at the high school and college levels, not specifically focused on statistics but still relevant to the goals of keeping students interested and active in a difficult course. For instance, the journal *College Teaching* provides examples of course activities that can be adapted to multiple disciplines.

11.5.4 Web sites

Some of the materials described in this book, including course packets for the statistical literacy projects in Chapter 6, data for the stat labs described in Chapter 16, and related interactive learning tools, are at our own Web sites, www.stat.columbia.edu/~gelman/, www.stat.berkeley.edu/users/nolan/.

The *Journal of Statistics Education* (www.amstat.org/publications/jse/) information service lists several relevant sites, including a regular "Teaching bits" column. Chance News is an always-interesting monthly newsletter of probability and statistics (www.dartmouth.edu/~chance/chance_news/news.html). The Teaching Resource Center at the University of Virginia has a good selection of general advice on college teaching at www.trc.virginia.edu/tips.htm/.

In addition, a Web search on any topic of interest is likely to yield some interesting statistical examples. For example, a search on palm beach voting yields the Web page madison.hss.cmu.edu/, which shows some linear regression analyses of the vote for Patrick Buchanan in the Presidential election of 2000, along with many references. The analysis was also published in *Chance* (Adams, 2001).

11.5.5 People

A big source of teaching tips is other teachers reporting what works and does not work for them. In addition, do not overlook the possibilities of input from your own students. One approach that has been recommended is to give students a minute at the end of each class to write the answers to the following questions:

- What was the most important point of the lecture?
- What would you like to know more about?
- What was the muddiest point of the lecture?

The instructor collects these, reads them, and then at the beginning of the next lecture tells the students what were the most common responses to the questions.

Of course, traditional midterm and end-of-term student evaluations can be useful too, but by the time you get these evaluations it can be too late to help.

12 Structuring an introductory statistics course

The demonstrations and examples in this book are presented as separate modules so that the reader can easily use any subset of them. This chapter illustrates how we integrate our demonstrations and other teaching material into a non-calculus-based semester-long introductory course. We outline the course material and the student activities for each of 26 lecture periods of 75 minutes each. We make no claims for the optimality of this syllabus; rather, we include it to show how class-participation activities can be inserted into a standard course. In addition, we sketch an alternative list of activities for each week of a 15-week semester.

12.1 Before the semester begins

We prepare ahead of time two packets for the students. For the first day of class, we prepare:

1. A short packet for the first day of class, containing:
   - A sheet with scheduling information for the class, office hours, due dates, and so forth
   - A summary of the material that the students will be expected to learn during the semester. This is a list of about 50 short exam-type problems, to which we refer throughout the course to emphasize the links between the concepts covered and the specific skills being taught.
2. A course packet of material that we make available through the copy center. It includes:
   - Copies of all handouts to be used during the semester
   - A detailed course syllabus with a paragraph on each lecture, along with a schedule of reading assignments from the textbook
   - All the homework assignments
   - Instructions and help for the computer assignments
   - Guidelines and suggested topics for the final projects
   - Old midterm and final exams with solutions.

   In addition to simplifying the mechanics of the course, the packet gives students a sense of what will be happening next in class.

Students are required to bring the following materials to every class:

- Their course packet
- A pocket calculator
- Their 20-sided die or personal sheet of random digits (see Section 7.2).

12.2 Finding time for student activities in class

A common response to this book is, "student activities are fine, but I don't have time for them in my class!" We attempt to answer this objection with a schedule in Section 12.3 of our introductory statistics class, indicating in each lecture where the demonstrations and examples fit in, along with the approximate amount of time taken by each activity. (Section 12.4 outlines an alternative schedule with fewer activities.) We usually complete the scheduled material within the 75-minute lecture period (the key is to keep things moving and not to spend too much time standing at the blackboard), but when we realize we will not get through the entire scheduled material, we pause about 10 or 15 minutes before the end of the lecture and explain what we will skip and the students must thus study on their own. We do not drag half-finished topics across lectures.

The course schedule gives all the topics that we cover in class; extra time is also allowed for discussing homework problems, projects, and student questions. (One of the goals of doing all these demonstrations is to make students more comfortable with speaking up in class!) In addition, we often refer back to examples we have covered in earlier lectures—although this is not always listed in the schedule, these backward references are helpful in maintaining continuity. Throughout, we follow the generally recommended approach of starting with an example, moving to the general principle, and then illustrating with another example. Especially in the later part of the class, where students are learning specific techniques for solving probability problems, constructing confidence intervals, and so forth, we reinforce the basic material with drills in class and recitation sections.

12.3 A detailed schedule for a semester-long course

Lecture 1: Introduction

Bring to class: enough first-day-of-class handouts for the entire class, index cards for all students, set of 10 photo cards and placards for age-guessing demonstration (Section 2.1), 10 copies of "estimating ages" form (page 12), transparencies of cancer maps (pages 14–15), transparency projector

1. (20 minutes) Estimating ages from photographs (Section 2.1)
2. (5 minutes) Introduction: why is statistics important? (Section 1.4)
3. (10 minutes) Some examples of things you'll be able to do when the course is over:
   (a) Sampling
   (b) Descriptive statistics
   (c) Inference
   (d) Probability

4. (10 minutes) Administration
   (a) Hand out cards. To put on card:
      i. Name, section, year, email, major, sketch/description of self
      ii. Total Left and Right scores on handedness inventory (on first page of their packet of handouts; see Section 2.5)
      iii. Other information?
   (b) Mechanics of course
   (c) Responsibilities of the instructor and the student. Ask questions, etc. Calculus is not a prerequisite. Class participation is important.
5. (15 minutes) Example of numerical thinking: order-of-magnitude estimation
   (a) How many school buses are there in the U.S? (Section 2.3)
   (b) Other order of magnitude questions can be done every once in a while throughout the term. Encourage students to bring them to class.
6. (5 minutes) Quantitative and categorical variables (Section 3.3)
7. (10 minutes) Example of statistics: map of kidney cancer rates (Section 2.2)

Lecture 2: Distributions and histograms

Bring to class: more handouts (information sheet, packet of handouts, index cards) for students who did not show up last time, transparency of world record times in the mile run (page 20), transparency projector, pad of graph paper, tabulated data from students' handedness scores

1. (10 minutes) What's in the news? (Section 2.4)
2. (5 minutes) Administration
3. (5 minutes) Basic graphical displays we are about to discuss: time plots, histograms, distributions
4. (10 minutes) Time plots. Example: world record times in the mile run (Section 3.2.1)
5. Histograms
   (a) (5 minutes) Simple example such as students' ages
   (b) (10 minutes) Example of a histogram: handedness scores (Section 3.3.2)
   (c) (10 minutes) Example of a distribution: soft drink consumption (Section 3.3.3)
6. Computing probabilities and averages from histograms
   (a) (5 minutes) Diagram with area under the curve
   (b) (5 minutes) Example: soft drink consumption (Section 3.3.3)
7. (5 minutes) Normal distribution, z-scores. Example: heights of adult women (Section 3.6.1)
8. (5 minutes) Linear transformations of one variable: $y = a + bx$

Lecture 3: Scatterplots and bivariate distributions

Bring to class: graph paper transparency (Section 3.1), transparency projector

1. Before the lecture begins, load the blackboard with a set of numerical data that will be displayed in a plot (for example, the guessed and actual exam scores for men and women on page 25).
2. More on linear transformations
   (a) (5 minutes) z-scores as a special case of linear transformation, setting the mean to 0 and the sd to 1)
   (b) (10 minutes) College admissions: an example of linear combination of two variables (Section 3.7.1)
   (c) (10 minutes) Drills with more examples
3. Plotting data, items, x- and y-variables.
   (a) (10 minutes) Guessed and actual exam scores for men and women (Section 3.5.1)
4. (5 minutes) Trace of conditional mean, median, and quantiles (illustrate on graph of exam scores)
5. (5 minutes) Concepts of "regression" and conditional distributions
6. (5 minutes) Statistical independence and dependence. Illustrate with error in guesses and time to finish exam (Fig. 3.6).
7. (10 minutes) Bivariate normality: scores on two exams (Section 3.6.3)

Lecture 4: Least-squares regression

Bring to class: graph paper transparency (Section 3.1), transparency projector

1. (10 minutes) Review of what we have covered so far
2. (10 minutes) Linear model: $y_i = a + bx_i + \text{error}_i$
   (a) Errors have mean 0 and constant sd of σ and are independent of x
   (b) Suppose $y = a + bx + \text{error}$. Then $\text{mean}(y|x) = a + bx$ (linear transformation) and $\text{sd}(y|x) = \text{sd}(\text{error}) = \sigma$.
3. (10 minutes) Example: actual and guessed exam scores (Fig. 3.5)
4. (15 minutes) Running a regression on the computer: tall people have higher incomes (Section 4.1.2)
5. (15 minutes) Understanding the least-squares line (Section 4.1.1)
6. (5 minutes) Regression residuals
7. (10 minutes) Predictions: mean and variation
8. (5 minutes) Interpolation and extrapolation: height and income example (Section 4.1.2)
9. (5 minutes) Summary of least-squares regression

Lecture 5: Log transformation

Bring to class: graph paper transparency (Section 3.1), transparency projector, data on world population (page 34)

1. Before the lecture begins, load the blackboard with the first two columns of the world population data (Fig. 3.12)
2. (10 minutes) Powers of 10, logarithms, exponential growth, amoebas (Section 3.8.1)

3. (10 minutes) Simple example of log-log transformation: cubes (Section 3.8.1)
4. (15 minutes) Example of log transformation: world population (Section 3.8.2 and 4.1.3)
5. (15 minutes) Example of log-log transformation: metabolic rates (Section 3.8.3)
6. (15 minutes) Discuss topics for group projects

Lecture 6: Correlation

Bring to class: yardsticks for body measurements (Section 4.2.1), lists of words for the memory quizzes (Section 4.3.1)

1. (5 minutes) Correlation: definition of correlation and explanation in terms of linear transformations (z-scores) of x and y (Section 4.2)
2. (5 minutes) Introduce the example of correlations of body measurements. Gather data while the class proceeds.
3. (5 minutes) Example: correlation of exam scores (Section 3.6.3)
4. (5 minutes) Theoretical properties of the correlation
5. (10 minutes) Discussion of examples where correlation is relevant and where it is irrelevant (Section 4.2)
6. (10 minutes) Example: correlations of body measurements (Section 4.2.1)
7. (10 minutes) Correlation and least-squares regression (Section 4.2)
8. (15 minutes) Demonstration of regression to the mean: a memory experiment (Section 4.3.1)
9. (5 minutes) Understanding regression to the mean (Section 4.3.2)

Lecture 7: Categorical data and lurking variables

1. (15 minutes) Example of a 2-way table: who opposed the Vietnam War? (Section 3.5.2)
2. 2-way and 3-way tables: use an example from the textbook
   (a) (10 minutes) Description of the 2-way table: marginal and conditional distributions and their interpretations
   (b) (20 minutes) Looking at a lurking variable using the 3-way table
3. More examples of lurking variables
   (a) (5 minutes) Tall people have higher incomes; is sex a lurking variable? (Section 4.1.2)
   (b) (10 minutes) Soliciting examples from students (Section 5.4)

Lecture 8: Experiments

Bring to class: U.N. questionnaire forms for all students in the class (page 67)

1. (5 minutes) Distinction between sampling and experimentation
2. (15 minutes) The "U.N. experiment" that looks like a survey (Section 5.3.1)
3. (25 minutes) Design of experiments: units, treatments, factors, control. Discuss in context of U.N. experiment and of a recent newspaper article

4. (10 minutes) Example: coaching programs for the Scholastic Assessment Test (Section 5.4.3)
5. (10 minutes) Problems with before–after studies, regression to the mean
6. (10 minutes) Some topics in experimentation: matching, randomization, statistical significance, replication, blindness

Lecture 9: Sampling

Bring to class: Random numbers (see Section 7.2), telephone book, telephone book rulers (see Section 5.1.1)

1. (15 minutes) Discussion of sampling in the context of a recent newspaper article. Structure of sampling: population, units, sample, measurements. Concept of a "representative sample" and its problems.
2. (10 minutes) Some topics in sampling: stratification, clustering, undercoverage, nonresponse, response bias
3. (5 minutes) Demonstration on question-wording bias
4. (10 minutes) Wacky surveys (Section 5.1.3)
5. (10 minutes) Survey of number of siblings in families (Section 5.1.6)
6. (15 minutes) Benford's law demonstration with telephone book pages (Section 5.1.2)

Lecture 10: Concepts of statistical inference

Bring to class: Scale, bag of candies

1. (5 minutes) Begin the candy-weighing demonstration (Section 8.1). The class proceeds with the weighing for the next 40 minutes or so while the lecture continues.
2. (10 minutes) Discuss examples of interesting research designs from recent newspaper articles
3. (10 minutes) Introduce bias and variability using the data from estimating ages (Section 2.1)
4. (10 minutes) Bias and variability: general discussion and simple examples
   (a) Zero bias, high or low variability
   (b) Moderate bias, high or low variability
   (c) High bias, high or low variability
5. (10 minutes) Randomization; discussion in the context of the U.N. survey/experiment (Section 5.3.1)
6. (10 minutes) Conclude candy-weighing demonstration (Section 8.1)

Lecture 11: Probability: introduction

Bring to class: Copies of coin-flipping handout (page 107)

1. (5 minutes) Probabilities and proportions
2. (5 minutes) Rules of probability, probabilities of compound events
3. (10 minutes) Probabilities of various sequences of boy and girl births (Section 7.3.1)

4. (15 minutes) Example with probabilities of more complicated events: length of the World Series of baseball (Section 7.4.1)
5. (5 minutes) Realism and approximations in probability models
6. (20 minutes) Real and fake sequences of coin flips (Section 7.3.2)
7. (10 minutes) Probabilities of rare events (Section 7.3.3)

Lecture 12: Probability: random variables

1. (5 minutes) Definition of random variable: examples of length of baseball World Series, number of children, height of a randomly sampled person
2. (10 minutes) Distribution of X, Y, $X+Y$ in a simple example with dice
3. (5 minutes) Expectations (means) and variances
4. (5 minutes) Linear transformations
5. (5 minutes) Probability trees for composite outcomes, for example, twins (Section 7.3.1)
6. (10 minutes) Rules for probability trees and another example
7. (35 minutes) Review for midterm exam
8. (15 minutes) (if there is time) Voting and coalitions (Section 7.4.2)

Lecture 13: Probability: joint distributions and conditional probability

Bring to class: packets of three cards (Section 7.5.1)

1. (15 minutes) Discuss midterm exams. If there was randomizing of the exam questions (Section 5.3.2), discuss whether it is fair to make an adjustment for the students who took the exam form with the lower average grade.
2. (15 minutes) What's the color on the other side of the card? (Section 7.5.1)
3. (10 minutes) Formal definition of conditional probability
4. (10 minutes) Lie detector demonstration (Section 7.5.2)
5. (10 minutes) Joint distributions, independence, conditional distributions
6. (10 minutes) Correlation and the mean and standard deviation of $aX+bY$

Lecture 14: Midterm exam (covering material up to and including lecture 12)

Lecture 15: Probability: binomial distribution and normal approximation

Bring to class: roulette set (Section 8.2.2)

1. (10 minutes) Example: number of girls in 100 births (Section 7.3.1)
2. (10 minutes) Exact form of the binomial distribution
3. (10 minutes) Derivation of mean and variance of binomial distribution
4. (5 minutes) Normal approximation and Central Limit Theorem
5. (5 minutes) Where are the missing girls? (Section 8.2.1)
6. (10 minutes) Rolling dice to get normally distributed "IQs" (Section 7.2.3)
7. (5 minutes) Mean and standard deviation of a sample proportion
8. (15 minutes) Real-time gambler's ruin (Section 8.2.2)

Lecture 16: Probability: sums of independent random variables

1. (5 minutes) Continuity correction for the normal approximation (example such as probability of exactly 50 heads in 100 coin flips)
2. (10 minutes) Sampling distribution of the sample mean
3. (10 minutes) Sums and differences of sample means
4. (10 minutes) Example: playing roulette once for $10 000 vs. placing 10 000 bets of $1 each
5. (15 minutes) Poll differentials: an example of a discrete distribution (Section 8.3.4)

Lecture 17: Parameters and estimates

1. (10 minutes) Example of a confidence interval: land or water? (Section 8.3.3)
2. (10 minutes) Definition of confidence intervals
3. (10 minutes) Confidence interval from poll differentials (Section 8.3.4)
4. (15 minutes) Demonstration of coverage of intervals (Section 8.4.1)
5. (10 minutes) Demonstration of *noncoverage* of intervals (Section 8.4.2)
6. (10 minutes) Discuss how to estimate how common is your name (Section 8.6.2); students will gather the data as a homework assignment

Lecture 18: Significance testing

1. (10 minutes) Example: analysis of the data from the U.N. survey/experiment: is there evidence for a real difference between the two groups? (Section 8.3.2)
2. (10 minutes) Example: are average scores on the two forms statistically significantly different? (Section 5.3.2)
3. (10 minutes) Definition of hypothesis testing
4. (10 minutes) "Statistical significance" vs. practical significance
5. (10 minutes) Discuss the example, "Praying for your health" or some other example from the news (Section 8.7.3)
6. (10 minutes) Do-it-yourself data dredging (Section 8.7.2)

Lecture 19: Inference using the t distribution

1. (10 minutes) Confidence intervals and hypothesis test for the U.N. survey/experiment (Section 8.3.2)
2. (10 minutes) The t distribution for confidence intervals and hypothesis tests: x scores and t scores
3. (10 minutes) Analyze data from some previous example such as the age-guessing demonstration (Section 2.1)
4. (10 minutes) Demonstration: coverage of confidence intervals (Section 8.4.1)
5. (10 minutes) Inference for linear combinations

Lecture 20: Inference for proportions

Bring to class: 10 tennis balls (Section 8.7.1)

1. (5 minutes) Margin of error in a sample survey

2. (10 minutes) Inference for proportions using the normal approximation
3. (5 minutes) Comparing two proportions
4. (10 minutes) Comparing abilities of basketball shooters (Section 8.7.1)
5. (10 minutes) Using an example from the news, discuss the relation between sample size and margin of error
6. (10 minutes) Example of several confidence intervals: how good is your memory? (Section 8.6.1)
7. (15 minutes) Using the binomial distribution to check the fit of the models for the World Series (Section 8.5.5) and Benford's law (Section 8.5.4) examples

Lecture 21: More examples of statistical inference

1. (10 minutes) Discussion of some recent newspaper article
2. (10 minutes) Sample size calculations; refer to the basketball shooting demonstration (Section 8.7.1)
3. (10 minutes) One-sample and two-sample problems (discuss $\frac{1}{n_1} + \frac{1}{n_2}$ effect)
4. (15 minutes) Matched pairs experiment vs. comparing two means: demonstration using data gathered from pairs of students

Lecture 22: Inference for simple linear regression, begin multiple regression

1. (15 minutes) Regression of income on height (Section 9.1.1)
2. (10 minutes) Statistical model for linear regression
3. (5 minutes) Confidence intervals for the regression coefficients
4. (5 minutes) Comparing coefficients from two regressions (using an example with data from two different surveys)
5. (15 minutes) Prediction intervals; illustration using scatterplots
6. (15 minutes) Multiple regression of income on height and sex (Section 9.1.2)

Lecture 23: Multiple linear regression

1. (10 minutes) The multiple linear regression model, referring to the regression of income on height and sex (Section 9.1.2)
2. (15 minutes) Another example, possibly from a social science journal
3. (10 minutes) Estimates and confidence intervals for multiple regression
4. (10 minutes) Prediction intervals
5. (15 minutes) Regression with interactions (Section 9.1.3)

Lecture 24: Data collection and inference

1. (20 minutes) Discussion (with examples) of general approaches to dealing with lurking variables: just use one condition of the lurking variable, control for the variable, or balance over the lurking variable
2. (10 minutes) Aggregation: scatterplots to show how it can increase, decrease, or make no change to correlation
3. (10 minutes) How to learn about causation? Discussion in the context of some recent newspaper article with a controlled comparison

4. (15 minutes) Causation: the fundamental problem of causal inference, comparison of treatment with control
5. When to do an experiment, survey, or observational study
   (a) (10 minutes) Coaching for the SAT (Section 5.4.3)
   (b) (10 minutes) Surgeon General's Report described (Section 5.4.1)

Lecture 25: Statistical communication

1. (25 minutes) Methods of lying with statistics (Section 10.1)
2. (20 minutes) Interpreting the semester's material from the perspective of statistical communication (Section 10.3)
3. (15 minutes) Ethics and statistics (Section 10.5)

Lecture 26: Conclusion

1. (30 minutes) Discussion of students' data collection/analysis projects, interspersed throughout the lecture
2. (20 minutes) Review. What can the students now do that they could not do before taking this class?
3. (20 minutes) More advanced topics in probability and statistics

12.4 Outline for an alternative schedule of activities

Here we provide a briefer outline showing how a few classroom activities can be used each week in an introductory statistics class. We recommend reading through this book, picking out the demonstrations that you like best, then developing your own personal versions as you try them out on your students.

Week 1: Distributions and histograms

1. What's in the news? (Section 2.4)
2. Collecting data from students (Section 2.5)
3. Soft drink consumption (Section 3.3.3)

Week 2: Descriptive statistics and the normal distribution

1. The average student (Section 3.4.2)
2. Heights of men and women (Section 3.6.1)
3. Heights of conscripts (Section 3.6.2)

Week 3: Observational studies and experiments

1. Taste testing (Section 5.3.3)
2. Surgeon General's report (Section 5.4.1)
3. Vietnam war opinions (Section 3.5.2)
4. Age adjustment (Section 3.7.3)
5. Literacy packets (Chapter 6)

Week 4: Scatterplots and correlation

1. Guessed and actual exam scores (Section 3.5.1)

2. Body measurements (Section 4.2.1)
3. Scores on two exams (Section 3.6.3)

Week 5: Linear regression and log transformation

1. Simple examples (Section 4.1.1)
2. World population (Sections 3.8.2 and 4.1.3)

Week 6: Regression effect, regression fallacy, and causation

1. Memory experiment (Section 4.3.1)
2. Height and income (Section 4.1.2)

Week 7: Midterm exam and introduction to probability

1. Real vs. fake coin flips (Section 7.3.2)
2. Modifying dice and coins (Section 7.6)

Week 8: Probability: compound events

1. Rare events (Section 7.3.3)
2. Color on the card (Section 7.5.1)
3. Lie detection (Section 7.5.2)

Week 9: Probability: Central Limit Theorem

1. Random digits via dice (Section 7.2)
2. Roulette (Section 8.2.2)
3. Weighing a random sample (Section 8.1)

Week 10: Sampling

1. Wacky surveys (Section 5.1.3)
2. Sampling entries from the phone book (Section 5.1.1)
3. Family size (Section 5.1.6)

Week 11: Confidence intervals

1. Coverage of confidence intervals (Section 8.4.1)
2. Weighing a random sample revisited (Section 8.1)

Week 12: Hypothesis testing

1. Taste tests and the hypergeometric distribution (Section 8.5.3)
2. Comparing two groups (Section 8.3.2)
3. Benford's law and the χ^2 test (Section 8.5.4)

Week 13: Inference: interpretation

1. Do-it-yourself data dredging (Section 8.7.2)
2. How good is your memory? (Section 8.6.1)
3. Praying for your health (Section 8.7.3)

Week 14: Statistical communication

1. Methods of lying with statistics (Section 10.1)
2. One in two marriages end in divorce (Section 10.4)
3. Breast cancer trials (Section 10.5)

Week 15: Review

Part III

More advanced courses

13

Decision theory and Bayesian statistics

We have taught several times an introductory course in decision theory and Bayesian statistics, requiring one term of probability as a prerequisite. In addition to the usual lectures, homework, and problem solving, we have found it useful to conduct frequent classroom demonstrations. Some of these are essentially lectures, but with extensive class participation, whereas others involve actions or calculations by the students. This chapter outlines some of our more effective demonstrations. Our contribution here is in the tricks used to involve students; the ideas behind most of the demonstrations are well known, and we refer instructors and students to textbooks in applied decision analysis and Bayesian statistics for further references (see the Notes at the end of the book).

The activities serve several purposes, including focusing student attention on difficult conceptual issues that are hard to learn in a lecture or by solving homework problems (for example, the principle of expected gain in Section 13.1.1, determining the value of a life in Section 13.1.4); alerting students to their cognitive illusions and that they are shared with others (for example, the incoherent utilities for money in Section 13.1.2 and the uncalibrated subjective probability intervals in Section 13.2.2); bringing personal issues into the class, thus allowing each student to make a personal contribution to the discussion (for example, different areas of knowledge in the subjective probability intervals, different preferences regarding the value of a life, and personal decision problems in Section 13.1.7); dramatizing counterintuitive results which a student might not realize as counterintuitive if he or she were not forced to guess out loud (Sections 13.1.1 and 13.2.1); and demonstrating the multiple levels of uncertainty in a Bayesian analysis, as well as the coverage property of posterior intervals. In addition, eliciting discussion in these demonstrations has been useful in introducing the students to the instructor and each other and has led to a high level of student participation.

We divide the activities in two sections: decision analysis in Section 13.1 and Bayesian statistics in Section 13.2, but the topics greatly overlap and we recommend you read the entire chapter if you are teaching a course on either topic. Table 13.1 lists the demonstrations, the concepts they are intended to convey, and the materials they require.

Table 13.1 Concepts that are intended to be conveyed and additional materials required to conduct the demonstrations on decision analysis and Bayesian statistics in Chapter 13.

Activity	Concepts covered	Materials required
13.1.1: How many quarters are in the jar?	Expected values, subjective probability, optimization	Jar filled with quarters, count of the number of quarters in the jar
13.1.2: Utility of money	Utilities, coherence, decision trees	none
13.1.3: Risk aversion	Cognitive illusions, coherence	none
13.1.4: What is the value of a life?	Utilities, calibration of low probabilities	none
13.1.5: Probabilistic answers to true–false questions	Proper scoring rules, Brier score	Special question on midterm exam
13.1.6: Evaluating real-life forecasts	Calibration, accuracy	none
13.1.7: Real decision problems	Multiattribute utility functions	none
13.2.1: Where are the cancers?	Adjustment of data, sampling variability, shrinkage	Handouts of Figs. 2.3, 2.4, and 13.4
13.2.2: Subjective probability intervals	Overconfidence, calibration	Handouts of Fig. 13.5, list of uncertain quantities and their true values
13.2.3: Drawing parameters out of a hat	Bayes inference (normal model), coverage of posterior intervals	Hat filled with draws from a normal distribution
13.2.4: Where are the cancers, a simulation	Bayes inference (Poisson model), shrinkage, prior distributions	List of counties with populations, envelope filled with draws from a gamma distribution
13.2.5: Hierarchical modeling and shrinkage	Bayes inference, validation	Material for quick quizzes, laptop computer set up to fit the model

13.1 Decision analysis

13.1.1 How many quarters are in the jar?

This demonstration, which we do on the first day of class, is one of our favorites, and we describe it in detail to show how we use it to convey concepts of decision analysis as well as mathematical techniques. Before performing this demonstration, we lead the class through the demonstration of subjective probability intervals described in Section 13.2.2. (Please go to page 216 and read that section before continuing.)

Constructing a consensus subjective probability interval for the number of quarters

Once the students have gone through the subjective probability demonstration, we pull out a glass jar full of quarters and let the students pass it around and examine it. The jar has previously been filled to a specified level, and the instructor does not know how many quarters are in the jar—the answer has been written on a sheet of paper and is in a sealed envelope, which the instructor places on a table. The students are asked how many quarters they think are in the jar. A few students will state guesses, and we then encourage them to explain and discuss the guesses (for example, "So, Ned, given that Louise guessed that the jar has 200 quarters, do you still want to guess 100? Is that your final answer?"). A student is then asked for a 50% subjective probability interval, so that the probability is 25% that the true value is below the low point of the interval and 25% that it is above the high point. The instructor and the class then prod the student (for example, "Your interval is $[125, 150]$. If I offered you an even-money bet, where you would win if the true number of quarters is between 125 and 150, and I would win if it is below 125 *or* above 150, would you take this bet?"). Once the student has settled on an interval, we ask if any students disagree with the interval. Someone will answer and give their interval, and the class is led in more discussion until they are brought to agreement on an interval that seems about right for everyone (for example, $[120, 200]$).

We then sketch a normal density on the blackboard and ask what the mean and standard deviation should be for the specified interval to contain 50% of the probability. The computation is easily done, using the fact that $[\mu - \frac{2}{3}\sigma, \mu + \frac{2}{3}\sigma]$ is an approximate 50% interval for the normal distribution with mean μ and standard deviation σ. (The students are already familiar with the normal distribution from their probability prerequisite.) The horizontal axis of the density is then labeled appropriately, and the students are asked if this seems to represent their uncertainty ("You are 90% sure the number of quarters is less than X?", etc.). We ask if their uncertainty can be expressed exactly by a normal distribution. Some students will realize the answer is no, because the true number of quarters must be (a) positive, and (b) an integer. We discuss how, with a distribution with mean 160 and standard deviation 60, for example, zero is far in the tail of the distribution, and the discreteness is a minor issue, so it can be reasonable to characterize the students' uncertainty by a normal distribution. At the end of the demonstration, we return to this issue and discuss why the results are basically valid for any unimodal distribution.

What is the optimal guess?

We then state the puzzle: you [the class] will be given *a single guess* as to the number of quarters in the jar. After the class submit its collective guess, we shall open the envelope to reveal the true value. If the guess turns out to be correct, the money will be given to the class (yes, we would really do this) and split equally among the students. If the guess is incorrect, the students get nothing. What should the class guess? In answering, assume that the normal distribution sketched on the board represents your true state of uncertainty about the number of quarters; thus, you believe that 160 (say) is the most likely value, 159 and 161 are the next most likely, there is only a 5% chance that it will be more than two standard deviations away, etc. Recall also that the jar was filled to a specified level—the number was not picked in advance. So you need not worry about psychological issues such as, "He wouldn't have picked 150, because it's a round number." It's just a problem of geometry—how many quarters are in the jar—and the distribution on the blackboard represents your uncertainty.

After a pause, a student speaks up and says they should guess 160, the mode of the distribution. Does everyone agree? Yes, they all agree, although some are wary, suspecting a trick. We state that, in fact, the "obvious" answer is wrong—and there is no trick! Why is this? We pause to let the students think.

Building up intuition on decision making under uncertainty

We then pull a quarter out of our pocket and shake it between our two hands, then hold out both fists. One fist contains the quarter. A student is asked to pick a hand, and then we say, "This hand contains 0 or 1 quarters. There's a 50% chance that this hand holds a quarter, and if you guess right, you will get all the quarters in the hand. Should you guess 0 or 1?" The students start to realize—if you guess 0 and you're right, you don't win anything anyway, so you might as well guess 1. "Suppose there's only a 10% chance that there's a quarter—what should you guess?" You should still guess 1—it can't hurt.

Now to a more complicated problem. Suppose the jar contains either 100 or 200 quarters, and you think the two possibilities are equally likely. Should you guess 100 or 200? What if there is a 51% chance of 100 quarters and a 49% chance of 200? Which should you guess? At this point, some students will choose 100 and some will choose 200. Which choice is better? We bring two students to the blackboard—suppose Ned would guess 100 and Louise would guess 200—and play the guessing game repeatedly. At each play, we choose a random number from 00 to 99 by rolling dice (see Section 7.2); if the outcome is in the range $[00, 48]$, we give Louise \$200, and if it is in the range $[49, 99]$, we give Ned \$100 (Monopoly money in both cases). After playing ten or so times, it becomes clear that Louise is doing better than Ned. We derive this on the blackboard by showing the expected value per play of Ned and Louise, and referring to the law of large numbers.

So Louise's strategy is better. But, a student asks, we are only playing the quarters game once, not playing repeatedly, so why are expected values and the law of large numbers relevant? Well, life is full of uncertainties—in a given week, you may buy insurance, bet on a football game, make a guess on an exam question, and so forth. As you add up the uncertainties in the events, the law of large numbers comes into play, and the expected value determines your long-run gain (just like Ned and Louise). As long as no single decision or small set of decisions are dominant (this condition can be made more precise in a more theoretical course on probability), you can go with the expected value. (We are ignoring nonmonetary gains such as the thrill of getting the correct guess.)

Maximizing the expected gain

Now back to the quarters. Should you guess 160, or something higher, or something lower? Yes, something higher. Let's work it out mathematically. Your goal is to maximize your expected return. Let x be your guess; then your expected gain is just x times the probability that the number of quarters is x. For our distribution with mean μ and standard deviation σ, that probability is approximately the normal density at x, so the expected gain is approximately $x\frac{1}{\sqrt{2\pi}\sigma}\exp\left(-\frac{1}{2\sigma^2}(x-\mu)^2\right)$. To find the maximum of the expected gain, differentiate with respect to x and set the derivative to zero; after some cancellation this yields a quadratic equation, with solution $x = \frac{1}{2}(\mu \pm \sqrt{\mu^2+4\sigma^2})$. The answer cannot be negative, so the $\pm$ must be $+$. We plug in μ and σ to compute an answer, rounding to the nearest integer (for example, in the above example, with $\mu = 160$ and $\sigma = 60$, the optimal guess is $x = 180$.) Are the students happy with this guess? It is useful to tell the story of the motorist who is stranded at night and is looking for his keys, not by his car (the most probable location), but near the street lamp (less probable, but more likely that he will find his keys, if that is where they are). We then open the envelope and find the true answer, paying out in the unlikely case that the guess is exactly correct.

An appealing feature of this demonstration is that the answer can be computed exactly, but it requires some nontrivial analysis (differentiation and the quadratic formula). This is probably the first problem they have ever seen in which the exact formula for the normal density is useful. We conclude by noting that their expected gain (and thus the instructor's expected monetary loss from doing the demonstration) equals the value of x times the probability that x is the correct guess. Computing this for the chosen x and the approximate normal distribution shows the instructor's expected loss to be reassuringly small (if $\mu = 160$ and $\sigma = 60$, an expected loss of 28.3 cents).

13.1.2 Utility of money

To introduce the concept of utility, we ask each student to write on a sheet of paper the probability p_1 for which they are indifferent between

- a certain gain of \$1, or
- a gain of \$1 000 000 with probability p_1 or \$0 with probability $1 - p_1$.

We write this formally as

$$U(\$1) = p_1 U(\$1\,000\,000) + (1 - p_1)U(\$0),$$

where U is the student's utility function for money.

The students are then asked to write, in sequence, the probabilities p_2, p_3, p_4, for which

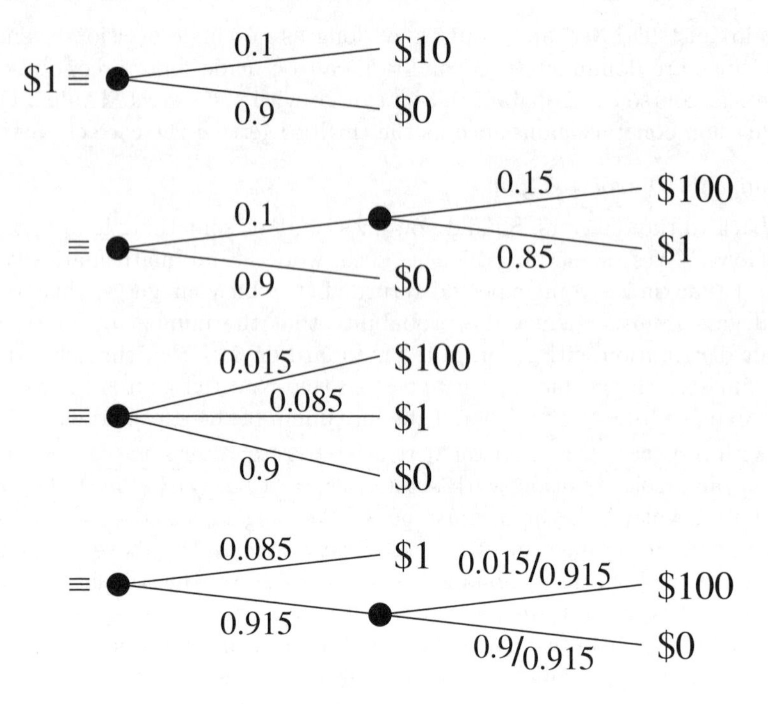

Fig. 13.1 A derivation of relative utilities using decision trees, replicating the steps of equation (1).

$$
\begin{aligned}
U(\$1) &= p_2 U(\$10) + (1 - p_2) U(\$0)\\
U(\$10) &= p_3 U(\$100) + (1 - p_3) U(\$1)\\
U(\$100) &= p_4 U(\$1000) + (1 - p_4) U(\$10)\\
U(\$1000) &= p_5 U(\$1\,000\,000) + (1 - p_5) U(\$100).
\end{aligned}
$$

One of the students is then brought to the blackboard to give his or her answers to the questions. The probabilities are checked for coherence (that is, the existence of a consistent set of utilities and preferences), as follows. First, the answers to the questions involving p_2 and p_3 are combined to yield a comparison between \$1, \$100, and \$0. For example, suppose $p_2 = 0.1$ and $p_3 = 0.15$ (when working through the example in class, using the student's actual numbers is clearer than working with the algebra of p_1, p_2, p_3, etc.). Then,

$$
\begin{aligned}
U(\$1) &= 0.1\,U(\$10) + 0.9\,U(\$0)\\
&= 0.1(0.15\,U(\$100) + 0.85\,U(\$1) + 0.9\,U(\$0)\\
&= 0.015\,U(\$100) + 0.085\,U(\$1) + 0.9\,U(\$0)\\
&= 0.085\,U(\$1) + 0.915\left(\frac{0.015}{0.915}U(\$100) + \frac{0.9}{0.915}U(\$0)\right).
\end{aligned}
\tag{1}
$$

This holds if and only if $\left(\frac{0.015}{0.915}U(\$100)+\frac{0.9}{0.915}U(\$0)\right)=U(\$1)$. (In class we work this and subsequent utility computations out using decision trees, as illustrated in Fig. 13.1, rather than equations.) Given this student's answers to the questions, we have deduced that $U(\$1)=0.0164\,U(\$100)+0.9836\,U(\$0)$. We then repeat this procedure, using the student's value of p_4, to determine the utility of \$1 relative to \$1000 and \$0, and then once again, using p_5, to determine the utility of \$1 relative to \$1 000 000 and \$0. Finally, this derived value is compared to the student's original value of p_1. These will disagree, meaning that the student's preferences are incoherent. The students in the class then discuss with the student at the blackboard how to change $p_1,\ldots,p_5$ to give coherent and reasonable answers. It may be necessary to remind the students that coherence does *not* require the utility for money to be linear. The student at the blackboard then is asked to sketch his or her utility function for money, as implied by the now-coherent indifference statements.

13.1.3 Risk aversion

A related demonstration focuses on the form of the utility function for money. We start by setting up the problem mathematically. Suppose a person is somewhat risk averse and is:

- indifferent between (a) a certain gain of \$10 and (b) a 55% chance of \$20 and a 45% chance of \$0;
- indifferent between (a) a certain gain of \$20 and (b) a 55% chance of \$30 and a 45% chance of \$10;
- indifferent between (a) a certain gain of $\$x$ and (b) a 55% chance of $\$(x+10)$ and a 45% chance of $\$(x-10)$; for $x=30,40,50,\ldots$.

Is this reasonable? The students assent.

Another way of eliciting these indifferences is to pick a student and ask for what value p is he or she neutral between the alternatives (a) \$10 and (b) a p probability of \$20 and a $(1-p)$ probability of \$0; then repeat this question for $U(\$20)=pU(\$30)+(1-p)U(\$10)$, $U(\$30)=pU(\$40)+(1-p)U(\$20)$; and so forth. The student will probably give a value of p that is near to or greater than 0.55.

We continue the example by asking the following question: for what dollar value y is this person indifferent between (a) a certain gain of $\$y$, and (b) a 50% chance of \$1 billion and a 50% chance of \$0? The answer, surprisingly, is that $\$y$ is between \$30 and \$40, as can be derived easily by mathematical induction. For example, using utility notation, the given indifferences can be written as $U(\$x)=0.55U(\$(x+10))+0.45U(\$(x-10))$ for each x, and thus $U(\$(x+10))-U(\$x)=\frac{0.45}{0.55}\left(U(\$x)-U(\$(x-10))\right)$. Setting $U(\$0)=0$ and $U(\$10)=1$ (the location and scale of the utility function can be set arbitrarily) and evaluating the expressions in order yields,

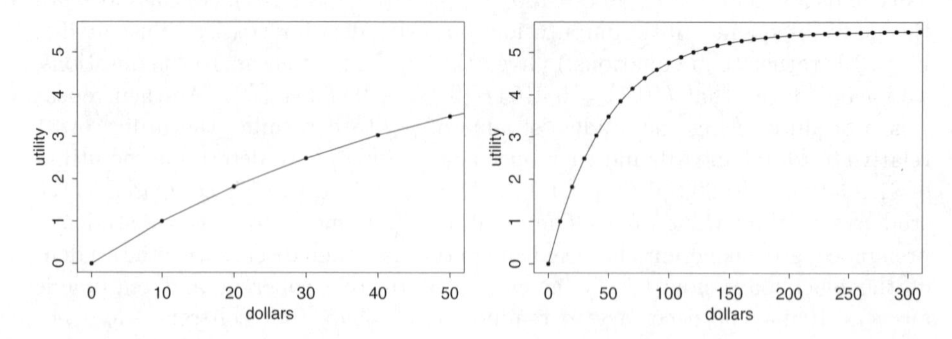

Fig. 13.2 Utility function for the example of Section 13.1.3: (a) After setting $U(\$0) = 0$ and $U(\$10) = 1$, we can derive $U(\$20), U(\$30), \ldots$. (b) The utility function has an asymptote of 5.5: under these assumptions, $U(\$1\text{ billion})$ is virtually the same as $U(\$300)$.

$$
\begin{aligned}
U(\$20) &= 1 + \frac{0.45}{0.55} = 1.818 \\
U(\$30) &= 1 + \frac{0.45}{0.55} + \left(\frac{0.45}{0.55}\right)^2 = 2.487 \\
&\cdots \\
U(\$1\text{ billion}) &= 1 + \frac{0.45}{0.55} + \left(\frac{0.45}{0.55}\right)^2 + \cdots + \left(\frac{0.45}{0.55}\right)^{999\,999\,999} \\
&\approx \frac{0.55}{0.55 - 0.45} = 5.5.
\end{aligned}
$$

Since

$$
\begin{aligned}
U(\$30) &< 0.5U(\$1\text{ billion}) + 0.5U(\$0), \\
U(\$40) &> 0.5U(\$1\text{ billion}) + 0.5U(\$0),
\end{aligned}
$$

and utility is an increasing function of money, $\$y$ must be between \$30 and \$40.

The derived utility function is displayed in Fig. 13.2.

The student believes each step of the argument but is unhappy with the conclusion. Where is the mistake? It is that fearing uncertainty ("risk aversion") cannot generally be explained as a rational response to a concave utility function for money. Rather, it implies behavior that is not consistent with any utility function. This is a good time to discuss cognitive illusions, many of which have been demonstrated in the context of monetary gains and losses.

Here are some questions to raise in class: Is decision theory descriptive? Is it normatively appropriate?

13.1.4 What is the value of a life?

We begin this demonstration by asking the students what is the dollar value of their lives—how much money would they accept in exchange for being killed?

They generally answer that they would not be killed for any amount of money. Now flip it around: suppose you have the choice of (a) your current situation, or (b) a probability p of dying and a probability $(1-p)$ of gaining \$1. For what value of p are you indifferent between (a) and (b)? Many students will answer that there is no value of p; they always prefer (a). What about $p = 10^{-12}$? If they still prefer (a), let them consider the following example.

To get a more precise value for p, it may be useful to consider a gain of \$1000 instead of \$1 in the above decision. To see that \$1000 is worth a nonnegligible fraction of a life, consider that people will not necessarily spend that much for air bags for their cars. Suppose a car will last for 10 years; the probability of dying in a car crash in that time is of the order of $10 \cdot 40\,000/280\,000\,000$ (the number of car crash deaths in ten years divided by the U.S. population), and if an air bag has a 50% chance of saving your life in such a crash, this gives a probability of about $7 \cdot 10^{-4}$ that the bag will save your life. Once you have modified this calculation to your satisfaction (for example, if you do not drive drunk, the probability of a crash should be adjusted downward) and determined how much you would pay for an air bag, you can put money and your life on a common utility scale. At this point, you can work your way down to the value of \$1 (as in the first demonstration in Section 13.1.2). This can all be done with a student volunteer working at the blackboard and the other students making comments and checking for coherence.

The student discussions can be enlightening. For example, one student, Julie, was highly risk averse: when given the choice between (a) the current situation, and (b) a 0.000 01 probability of dying and a 0.999 99 of gaining \$10 000, she preferred (a). Another student in the class pointed out that 0.000 01 is approximately the probability of dying in a car crash in any given three-week period. After correcting for the fact that Julie does not drive drunk, and that she drives less than the average American, perhaps this is her probability of dying in a car crash, with herself as a driver, in the next six months. By driving, she is accepting this risk; is the convenience of being able to drive for six months worth \$10 000 to her? This demonstration is especially interesting to students because it shows that they really do put money and lives on a common scale, whether they like it or not. There is a vast literature on the practical, political, and moral issues involved in equating dollars and lives.

In addition, people are often ignorant of the magnitudes of various risks, a point we illustrate by giving students the handout shown in Fig. 13.3.

13.1.5 Probabilistic answers to true–false questions

In discussing calibration and accuracy of probability forecasts, we introduce the Brier score for evaluating probabilistic forecasts of binary outcomes. If a forecaster assigns the probability p to an event, the forecaster's Brier score is defined as $1-(1-p)^2$ if the event occurs, or $1-p^2$ if the outcome does not occur. This scoring system is designed to give an advantage to forecasters who are calibrated (given that the forecast probability is p, the event should actually occur with frequency p) and precise (p should be as close as possible to zero or one, while

Cause of death	Estimated deaths per year
Botulism	
Flood	
Heart disease	
Homicide	
Motor vehicle accidents	40 000
Pregnancy	
Stomach Cancer	
Tornado	

Estimate the frequency of deaths (number of deaths per year in the United States) from each of the causes of death listed alphabetically above. As a reference point, about 40 000 persons die each year in the United States from motor vehicle accidents.

Fig. 13.3 Example of a handout used to demonstrate misunderstandings about risks. Try this out yourself; the true risks appear on page 274. There is nothing special about this list; we encourage you to develop your own list that will interest your students.

remaining calibrated). If a forecaster has a subjective probability π that an event will occur, the expected Brier score will be maximized by setting $p = \pi$; that is, it is a "proper" scoring rule.

We cover the Brier score extensively in class, using examples such as weather forecasting (the original motivation for the method). But we really bring the subject to life by including on the midterm exam several true–false questions, for which each student is asked to give a subjective probability p that the correct answer is "True." Their score for each question is five times the Brier score. We have found that students tend to be overconfident in their answers, frequently assigning probabilities of 0 or 1 (indicating certainty that the answer is "False" or "True," respectively) but being wrong. They have not internalized the mathematics of the Brier score: for example, suppose you think that the correct answer to a question is "True," but you are not *completely* sure. If you write "0.8," you will receive 4.8 points (out of a possible 5) if you are correct and 1.8 points if you are wrong. Even a blind guess of "0.5" nets you a certain 3.75 points. Students have a greater appreciation of calibration of forecasts after losing exam points from overconfident guessing.

13.1.6 Homework project: evaluating real-life forecasts

There are many examples of actual forecasts that students can evaluate for calibration and accuracy. Examples include point spreads and betting odds in sporting events, weather forecasts, and predictions of corporate earnings in business. For a long-form homework assignment, students can divide themselves into groups, with each group responsible for finding a series of forecasts from the newspaper or some other source.

These forecasts can then be checked for calibration and accuracy, which can be evaluated using expected value and mean squared error. For continuous outcomes, these concepts are straightforward (see, for example, the age-guessing data in Figure 2.2 on page 13). For probability forecasts of binary events, mean squared error is equivalent to the Brier score described in Section 13.1.5. For each trial $i = 1, \dots, n$, the prediction p_i is the forecast probability that the specified event will occur, and the outcome is labeled $y_i = 1$ if the event occurs or $y_i = 0$ if the event does not occur. The mean squared error is simply $\sum_{i=1}^{n}(y_i - p_i)^2/n$.

Calibration and accuracy have been studied in a variety of forecasting settings. For example, Las Vegas oddsmakers' point spreads for professional football games have been studied and found to be calibrated. That is, if y_i is the difference between the scores of the two teams in football game i, and x_i is the point spread (which can be interpreted as a predicted value for y_i), then $\mathrm{E}(y|x) = x$ for any value of x. More precisely, $\mathrm{E}(y|x) - x$ has been estimated empirically, from a dataset of actual football games, and found to be statistically indistinguishable from zero. The point spreads have an accuracy of about 14 points; that is, $\mathrm{sd}(y|x) \approx 14$. Professional basketball games can be predicted to an accuracy of about 12 points.

For a more serious example, economists have studied the calibration and accuracy of forecasts of inflation by the Federal Reserve Board and by private companies. Another active area of research is the information available from commercial weather forecasts one or two weeks ahead.

13.1.7 Real decision problems

A standard assignment in any decision analysis class is for students to formally analyze their own decision problems. Because this can be difficult for students, one approach is to break the problem into parts. We follow treatments of business and social decision-making, which emphasize that the key part of setting up the problem is identifying goals and values, after which the formal steps of constructing and evaluating the decision tree are relatively straightforward.

One plan is for students each to consider a personal goal and then various strategies he or she might use to achieve this goal. Familiar examples of goals (and decision options) include: finding an affordable and convenient place to live (deciding among different housing options); achieving academic success and a happy social life (deciding how many hours to spend studying and socializing each week); having a useful and affordable consumer product (deciding how much to pay for different features); and having an interesting and socially useful career (deciding what sort of job or study to take after graduation).

An alternative project is to study a public or social decision problem. For example: how much should the university charge in tuition and offer for financial aid; what hours should the university library be open; what frequency should the city run buses on a particular route; how much money should a particular local company spend on advertising; how much should the government tax gasoline.

Multiattribute value functions

All of these examples feature tradeoffs between different goals, and to study these tradeoffs using decision analysis it is necessary to set up a *value* or *utility* function. Before getting to the personal decision problems, we conduct a class exercise on setting up multiattribute value functions. Section 13.1.2 describes a demonstration of a one-dimensional value function (for money), and in Section 13.1.4, we consider a very specific two-dimensional value function for money and life.

Here, we pick two continuous attributes and set up a value function—that is, we define a utility for any combination of the two outcomes. We typically choose a simple consumer-product example such as weight and battery-life of a laptop computer, or carrying capacity and gas mileage of a car, or cost and location of seats at a popular concert.

Setting up a utility function is difficult and is discussed in modern textbooks on applied decision analysis. One recommended strategy is to first set up indifference curves and then assign numerical utilities to the curves. In any case, this must be done carefully in class; we have found that students are generally unable to construct a two-dimensional utility function as a homework assignment without supervised practice first.

Setting up a decision problem

The next step is for students to attack a real-life decision problem of the sort described at the beginning of this section. Personal decision problems can be worked on in pairs, or the entire class can study a social decision problem, dividing the tasks among groups of students. If students are to work on real-life decision problems for their final projects, it makes sense to first do a practice problem together in class. Here we briefly lay out the steps required to set up and formally "solve" a problem.

1. *Goals and decision options.* The first step—before setting up a decision tree—is to state the decision-maker's goals, which in turn can suggest strategies to achieve them.
2. *Values and utilities.* Decision analysis then requires that a utility function be defined on the space of all possible outcomes, as in the activity described earlier in this section.
3. *The decision tree.* The students can then set up a tree identifying the structure of decision options and uncertainties.
4. *Probabilities.* Probability distributions must then be assigned to all uncertainty nodes in the tree. For a classroom activity, these probabilities can be guessed at, but the students should think about whether they can be estimated quantitatively using reliable data.
5. *Evaluating the tree.* At this point, the decision problem can be formally solved, averaging utilities over uncertainty nodes and maximizing over decision nodes to determine the optimal decision at each point in the tree.

6. *Understanding the results.* Does the recommended decision make sense? If not, is it possible that aspects of the decision tree, value function, or problem definition have not been specified correctly? Like a computer program, the decision analysis is only as good as its inputs.

Ideally, the discussions surrounding each step should bring out the connections between the real-world problem being studied, the concept of decision analysis, and the technical steps of specifying and averaging over value functions and probability distributions.

13.2 Bayesian statistics

13.2.1 Where are the cancers?

On the first day of our Bayesian statistics class we go over the example of the cancer map, as discussed in Section 2.2, including handing out copies of Figs. 2.3–2.4. (Please go back to page 13 and reread before proceeding.)

We then continue the example by handing out copies of Fig. 13.4, a map that shades the counties with the highest *Bayes-estimated* kidney cancer death rates, where the Bayes-estimated rate for each county is a weighted average of (a) the observed rate in the county and (b) the national average rate. In this weighted average, the relative weight attached to the observed rate is approximately proportional to the population of the county, so that in counties with extremely small population the rate is shrunk virtually all the way to the national average, in counties with moderate population the rate is shrunk part way toward the national average, and in very large counties the adjusted rate is essentially equivalent to the observed rate. Figure 13.4 looks much different from Fig. 2.3, with much more of the shading appearing in populous counties. Most of the extreme rates in Fig. 2.3 occurred in low-population counties, and they got shrunk so much toward the national average that they were no longer extreme in Fig. 13.4. (In fact, the shading in Fig. 13.4 overemphasizes the high-population counties, but this is not an issue we discuss in our introductory class.)

If the raw rates in Fig. 2.3 are y_j/n_j (y_j cancer deaths out of n_j white males in county j), and the true rates have a Beta(α, β) distribution, then the Bayes-estimated rates in Fig. 13.4 are $(y_j + \alpha)/(n_j + \beta)$. In this case, we can estimate $\alpha = 27$ and $\beta = 58\,000$, and $\alpha/\beta = 4.65 \cdot 10^{-4}$ is the average kidney-cancer death rate among all U.S. counties for white males in the decade 1980–1989. The actual adjustment is more complicated because the values are age-adjusted, but this is the general idea.

This example is a good introduction to applied statistics because it shows a case where statistical adjustment is clearly appropriate and is, in fact, a standard tool in epidemiology. but we can explain it without worrying about prior distributions or probability theory. We return to this example later in the semester to illustrate Bayesian inference (see Section 13.2.4).

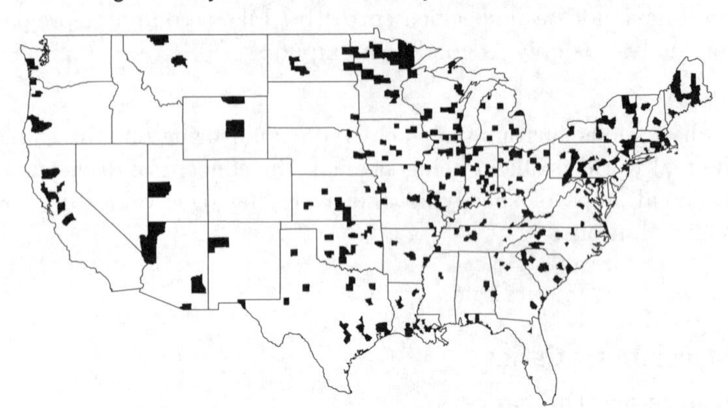

Fig. 13.4 The counties of the United States with the highest 10% *Bayes-estimated* age-standardized death rates for cancer of kidney/ureter for U.S. white males, 1980–1989. See Section 13.2.1 for discussion.

13.2.2 Subjective probability intervals and calibration

Subjective interval for a single unknown quantity

A well-known, and very useful, demonstration involves the calibration of probability intervals. We start by asking a student to give his or her guess at some uncertain quantity. For example, when performing this demonstration at Smith College, we asked the students to guess out loud the number of listings of persons named Smith in the local telephone directory. We wrote the first several guesses on the blackboard: 156, 250, 72, 210, 150, 120, 200, 35, 76, 49, 50. Substantial differences of opinion have been revealed.

We then ask one student to give a 50% probability interval (more precisely, the 25% and 75% quantiles) for the uncertain quantity—as we explain, this is an interval for which the student believes there is a 1/2 chance the true value is inside the interval, a 1/4 chance the interval is too high, and 1/4 chance the interval is too low. This interval is intended to summarize the student's uncertainty. For example, when we did the demonstration at Smith, a student volunteered the interval $[100, 200]$.

At this point we invite the other students to comment. If a student in the class would place more or less than 50% probability on the stated interval, we point out the opportunity for a bet that both parties should accept. We discuss the idea of a subjective interval with the class and alter the interval on the blackboard until the students generally consider it reasonable. For example, most of the students in the class at Smith thought $[100, 200]$ was too high, so we considered $[75, 175]$. There was a general consensus that 75 was too low—that is, the class judged that there was less than a 1/4 chance that the true value was below 75—so we

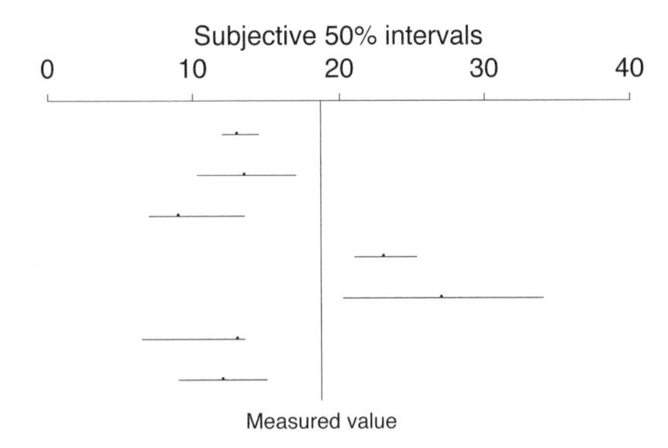

Fig. 13.5 Experts' predictions and 50% predictive intervals of the height at which an embankment would fail, along with the true value. None of the predictions included the true value.

adjusted again to [85, 175], which most of the class found reasonable.

We then revealed that the local telephone book had 458 listings of persons named Smith. This result—a true value that was well outside the 50% interval—is typical. People's uncertainty intervals tend to be too narrow—that is, they are *overconfident*. The students may be reassured to see the well-known example displayed in Fig. 13.5, in which a set of internationally known geotechnical engineers show overconfidence (in retrospect) in their probabilistic forecasts of the failure height of an embankment.

Several unknowns and calibration

We follow up with a written exercise in which students are divided into pairs and each pair is given a list of several unknown quantities and asked to write 50% intervals for each, which takes about 10 minutes. To keep students interested, we include questions on a range of topics. Figure 13.6 shows some questions we have used in class; we encourage you to develop your own list with the cooperation of your students. We perform the discussion before the written exercise because we have found that many students do not understand the concept of a probabilistic forecast until we have discussed it in class together.

We then write the questions on the blackboard and, for each question, reveal the true value (which is often surprising enough to elicit gasps from many of the students), then ask the pairs of students to raise their hands, first if their interval contains the true value, then if their interval does not. We write on the blackboard the coverage rate of the interval (for example, 8 intervals out of 25 containing the true value), and repeat for each of the 10 questions. We have performed this demonstration many times, and the students are invariably overconfident (even though we warn them about this in the preceding class discussion). Far fewer than half of the 50% intervals contain the true value. (A typical result for the questions shown in Fig. 13.6 is for the coverage to be less than 50% for all the

Uncertain quantity	25% lower bound	75% upper bound
% black		
# eggs		
# airline deaths		
% girl births		
% freshmen in phys sciences		
# French speakers		
# Super Bowl watchers		
# babies born		
# abortions		
$ median income		

Give 25% and 75% probability bounds for each of these quantities. You should specify the bounds so that, for an unknown quantity x, there should be a 50% chance that x is between your upper and lower bounds. Fill in all the blanks on the table. You will then be told the true values of these quantities.

1. The percentage of people in the United States who are "black" (from the 1990 Census).
2. The total egg production in the United States in 1965 (in number of eggs).
3. The number of airline passengers worldwide who died in plane crashes in 1980.
4. The percentage of babies born in the United States that are girls.
5. The percentage of entering college freshmen in the United States in 1990 whose probable field of study was physical sciences.
6. The number of native French speakers in Canada in 1981.
7. The number of people in the United States who watched the Super Bowl in 1995.
8. The number of babies born in the United States in 1992.
9. The number of abortions in the United States in 1992.
10. The median household income in the United States in 1996.

Fig. 13.6 Example of a handout we have used to demonstrate the difficulty of calibrating subjective probability intervals. Try this out yourself; the true values of the ten unknown quantities are on page 274. There is nothing special about this list; we encourage you to develop your own list that will interest your students.

questions except the probability of a girl birth, for which students tend to be slightly underconfident.)

After assessing the calibration of the students' intervals, we go on to explain an easy method for ensuring that your 50% intervals have perfect calibration. If you simply set half of your intervals to $(-\infty, \infty)$ and the other half to be empty, then you are automatically calibrated! Clearly, calibration is not the whole story; it is also important for the uncertainty intervals to be informative. Students can discuss how they would balance the goals of calibration and accuracy (see also Section 13.1.5).

For a more advanced exercise, students can discuss methods of calibrating subjective estimates in a setting such as the age-guessing example described in Section 2.1. The students in the Bayesian statistics class can be given data students' age guesses (see, for example, Fig. 2.2) and asked to come up with a procedure to get calibrated subjective intervals from these guesses.

13.2.3 Drawing parameters out of a hat

When introducing Bayesian statistics, a vivid way to illustrate population (or prior) and data distributions is by physical sampling. The demonstration goes as follows. Two students out of the class are picked to be "statisticians" and are taken out of the room. Each of the remaining students then draws a slip of paper from a hat, which, before the lecture, the instructor filled with random samples from a normal distribution with mean 100 and standard deviation 15. This slip of paper represents θ_j, the "true IQ" of student j. (IQ test scores are scaled so that their distribution is approximately normal; see Section 8.7.2.) Each of these students rolls a die several times and performs the appropriate linear transformation (see Section 7.2.3) to create a random variable with mean 0, standard deviation 10, and an approximate normal distribution to represent "measurement error," and then adds it to his or her true IQ to obtain a "measured IQ," y_j.

The two "statisticians" are then brought back into the room. They are told the population distribution of "true IQs," the distribution of measurement error, and the "measured IQs" y_j, and are asked to estimate the "true IQ" θ_j for each student in the class and to supply 90% posterior intervals. The length and coverage of these intervals are compared to the classical 90% intervals obtained from the "measurements" alone. We find that both sorts of intervals have the correct coverage properties (on average), but the Bayesian intervals are shorter, which makes sense since the Bayesian intervals make use of the known population distribution.

This example can be stretched out further by discussions of the prior distribution, the likelihood, and so forth. As in Section 8.7.2, the "IQ" context is a hook to get students involved.

13.2.4 Where are the cancers? A simulation

Near the end of the course, we work through the basics of Bayesian inference, including results for normal, binomial, and Poisson models. Where possible, we use prior distributions that correspond to actual populations, thus treating all Bayesian models as implicitly hierarchical (that is, with a prior distribution that represents the distribution of an actual population of parameters). As an example, we adapt the "drawing parameters out of a hat" demonstration to the kidney cancer mortality rates in U.S. counties (see Sections 2.2 and 13.2.1). To do this requires first setting up a probability model for the parameters θ_j (the underlying 10-year kidney cancer death rates in U.S. counties j) and the data y_j (the observed number of deaths out of a population n_j in each county j). We assume a Poisson distribution for each y_j with mean $n_j\theta_j$ (ignoring the age adjustment) and a conjugate gamma population distribution for the θ_j's, with hyperparameters set by matching moments (itself an interesting discussion topic), which has parameters $\alpha = 27$ and $\beta = 58\,000$. A student in the class is asked to compute the mean and standard deviation of this gamma distributionand to sketch its density function on the blackboard. We interpret the mean of the distribution, $\alpha/\beta = 4.65 \cdot 10^{-4}$, as a nationally averaged rate of $4.65 \cdot 10^{-5}$ kidney cancer deaths per person-year (since the data cover a 10-year period).

The demonstration now begins. Two students out of the class are picked to be "public health officials" and are taken out of the room. Each remaining student in the class is assigned a county j (identified by its name and population), taken from a list chosen to include a wide range of populations, ranging from about a thousand to over seven million (Los Angeles), and also, to keep the students' interest, counties that are well known (for example, New York), or with amusing names (for example, Jim Hogg County, Texas). Each student then selects a true (underlying) kidney cancer rate θ_j for his or her county by drawing from an envelope that we had previously stuffed with random simulations from the gamma distribution with parameters 27 and 58 000. The physical sampling brings an immediacy to the meaning of the prior distribution in a hierarchical model.

Each student then multiplies the county population n_j by the underlying rate θ_j to get an expected number of kidney cancer deaths in a 10-year period. The student then draws a random number, y_j, from the Poisson distribution with this mean (see Section 7.2).

All this is written on the blackboard. Then we erase the true rates and the expected rates, leaving only the county names, populations, "observed" deaths y_j, and "observed" rates y_j/n_j. The "public health officials" are then brought back into the room and asked to estimate the ranking of the counties in order of the true death rates θ_j. The highest and lowest observed rates, of course, will tend to be in low-population counties.

The class is then led through the Bayesian analysis, which yields the posterior mean and standard deviation of θ_j, conditional on y_j, for each county in the table. We conclude the demonstration by writing the true values of θ_j back on the blackboard, checking the confidence interval coverage, and comparing the underlying, observed, and posterior mean death rates.

13.2.5 Hierarchical modeling and shrinkage

A central topic in modern Bayesian statistics is hierarchical (multilevel) modeling. We have already considered one such example—the cancer rates as discussed in Section 13.2.4 are hierarchical in that there is a separate parameter for each county, and the counties themselves are part of a larger distribution.

It should be possible to develop a demonstration of similar concepts using data gathered directly from students. For example, suppose that each student in class is given a simple memory quiz twice (as in Section 4.3.1). It is then possible to estimate each student's ability with a hierarchical model, using the data y_{jt}, $j = 1, \ldots, J$, $t = 1, 2$, where j indexes the students in the class. The computer program to estimate the model can be set up ahead of time, and then the inference can be done immediately in class. Each student then gets an estimated true ability as compared to his or her raw score. The demonstration can be continued by giving each student a third memory quiz, and checking to see which estimate—the raw score or the Bayes estimate—better predicts the actual score, y_{j3}.

14
Student activities in survey sampling

The principles of active student participation that we have discussed in the context of an introductory statistics course can also be used in more advanced classes. Part I of this book contains many activities on sampling that we also use in our more advanced courses. For example, Section 8.1 describes a demonstration in which students draw samples of candies from a bag and weigh them. The candies are of different shapes and sizes; the larger ones are easier to grab, which leads to sampling bias. The activity on family size (Section 5.1.6) demonstrates size-biased sampling, and the examples of wacky surveys in the news (Section 5.1.3) offer a variety of examples of bias. Section 10.2 has many examples of selection bias.

This chapter describes student activities we have developed for undergraduate survey sampling classes of 20 to 40 students. In addition to demonstrations and examples, we describe our use of handouts, techniques for encouraging student participation, the use of news clippings and external reading materials, and ways to organize student projects.

14.1 First week of class

At the beginning of the course, we schedule activities for students to meet each other and get comfortable speaking in class. These activities help the class establish as a group, and they help students to choose partners for later projects. As the semester progresses we spend less class time on these sorts of activities and more time organizing and tracking projects. We do, however, continue to have the class work in small groups on worksheets and handouts that are focused on specific techniques (cluster sampling, sampling with probability proportional to size, network sampling, imputation, weighting, and so forth).

14.1.1 News clippings

The newspapers are full of reports on the latest survey results. These clippings can quickly generate discussion among students at the start of a course. For one of our first activities, we hand out a newspaper clipping, such as "1 in 4 youths abused, survey finds" (see page 16). Any news clipping that reports on a sample survey with the potential for several types of bias would work well here. After reading the article to the class, students work in pairs and write one criticism and one positive comment about the study. We use these papers to begin defining various types of bias in survey sampling—nonresponse, measurement, selection, question-wording, and so forth. Students hand in their comments with no names on them, and without sorting or looking through the papers, we take the top sheet from the stack, read it aloud, and discuss with the class. We continue reading papers, taking them one by one from the stack until we have heard and discussed a variety of ideas.

After class, we read through the remainder of the papers and select those with ideas that we did not cover and think would be good to address. We also make a pile of papers that contain incorrect statements or that indicate basic misconceptions. We bring these two sets of papers to the next class and continue our discussion. We have found it helpful to identify and address basic misunderstandings as quickly as possible, and we continue this process throughout the semester (see Sections 14.4 and 14.5 for alternative ways of generating discussion and identifying problem areas). Also, we have found that maintaining anonymity of the authors of the papers is crucial for guaranteeing complete student participation. And, as always, having students work in pairs reduces the pressure of performance and makes this a fun activity.

14.1.2 Class survey

In the first week of the course, we collect data on the students in our class. This information forms the basis of examples and demonstrations used in the first couple of weeks of the semester. The results of the survey are treated as a mini-population, and we draw samples to demonstrate many of the basic ideas behind simple random sampling.

We survey the class only once in the semester because after the first few weeks of the course we use examples with real survey data. Treating the students in our class as sampling units oversimplifies the issues in survey sampling. We do not treat the data collected from the class as a sample from a larger population, except to hold it up as a poor example of sampling undergraduate students because a probability method was not used to choose the sample.

To carry out our survey, we bring to class numbered index cards, one for each student. We mix them up, hand them out to the students, and ask them to write on the cards their responses to our questions, which we write on the board. Mixing the cards preserves anonymity, and we emphasize this point to the students. The information we request is often something impersonal such as the number of minutes they spent watching television the day before the survey. Other times we have passed out sheets of paper with a ruler photocopied on it and asked students to measure the span of their left hand (see Section 4.2.1). See Section 2.5 for more examples of questions to ask your students.

The cards serve as units in the population. We turn them face down, mix them up on the tabletop, and sample from our mini-population. It is easy in this way to take multiple samples, which helps students see the randomness in the sampling process. For each sample, we compute the average and standard deviation so students see how these statistics have a distribution. We also discuss various ways of picking the cards. We ask them if it matters if we stack the cards, shuffle them, and choose the top five for our sample. Or, is it acceptable to simply take the bottom five cards?

In a later class, we present a tally of the population values. The distribution of time spent watching television has a peak at 0 and a long right tail. We can compute the sampling distribution of the average via simulation, and the expectation and variance of the sample average can be computed exactly (see Section 14.4).

14.2 Random number generation

14.2.1 What do random numbers look like?

The question, What is a random number?, is important for understanding how sampling is done in practice. Although we use the computer to generate random numbers, the numbers it generates are not really random. They are pseudo-random numbers. We introduce one of the simple algorithms for generating pseudo-random numbers, and demonstrate how it is deterministic and eventually repeats itself. However, if started properly the numbers appear random-like. That raises the questions, What do random numbers look like? and Why not skip the computer and just generate our own numbers "at random" out of our heads? To answer this question we use the demonstration on real versus fake coin flips (see Section 7.3.2), where we show how we can tell the difference between a sequence of 100 real coin flips and a sequence of 100 coin flips made up by the students. If there is further interest in the topic, we refer students to Web sites in which pseudo-random numbers or physically generated random numbers are available (see Section 7.2.2).

14.2.2 Random numbers from coin flips

To see how a random number can be created, we use the randomness in coin flipping. This demonstration also serves as a review of some of the basic ideas in probability. We divide the class into groups of two to three students, give each group a coin, and ask them to come up with a procedure for generating a simple random sample of size 2—two numbers at random without replacement—from the integers 1 to 20. We hand out instructions on how to approach the problem. The instructions divide the task into four parts:

1. Develop a method to generate either the number 1 or 2 at random from one flip of the coin.
2. Extend your method to generate a number from 1, 2, 3, 4 using two flips of a coin. Each number should have an equal chance of being generated.
3. Now generate two numbers between 1 and 4 without replacement.
4. Finally, extend the technique for generating a number between 1 and 4 to the problem of generating a number between 1 and 20, inclusive. Be sure that each number has an equal chance of being produced from your procedure.

The last two parts can be hard for students to figure out. We write hints on the board to keep groups from getting discouraged. For example, we write the question, "Can we generate numbers with replacement and ignore those that have already appeared to get a sample without replacement?" In our experience, not all groups will find solutions to all four problems in the allotted time. We conclude the demonstration by choosing four groups to write up a solution to one of the four parts of the task. Rather than ask for volunteers, we approach groups and request them to write their results on the board. By controlling who writes solutions on the board, we ensure equal participation from all the students over the course of the semester. We make sure to leave time at the end of the class to go over the solution, elaborate on the students' solutions, correct mistakes, and tie up any loose ends. In our experience, students don't like a delayed punch line.

In addition to this activity, or as an alternative to it, we have students sample entries from a page in the telephone book (Section 5.1.1). Be sure to use 20-sided dice or handouts to generate random digits (see Section 7.2.1). Six-sided dice are too difficult to work with.

14.3 Estimation and confidence intervals

The candy-weighing example (Sections 8.1 and 14.3) can be adapted to use for introducing standard errors and confidence intervals. For this activity, we use candies that are all the same size and shape, so no sampling bias. The candies differ only by the color of the wrapper. Foil-wrapped chocolate candies work well for this activity. We fill a sack with 120 candies, say 72 gold and the rest silver. Each student draws 16 candies from the bag, counts the number of gold ones, puts them back, and passes the bag to the next student. While this is happening, we set up the blackboard for data collection, and we lecture on the standard error of a sample proportion. After the students have taken their samples, we go around the classroom, and each student shouts out his or her sample proportion. We mark the students' sample proportions and confidence intervals on the board (see Fig. 8.3). While we are drawing these confidence intervals, we pass the bag of candy around the classroom for consumption. When we reveal the true proportion, the visual image of the varying intervals makes the point that some intervals cover the true proportion and some do not.

Since we are estimating proportions, and the sample size is small, we can prepare in advance a list of all possible proportions, standard errors, and confidence intervals. This makes it easy to display the results quickly. As we go around the classroom soliciting proportions and intervals from the students, we write their names next to the intervals we draw on the board. Then in the subsequent discussion we identify unusual intervals that do not cover the true proportion by the sampler's name. Further tips on how to operate this demonstration can be found in Section 8.1.

For another example, we have developed a triptych template (Fig. 14.1) to help students differentiate between a population and its parameter, a sample and its statistic, and the sampling distribution.

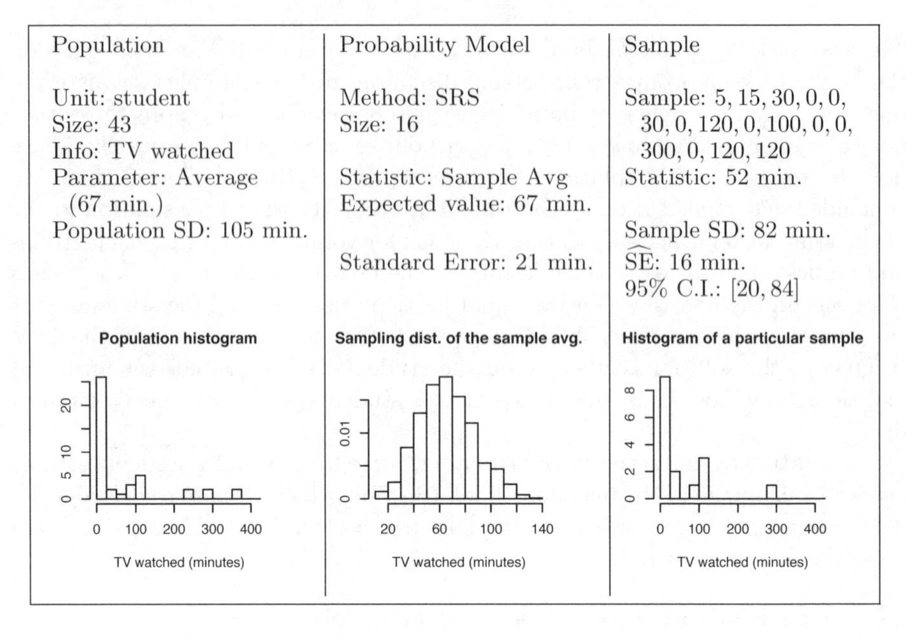

Fig. 14.1 Triptych template for a simple random sample from the class survey. Each member of the class reported the number of minutes he or she watched television the night before the survey. In this example we know all the population values, which are displayed at the bottom left panel. We simulated the sampling distribution by taking 1000 samples from the population. The resulting histogram of 1000 sample averages appears in the center panel. The right panel contains the values for the sample that we take in class. We hand out the triptych as a template and have students fill in the blanks. With examples where we do not know all of the information about the population, we draw in the left panel a sketchy outline that resembles the sample histogram, and we place a question mark over it to emphasize that we do not know what it looks like but it should resemble the sample histogram.

14.4 A visit to Clusterville

Figure 5.6 on page 54 shows a template that students can fill out to isloate the steps of a simple survey. In a course on survey sampling, students are introduced to more elaborate sampling methods, for example, stratified, cluster, block, and network sampling. These various methods are easier to understand when students discover the main concept behind the sampling technique on their own through activities or worksheets. The name of the sampling method often suggests how to do it. When we give simple examples with diagrams that suggest how to apply the technique, students can often derive the basic approach to the sampling method and develop an estimator using their data.

For example, "cluster sampling" suggests sampling groups or clusters of units. With the diagram and simple instructions found in Fig. 14.2, we have students work in small groups for about 10 to 15 minutes to figure out how to take a cluster sample from Clusterville. We also have them develop notation that is

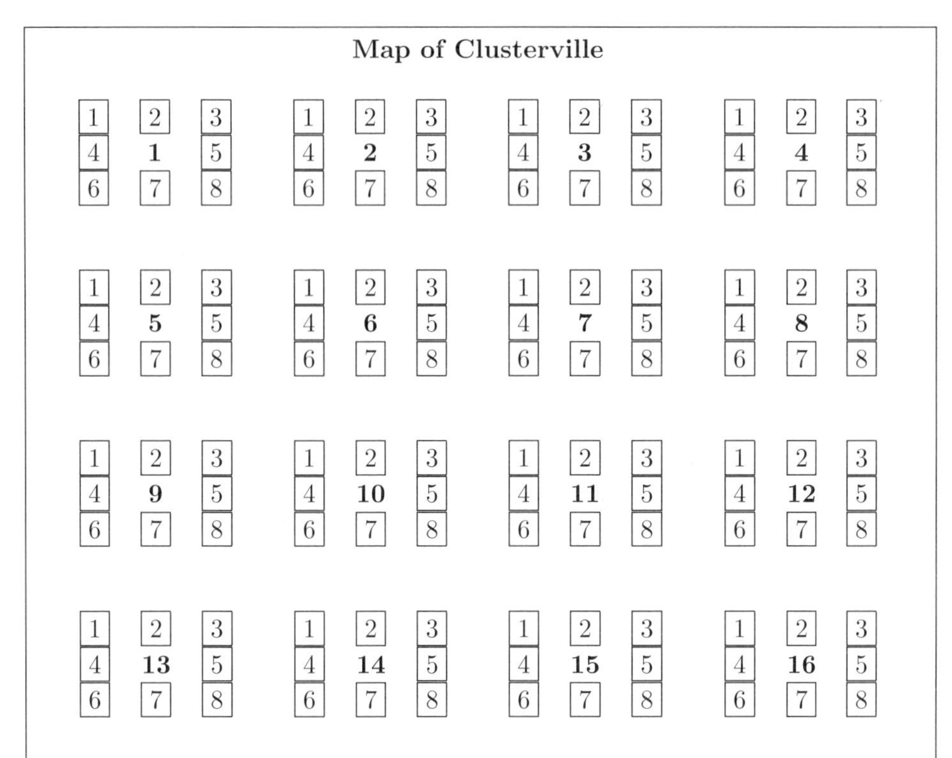

1. Take a "cluster sample" of 16 units from Clusterville. Generate your sample with coin flips and shade the selected households on the map.
2. Describe in words the cluster sampling technique that you have developed. Can you generalize it to other populations with a different number of clusters and a different cluster size?
3. What do you think are the pros and cons of your new sampling design? There is no need to make a calculation. Explain why this design might be preferred to the simple random sample or the stratified random sample.
4. How would you estimate the population average based on your sample data? Use the notation provided in the map to define your estimator.
5. Do you think your estimator is unbiased?

Fig. 14.2 Example of a worksheet for learning cluster sampling (see Section 14.4). In this map of Clusterville, each square represents a household. There are eight houses to a block, and one household in each house. We sometimes label the map with street and avenue names to make it appear more like a town. Students are asked to draw a random cluster sample of 16 households from Clusterville. For example, with four flips of a coin, we can choose a block of eight houses (TTTT=block #1, TTTH= block #2, and so forth).

rich enough to support their methodology.

In implementing this activity, it is important to keep it on course so that it does not become too time-consuming for the amount the students learn. For example, we have not had much luck with simulating capture/recapture sampling. In our activity, students rolled dice to move markers, which represented animals, around a game board that was marked with traps and free spaces. We found there were too many steps involved in simulating the population movements. In general, we bring a handout or template to each class meeting. Students use the handouts to take notes or to practice a technique discussed in class, as with the triptych template in Fig. 14.1. We also provide charts such as one on various ways of imputing missing data and the one in Fig. 5.6 on page 54 where the student must identify the target population, sampled population, and sampling frame of a survey. We think these handouts help students process the material at a deeper level by applying it to simple examples as they learn it.

14.5 Statistical literacy and discussion topics

Real examples of surveys make the important connection between statistics and the outside world. Articles from the newspaper and not straight from the book provide additional class examples and they also address statistical literacy.

We adapt the approach presented in Chapter 6 and expose the students to a variety of surveys in the first several weeks of class. We set aside time each week to discuss a survey. We vary the presentation somewhat from survey to survey, but we always provide a set of questions, excerpts from a report on the study, and sometimes a newspaper clipping. When the research report is long, we hand the report out in advance, and ask students to come to class prepared to discuss it.

Figures 14.3 and 14.4 give an example of a news story and accompanying assignment. The survey under discussion was conducted by the U.S. Department of Labor. In fact, two samples were taken: a popular sample was obtained by distributing questionnaires to employers, labor unions, and women's organizations; and a scientific, or probability, sample was collected using a telephone survey. The main task for the students is to compare the two samples. For this assignment we also hand out copies of excerpts from a government report on the survey.

Students work in pairs to answer the questions. Everyone answers the first question to get at the basic differences between how the two samples were conducted. Each group is also assigned one of the remaining four questions. We give them about ten minutes to answer both questions, and for each question, we ask two groups to write their answers on the board. We lead a class discussion on the questions, and we ask the groups to explain their responses. As an alternative to this type of discussion, we sometimes have the groups write up their responses to be turned in for grading. We encourage class participation and attendance by including questions pertaining to the surveys discussed in class on exams and homeworks.

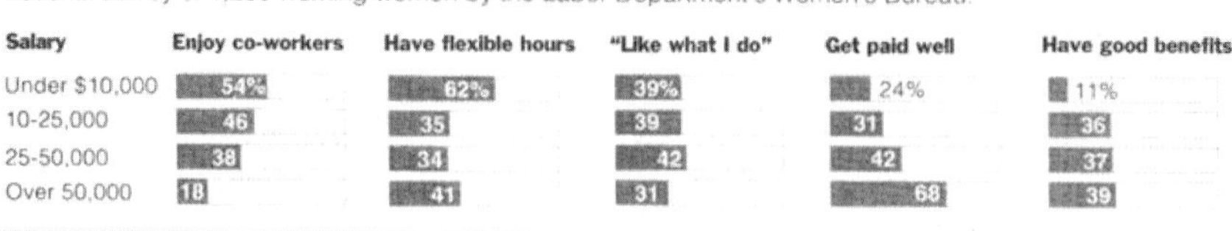

OVERVIEW

What Women Say About Their Jobs

National survey of 1,200 working women by the Labor Department's Women's Bureau.

Salary	Enjoy co-workers	Have flexible hours	"Like what I do"	Get paid well	Have good benefits
Under $10,000	54%	62%	39%	24%	11%
10-25,000	46	35	39	31	36
25-50,000	38	34	42	42	37
Over 50,000	18	41	31	68	39

Working Woman Say Bias Persists

By TAMAR LEWIN

Many working women still feel that they are not getting the pay, benefits or recognition they deserve, according to a new survey conducted by the Federal Government.

The survey, based on more than 250,000 questionnaires received in the four-month survey by the Women's Bureau of the Department of Labor, also found that working women mentioned stress as a serious problem more than any other.

And many women, in very different fields, complained of discrimination. An Alabama coal miner, who said she was the first woman to be hired in the mine where she worked, reported that she earned about $20,000 a year less than the men with comparable experience.

A Milwaukee woman with a bachelor's degree in electrical engineering said male co-workers with less education got the challenging work, and a Maryland woman in a three-person shipping and receiving department told how two male co-workers were paid more but had less accountability.

"The concern about discrimination and equal pay surprised me," said Karen Nussbaum, director of the Women's Bureau. "The popular wisdom had been that we don't talk about things that way anymore, but clearly it is the way women talk about it.

"I was also surprised by the consensus that emerged. We tend to think of training as a blue-collar issue, child care as a low-income issue, the glass ceiling as something professional women care about and discrimination as a concern for women of color, but each of these issues cut across all the lines."

In 1993, Government figures show, women earned 71 cents for every dollar earned by a man, up from 61 cents in 1978.

Still, nearly four out of five women said that they liked or loved their jobs — and only 4 percent said they disliked their work or found it "totally miserable."

For the Women's Bureau survey, known as Working Women Count!, questionnaires were distributed through more than a thousand businesses, community organizations, labor unions, newspapers and magazines, and more than 250,000 women returned them from May to August.

Discrimination cuts across class and racial lines, a survey finds.

Because that was not a scientific sample, the Women's Bureau also conducted a telephone survey in June, asking a nationally representative random sample of 1,200 women the same questions, to provide a benchmark for evaluating the overall responses. The margin of sampling error was given as plus or minus five percentage points.

Women in both the scientific and popular samples said their priorities for changes in the workplace were improved pay and health insurance. And the No. 1 issue that women said they wanted to bring to President Clinton's attention was the difficulty of balancing work and family.

Many women mentioned a "time crunch" and exhaustion, using adjectives like "hectic" and "hard" to describe their lives.

Among the findings from the scientific sample were these:

¶Forty-three percent of the women who worked part time, and 34 percent of those over 55 years old, lacked health insurance, compared with 18 percent of the general population.

¶Nearly half of the women said they were not paid what they deserved.

¶Stress was identified as a serious problem by almost 60 percent of the women, but it was particularly acute for single mothers and women in their 40's who held professional or managerial jobs.

¶Fourteen percent of white women and 26 percent of minority women reported losing a job or promotion because of sex or race.

¶Three out of five women said they had little or no likelihood of advancement.

¶Nearly a quarter of the women said they had no pension plan.

¶More than half the women with children 5 years old and under said that affordable child care was a serious problem.

¶Fourteen percent of the women — including almost a third of those earning less than $10,000 — said they had no sick leave.

The report acknowledged that while only women were surveyed, many of the problems described could also apply to men. The concerns, the report said, reflected "the trend toward a work force anxious about job insecurity, declining benefits and stagnant wages."

Fig. 14.3 This news article reports on two surveys of working women conducted by the U.S. Labor Department. We use this article and excerpts from the Labor Department's publication to promote discussion of the pros and cons of popular vs. scientific surveys. We promote the discussion using the questions appearing in Fig. 14.4.

1. The survey included "scientific" and "popular" samples. Compare the sampling methods with respect to the following methodological issues. Be as specific as possible.
    - Target population
    - Sampled population
    - Sampling frame
    - Sample size
    - Nonresponse size
    - Sampling method
2. What are three of the biggest difference between the samples that you think would affect women's attitudes toward their jobs?
3. What table in the report did the bar chart in the article come from? Describe what information was used in creating the bar chart. What are the advantages of the bar chart over the table?
4. Name three of the most important things in the report that are not in the article.
5. For each of the following items in the newspaper article, determine whether it came from the scientific sample or popular sample.
    - bar chart
    - 4% said they dislike their work
    - stress was identified as a serious problem for 60% of the women
    - an Alabama coal miner, who said she was the first woman ...

    Which results (scientific or popular) do you think the *New York Times* trusts more?

Fig. 14.4 Worksheet on which students work in groups for the statistical literacy assignment comparing a scientific sample and a popular sample used in a survey of working women. See Fig. 14.3 for the accompanying news story. Students are also given excerpts from a government publication describing the survey protocol and results (see Section 14.5).

14.6 Projects

We use four types of projects in our sampling course. We ask students to research and report on a complex survey; simulate and apply, using the computer, the techniques learned in the classroom; study a special topic in depth, which includes extra reading materials and homework assignments; or design and carry out a survey. In any semester, we assign only two projects, typically the research report and one of the other three types of projects. The project where the class carries out a survey of its own design is described in Section 5.2. The other three are described here.

Survey	Organization
National Survey of Family Growth	National Center for Health Statistics
Youth Risk Behavior Surveillance	Centers for Disease Control
National Health Information Survey	National Institutes of Health
Consumer Expenditure Survey	Bureau of Labor Statistics
Current Population Survey	Bureau of Labor Statistics
American Housing Survey	Dept of Housing and Urban Development
Commodity Flow Survey	Census Bureau
National Survey of Speeding and Other Unsafe Driving Actions	Department of Transportation
Survey of Income and Program Participation	Census Bureau
National Crime Victimization Survey	Department of Justice

Fig. 14.5 Several examples of complex U.S. government surveys and the organizations in charge of administering them. Our students have used these surveys for projects. Other class projects have involved complex surveys conducted by private and academic research groups. Titles of some student presentations of these surveys also appear in Section 14.6.1.

14.6.1 Research papers on complex surveys

Three weeks into the semester, students begin work on the first course project, researching a complex survey. They work in groups of 3–5 to prepare a 10–15 page report describing a complex survey. Each group works on a different survey. Figure 14.5 contains a list of surveys students have studied in the past.

The first time we assigned this project, the results were disappointing. Too many papers read like abridged government documents. Most often, students had read through an official government publication that described the survey, and condensed it into a 10-page paper. They did not write the description in their own words, and they did not use the notation and conventions they had learned in the course, so it was unclear how much they got out of the project. After that experience, we provided much closer supervision, and the assignment became quite successful. We found that with closer supervision, students hone their research and writing skills, and the projects are more ambitious and more rewarding for them.

To get the students started, we provide them with a list of possible surveys to study (Fig. 14.5), and we spend part of one lecture period reviewing the topic of each survey. To help students form groups, we write the main subject areas on the board (politics, sociology, health, economics, and so forth) and have students stand near the area that interests them. Then they divide into groups and decide which of the complex surveys they wish to study. Or, if they want, they may choose to study a survey they have found on their own. At the time the groups form, they exchange contact information, and sign up for the first of two 15-minute appointments to discuss the project with us.

At the first appointment, students bring a completed template (see Fig. 6.1 on page 79), a copy of the main source, and their draft description of the survey. We use the main source to check their work to see if they have missed an important aspect of the survey design. At the meeting, we identify a few important aspects of the survey for the group to study in more depth, and we discuss strategies for investigating these additional topics. These topics are typically the subject of oral presentations. For example, here are the titles of some student presentations:

- Public opinion polls and probability weighting
- Random digit dialing and political polls
- Binge drinking on college campuses
- Oak tree regeneration and the Sen–Yates–Grundy method of variance estimation
- Customer satisfaction with public transit and nonavailability bias
- How is unemployment measured? The Current Population Survey
- The Consumer Expenditure Survey—time lines, diaries, and interviews
- The evolution of the National Survey of Family Growth
- Obtaining accurate information on pregnancy rates
- Commodity flow survey—design and measurement.

At the second appointment, the students in the group bring their nearly complete version of the paper. The purpose of this meeting is to answer questions that have arisen in addressing the additional topic and to focus on the presentation of the written report. Also, if oral presentations are planned, we discuss which aspects of the report they will present in class. Finally, we also offer to review and comment on their draft of the paper, and we require in advance of the oral presentation copies of handouts and an outline of the presentation.

14.6.2 Sampling and inference in StatCity

Many of the methods studied in an advanced survey sampling course, such as weighting, imputation, and variance estimation, are quite complex. Often these methods rely on the computer for implementation, and to get a fuller understanding of how they work takes practice. We have developed a computer project where students practice these methodologies.

For our project, we use census data to make up a population that we call StatCity. Students select a stratified cluster sample from the city, compute weights to match city totals, impute missing values using the hot deck technique, and construct a confidence interval for median family income. They use the statistical package `R` to do this.

Our handout outlines the project. For each step, the students write a description of what they did, and they must include evidence that they have done it properly along with their code. For this evidence, we request specific sorts of information, such as a table of weights or a histogram of bootstrap values. As students progress on the project, we receive many questions in class, in office hours, and in email to us. It helps all the students, and reduces our work, if we put their questions with our answers to them on the course Web site.

StatCity project

StatCity is divided into two districts: Northeast and North Central. Imagine that you are the city Census Director and that you have been asked to take a sample of the people of StatCity in order to estimate the median family income.

Proceed as follows:

1. Your sample should be a stratified cluster sample, where the strata are the city districts and the clusters are families. The families should be selected proportional to the number of persons in the family. You may ignore the difference between sampling with and without replacement. One person is to be sampled from each family chosen for the sample. The sample size is 1000.
   Provide the following:
   (a) A description of the method used to select the sample.
   (b) A subset of the random indices of the persons chosen for the sample.
2. Weight your sample to:
   (a) Account for the probability that the family is included in the sample. Describe these weights.
   (b) Adjust for nonresponse using ethnicity (white non-Hispanic, black non-Hispanic, Hispanic, other) and sex. Use the method of raking. Match the city totals:

White non-Hispanic	42790	Males	26100
Black non-Hispanic	5022	Females	28638
Hispanic	5043		
Other	1883		

   Provide a table of the weights to be used.
3. Impute any missing values of education level using the hot deck technique with ethnicity and sex. Provide a list of the persons that had their educational items imputed, and provide the persons that were used to impute the missing items. Estimate the proportion of adults with a college education. Compare your estimate to that obtained without imputation.
4. Graph the distribution function for family income for those in your sample. Be sure to use the weights. Use the distribution function to estimate the median family income.
5. Use the bootstrap to determine a 95% confidence interval for median family income. Provide a histogram of the bootstrap values for median family income.

Frequently asked questions for the StatCity project

1. Is each record in the dataset one family?
   No, each record represents one person. If there are three people in a family then there will be three records in the file, one for each. There is no family identifier in the file, so families cannot be exactly matched up.
2. Do I take a SRS of 1000 from the whole population?
   No, you are to take a stratified cluster sample. To do this you take a sample from each of the regions in the city separately. You need to figure out how many persons you will sample from each region. The total sample size should be 1000.
3. How do I take a cluster sample when the dataset is not in clusters?
   Look at the random digit dialing example in your text. It shows how a SRS of phone numbers can be used to sample banks of 100 phone numbers where the probability is proportional to the number of residential phones.
4. What if I wind up with two people from the same family?
   This should be a rare event; ignore it.
5. How do I rake when the ethnicity codes in the dataset do not match the categories supplied (white non-Hispanic, etc)?
   You will need to create these categories by using two variables in the original data: `race` and `ethnicity`. For example, white non-Hispanic are those with `race` = 1 and `ethnicity` = 8, and Hispanics are those with `ethnicity` between 1 and 7. The other category should be the remainder (you can include the NA and DKs in this group).

6. When I rake, do I need to weight the data according to the probability of picking a family?
   No, the raking should be done at the individual person level. Make a table of the counts in your sample of the eight categories: male white non-Hispanic, female white non-Hispanic, ..., male other, and female other. Use this table to find the weights using the raking methods and the population totals that you have been given.
7. What do I do with the records that have NA in them for everything but the region?
   You just drop these records from your sample. They are your nonrespondents. Your total "working sample size" will be less than 1000.
8. When I rake, do I need to worry about the way that I allocated the sample sizes to the strata?
   Yes, if you have allocated your sample proportionally to the population totals for the strata then you need not worry. Otherwise, the counts in your table should take into consideration the allocation.
9. When I impute the education level for those with missing values, do I need to use the weighting scheme in selecting the records at random in the hot deck procedure?
   No, you can simply choose at random with replacement from the group of persons with sex and ethnicity that match the record with the missing education. (Ethnicity is defined according to the variable you created in 2b: white non-Hispanic, black non-Hispanic, Hispanic, and other.)
10. What should my estimator for the proportion of college educated adults look like?
    (a) Start by creating a new variable that indicates whether a person is college educated or not: y_i is 1 if the person has at least a college education, and 0 otherwise. Also create an indicator for whether the person is an adult or not: x_i is 1 if the person is 18 or over and 0 otherwise.
    (b) Use the weights from step 2b in creating your estimator. Call these rates r_i for raking weights. These are the only weights that you will need, because your estimator is at the person level, not family level.
    (c) Find the proportion of people in each stratum that are college educated. Be sure to use your weights. Your estimator should then be $(\sum_i y_i r_i)/(\sum_i x_i r_i)$, where the sum is over those individuals sampled from the first (or second) stratum.
    (d) Use the population totals to combine your two proportions into a single estimate.
11. What weights do I need to use in finding the family income?
    You will need to take the product of the family weight (call it f_i) and the raking weight: $w_i = f_i r_i$.
    Also if you have not proportionally allocated then you will need to adjust for the allocation method. (It's probably a good idea to proportionally allocate your sample.) You should probably check to see if $\sum_i r_i$ for stratum 1 is reasonable close to the population total for stratum 1. If not, you may want to include a third weight to adjust for this difference.
12. Do I compute the median separately for each stratum and then combine them as I did with the estimate for proportion of college educated adults?
    Unfortunately that won't work. Instead, you need to find that income I^* such that $\sum_i w_i$ for those with incomes less than or equal to I^* is $\frac{1}{2}\sum_i w_i$ for all those in the sample.
13. How do I compute the median using weights?
    Try the following: sort your weights according to salary. Then use the R commands,

```
htot <- sum(wt)*.5
indexmedian <- min ((1:length(wt))[cumsum(wt)>htot]
```

Judges Bar Sampling For Census

Survey intended to correct undercounting of minorities

By Ramon G. McLeoa
Chronicle Staff Writer

A three-judge panel in Washington, D.C., dealt the Clinton administration another major defeat yesterday in its plan to use statistical methods in the 2000 census to correct the undercounting of minorities.

The panel ruled unanimously in favor of a suit filed by congressional Republicans that sought to stop plans by the Census Bureau to use a post-census survey to account for missing minorities.

HOW SAMPLING WORKS

The technique of sampling is being tested now in Sacramento, which participated in a test census in April.

Here's how it works:

In the first phase of a census, residents are asked to mail back census forms, the standard procedure now used as the first way to get compliance. In the Sacramento test, only about 50 percent of the forms were returned.

After the mail-in phase ends, census-takers go out in the field, knocking on doors and try to interview people who live at addresses that have not returned their form.

When this phase ends, the sampling phase starts.

Census-takers will go to selected areas and knock on doors to get a sample of the population in a particular kind of neighborhood. Ultimately, the census-takers determine which specific kinds of people are typically not being found by the other approaches.

Those results would then be applied to other neighborhoods thought to have similar characteristics.

Fig. 14.6 A few years ago, the newspapers were full of stories about the U.S. Census and the debate over whether sample survey results should be used to correct it. This story appeared in the *San Francisco Chronicle* on August 25, 1998. It reports on the decision to not adjust the Census and the possible implications of this decision for the state of California. The inset describes briefly how a probability sample was to be used to adjust the census for possible undercount.

14.6.3 A special topic in sampling

We like to conclude the class with an example that students can research in depth, for example the high-profile controversy over whether a sample should be used to adjust the U.S. Census for possible undercounting. There is an abundance of literature on this topic that is accessible (with our help) to undergraduates who have studied sampling surveys.

We begin this project with a discussion of the importance of the topic, and we bring to class several news stories and advertisements about the Census (see Fig. 14.6, for example). However, the main task here is for the students to work through several articles and research papers on the census, which we place on reserve at the library.

For sources, we use the Census Web site, proceedings of the American Statistical Association and articles in statistical journals. In particular, the Census 2000 Operating Plan is available on the Web. This 150-page document provides a broad overview of the census, covering topics such as questionnaire content, long form sampling plan, address list development, enumeration strategies, nonresponse followup, and the accuracy and coverage evaluation. Papers in the American Statistical Association Proceedings on Survey Research Methods, 1991, contain detailed descriptions of some aspects of the census. For example, we found "The 1990 Post-Enumeration Survey: an overview" (Hogan, pp. 518–523) and "Address reporting error in the 1990 Post-Enumeration Survey" (West, Mulry, Parmer, and Petrik, pp. 236–241) to be quite readable. Finally, the main ideas behind the debate over adjustment can be found in articles in *Statistical Science*, *Chance*, *Evaluation Review*, and *Jurimetrics*. For example, the Fall, 1993, issue of *Jurimetrics*, "Adjusting the Census of 1990: an exchange," is dedicated to the topic.

To help the students read and digest this material, we provide several extensive homework assignments that require going through the papers, filling in holes, working out special cases, defining terms, and proving results related to the published work. After careful completion of these assignments, the students should have an understanding of the main points of the articles. When the students are working on this project, we dedicate class-time to covering the basics—discussing the main ideas and preparing them to carry out the assignments. We have found that this project provides an excellent capstone experience for our students because it builds on what they have learned in the semester, and because it is a highly publicized political debate surrounding a statistical controversy.

15 Problems and projects in probability

When probability is taught as a separate unit, not as part of a statistics course, then it makes sense to include exercises and examples that are motivated from mathematics rather than from applications. This chapter illustrates one way to promote active student involvement in a more advanced class. The material is designed to accommodate various mathematical backgrounds and to challenge all students.

There are essentially three basic aspects to our mathematical probability course. We teach in an interactive seminar style; we give students many challenging problems; and we require students to work on longer projects where they derive complicated results step by step.

15.1 Setting up a probability course as a seminar

Our interactions with students in the first several weeks of the semester centers on group problem-solving. We begin each class with a quick review of the material covered in the previous lecture; then we introduce new concepts via an example that the class works on with the instructor. After completing the example, the students split into groups of two to four to work on problems that further develop the material just introduced. Typically we set a time limit for the groups to work on the problems, say 10 to 15 minutes, and we circulate among the students to keep them on track and offer advice. When many groups encounter the same point of confusion, we write hints or points of clarification on the board. In the last part of the class period, we ask each group to write up a solution to one problem on the board. When there are more groups than problems, we sometimes have more than one group write a solution. The groups often work at different speeds; some will not have completed all the problems in the allotted time, and others will finish quickly. We bring extra problems for the rapid learners, and those still working on the first set of problems are asked to write up a problem that they have finished. At the end of the "lecture" period, we bring everyone together to review the solutions, make corrections if needed, tie up loose ends, and draw connections to the big picture.

This kind of group problem-solving also works well when each group works on a different problem. For example, to introduce the standard discrete distributions (binomial, geometric, hypergeometric, and negative binomial), each group of students works with a different distribution. They first find a numerical expression for a specific probability; for example, the binomial group finds the probability of three aces in seven rolls of a fair die. When they finish this exercise, they generalize their result to find the probability of exactly k aces in n rolls of a die, where the die lands ace with probability p. In these exercises, students gain practice formulating a parametric probability model and they are introduced to many of the standard discrete probability distributions via the wrap-up at the end of the class period.

With group work we can cover a lot of basic material quickly. Indeed, we have found it crucial to set a rapid pace from the start and get students accustomed to the idea of discovering results and proofs for themselves, without the aid of texts. We have developed a series of problems, which we provide in weekly handouts. The handouts include problems that illustrate the main concepts and just-for-fun, difficult problems about sleazy gambling joints, winning the lottery twice, and random cuts of spaghetti. By the end of the semester, each student has compiled a set of solutions to approximately 100 problems covering all of the course material. These problems are in addition to their homework assignments. Section 15.2 provides examples of the problems handed out to students. They are designed to help students discover fundamental properties of probability and expectation. The students' eagerness and interest leads them to demolish these problems in two weeks, whereas a typical undergraduate class would take almost double that time to get this far, even with the aid of detailed lectures.

15.2 Introductory problems

The following are two sets of simple problems that we give at the beginning of the course to introduce the students to mathematical probability. We include them here not because there is anything special about our presentation but rather to illustrate how we construct assignments for these students. More probability examples appear in Chapter 7.

We demonstrate some of these problems in class before letting the students loose on them. For example, with the birthday example (problem 3 in Section 15.2.1 and problem 6 in Section 15.2.2), we go around the class looking for same-birthday pairs. We write students' birthdays on the board and count the matches. Once when we did this with a class of 60 students, we found we had twins in the class (we thought they were just brothers until then), and another student shared the twins' birthday! All together we had five matches. After we talked about the assumptions we made in computing the probability of a same-birthday pair, we computed the expected number of pairs, and found it to be $60 \cdot 59/(2 \cdot 365) = 4.8$, another surprise for the students.

For many of our demonstrations, we bring props to class. We collect them and keep them on hand for our demonstrations. We have jumbo red and blue dice, a set of lockable boxes (problem 9 in Section 15.2.2), oversized playing cards (problem 1 in Section 15.2.1 and problem 7 in Section 15.2.2), sets of double-sided colored cards (Section 7.5.1), and a roulette wheel (Section 8.2.2). Using props adds variety to our lectures, makes problems more concrete when we describe them to the class, and takes little additional time over simply describing the problems in words.

15.2.1 Probabilities of compound events

Here is a set of problems that we give to students to start them thinking about the probabilities of complex events in abstract settings. These complement the more specific demonstrations and examples given in Chapter 7.

1. *Symmetry in card shuffling.* A deck of cards is well-shuffled, and the cards are dealt one by one. What is the probability that the second card is an ace? What is the probability that the seventeenth card is an ace? What is the probability that the second and third cards are both aces? What is the probability that the seventeenth and fiftieth cards are both aces?
The symmetry in these answers points to a theorem about random permutations. Figure out what the theorem is; then state and prove it.

2. *The gambler's rule.* Many questions in elementary probability theory were originally posed by gamblers some centuries ago. Here is one of them. A gambler bets repeatedly on an event that occurs with probability $1/N$, where N is a fixed positive integer, usually thought of as large. So for example, he or she could be betting on the number 17 at roulette with a probability $1/38$ of winning. Suppose successive bets are independent of each other. How many times should the gambler bet so that the probability of winning at least once is greater than $1/2$? Gambling experience suggests that the answer is about $2/3$ of N. Is this consistent with your calculation?

3. *The birthday problem.* There are n people in a class. What is the probability that at least two of them have the same birthday? (What assumptions are you making?) Roughly how big is this probability if $n = 40$? About how many people should there be in the class to make the probability at least 0.5?

4. *Another birthday problem.* There are n students in your class. What is the probability that at least one of them has the same birthday as yours? Is this problem the same as the birthday problem above?

5. *Binomial distribution with parameters n and $p = 1/2$.* A coin is tossed n times. What is the probability of getting exactly k heads? What is the most likely number of heads? Roughly what is the probability of getting exactly 1000 heads in 2000 tosses?

6. *Binomial distribution with parameters n and p.* Consider n independent trials, each of which results in a success with probability p. What is the probability of getting exactly k successes? What is the most likely number of successes?

7. *Negative binomial distribution with parameters k and p.* A gambler bets repeatedly on an event that has probability p. The bets are independent of each other, but the gambler decides to stop betting as soon as he or she has won k times. What is the probability that she stops immediately after the nth bet? (The special case $k = 1$ gives rise to the *geometric* distribution with parameter p.)

8. *The symmetric gambler's ruin problem.* A gambler bets repeatedly on tosses of a coin, on each bet winning \$1 if the coin lands heads or losing \$1 if the coin lands tails. The gambler starts with a dollars and decides to keep betting until he or she has won an additional b dollars or is broke. What is the probability that he or she ends up broke? (Assume a and b are positive integers.)
9. *The asymmetric gambler's ruin problem.* Do the gambler's ruin problem again, replacing the coin toss (which has probability 0.5 of heads; see Section 7.6) with a randomizer that yields "heads" with probability p .

15.2.2 Introducing the concept of expectation

We introduce the mathematics of expected values through the following series of problems for students to work on in groups.

1. *Calculating* $E(X)$ *from the distribution of* X. Show that

$$\mathrm{E}(X) = \sum_x x\mathrm{Pr}(X = x),$$

where the sum is over all x in the range of X. Most texts use this formula to define $\mathrm{E}(X)$, by saying something like: Let X be a random variable with possible values $x_1, x_2, \ldots$. The *expectation* $\mathrm{E}(X)$ is defined as,

$$\mathrm{E}(X) = \sum_i x_i\mathrm{Pr}(X = x_i).$$

This is fine, and usually all that's needed in most elementary calculations. But to understand and prove linearity, it is essential to think of the more formal definition: let X be a random variable defined on a countable outcome space Ω, on which there is a probability distribution p. The expectation of X is defined as,

$$\mathrm{E}(X) = \sum_{\omega\in\Omega} X(\omega)p(\omega),$$

provided the series is absolutely convergent; if not, the expectation does not exist.
2. *Linearity of expectation.* Let X and Y be random variables defined on Ω, and assume that $\mathrm{E}(X)$ and $\mathrm{E}(Y)$ exist. For constants a and b, show that

$$\mathrm{E}(aX + bY) = a\mathrm{E}(X) + b\mathrm{E}(Y).$$

Though it might seem simple, this property of expectation is of fundamental importance, as you shall see in the problems below.
3. *The method of indicators.* Let I_A be the *indicator of the event* A; that is, $I_A = 1$ if A occurs, and $I_A = 0$ if A does not occur. Find $\mathrm{E}(I_A)$.

a) A sequence of independent events, each with probability p of success, occurs. Use indicators to compute the expected number of successes in n trials.

b) A die is rolled repeatedly. Find the expected number of different faces that appear in the first n rolls.

4. *Tail sums.* Suppose the possible values of X are $0, 1, 2, 3, \ldots$ Show that

$$\mathrm{E}(X) = \sum_{x=0}^{\infty} \Pr(X > x).$$

Now suppose a sequence of independent events, each with probability p of success, occurs. Let k be a fixed positive integer. Find the expected number of trials until the kth success occurs. (It's a good idea to think first about the case $k = 1$.)
5. *The coupon collector's problem.* A cereal company puts one coupon in each of its cereal boxes. There are n distinct coupons. If you collect a complete set of these coupons, you get a grand prize. How many cereal boxes do you expect to buy to get a complete collection? (What assumptions are you making?) Get a good approximation to this expectation for large n.
6. *The birthday problem, revisited.* In a classroom with n students, find the expected number (and variance) of same-birthday pairs.
7. *Symmetry in card shuffling, revisited.* You deal cards one by one from a well-shuffled deck of cards, until all the aces have been dealt. How many cards do you expect to deal?
8. *Fixed points: the matching problem.* There are n letters, labeled 1 through n. And there are n envelopes, also labeled 1 through n. The letters are distributed randomly into the envelopes, one letter per envelope, so that all $n!$ possible arrangements of letters in envelopes are equally likely. Say that a match occurs in envelope i if the letter labeled i falls into the envelope labeled i. Find the expected number of matches.
9. *Cycles: the locked boxes problem.* There are n boxes. Each box locks itself when slammed shut, and must be opened with its own special key. An annoying person permutes the keys randomly and throws them into the boxes, one key per box, and then slams all the boxes shut. You want to open the boxes but are willing to break open only one box. This will yield a key that may open another box, which will yield another key, and so on. Find the expected number of boxes you can open. (Hint: You have solved part of this problem in Exercise 7 in Section 15.2.1.)

15.3 Challenging problems

In addition to problems that reinforce basic concepts, we hand out additional challenging problems on a regular basis. Some of the problems relate to a group research project.

The problems may vary in difficulty. Students work those that interest them, and divide up the task of writing solutions. They enjoy the freedom of choosing their assignment. Some work nearly all the problems, while others work in groups on a subset. Occasionally, the class period is dedicated to presenting solutions.

We include here examples of handouts for two kinds of challenging problems. The first handout (see Fig. 15.1) presents a result in number theory. Number theory is a popular subject among math undergraduates, and the following problems were designed to demonstrate a connection between probability and

Euler's formula

The *Riemann zeta function* ζ is defined by

$$\zeta(s) = \sum_{n=1}^{\infty} n^{-s}, \quad s > 1.$$

Consider a positive integer-valued random variable X with distribution given by

$$\Pr(X = n) = \frac{n^{-s}}{\zeta(s)}, \quad n \geq 1.$$

1. Let p be a prime number. Find $\Pr(X \text{ is divisible by } p)$.
2. First, a definition. Recall that two events A_1 and A_2 are independent if $\Pr(A_1 A_2) = \Pr(A_1)\Pr(A_2)$. Events $A_1, A_2, A_3, \ldots$ are called independent if, for every n and every n-tuple of indices $i_1, i_2, \ldots, i_n$,

$$\text{independence:} \quad \Pr(A_{i_1} A_{i_2} \cdots A_{i_n}) = \Pr(A_{i_1})\Pr(A_{i_2}) \cdots \Pr(A_{i_n}).$$

Now back to the problem. Let D_p be the event that X is divisible by p. Show that the events $\{D_p : p \text{ a prime}\}$ are independent.
3. Here is a general result about the monotonicity of probabilities. Let $A_1, A_2, A_3, \ldots$ be a decreasing sequence of events, that is, $A_1 \supseteq A_2 \supseteq A_3 \supseteq \cdots$. Show that

$$\Pr(A_n) \downarrow \Pr\left(\bigcap_{i=1}^{\infty} A_i\right) \quad \text{as } n \uparrow \infty.$$

4. Use the previous exercises to prove *Euler's formula*: for every $s > 1$,

$$\frac{1}{\zeta(s)} = \prod_p \left(1 - \frac{1}{p^s}\right),$$

where the product is over all prime numbers.

Fig. 15.1 Occasionally, we assign students a sequence of related problems such as this one where they use probability to establish Euler's formula, a result in number theory.

number theory. Students enjoyed using their new probabilistic tools to establish a well-known result in another branch of mathematics.

The second problem (see the handout in Fig. 15.2) gives an example of how a classic problem can be generalized and extended in a variety of ways. The Buffon needle problem was solved by the class as a whole. Then variations were given to different groups to tackle and present to each other. The answers to all of the variations are the same: $2L/\pi d$.

Buffon's noodle

1. A needle of length l is dropped at random onto a floor marked with parallel lines d apart ($l < d$). What is the probability that the needle crosses the line?
2. What is the expected number of lines crossed if the needle is of length L, where $L > d$? View the long needle as k short needles each of length l, where $kl = L$ and $l < d$. Then use indicator functions to find the expectation.
3. What is the expected number of lines crossed if the needle is bent?
4. A plane curve of length L is inside a circle of diameter d, and a cut is made along a randomly chosen chord as follows: select a distance uniformly from $-d$ to d and an angle uniformly between 0 and π; take the chord to be that which is perpendicular to the diameter corresponding to the chosen angle and which is the chosen distance from the origin. What is the expected number of cuts to the curve? (Hint: Try an infinitesimal calculus approach similar to that in the previous problem.)
5. What if 20 pieces of spaghetti, each 10 inches in length, are tossed on a plate, and six cuts are made along randomly chosen chords as above? How many pieces of spaghetti do you expect to get?
6. Two- and three-dimensional generalizations of these problems from stereology can also be solved in groups. For example, the area of a two-dimensional region inside a circle of diameter d can be estimated from the ratio of the length of the region along a randomly chosen chord to the total chord length. In this case, the chord is not generated as above. Instead, an angle is uniformly chosen between 0 and π; a point is chosen at random from the interior of the circle; and the random chord is that which passes through the chosen point along the chosen angle.

Fig. 15.2 In this sequence of problems we start with the classic Buffon needle problem of estimating the length of a needle thrown on a floor and generalize the result to throwing several pieces of cooked spaghetti on a plate.

15.4 Does the Poisson distribution fit real data?

We have also tried open-ended homework problems, for example, to find a set of real data that could be modeled by the Poisson distribution. For this homework assignment, we require at least 25 data points (as we tell the students, this could be data from 25 states, or 25 countries, or 25 years). After collecting the data, the students are instructed to fit a Poisson distribution, plot the data and the fitted distribution, and comment on aspects of lack of fit. Most real count data are overdispersed, and this was the point we wanted the students to learn.

However, in assigning the homework we found an additional problem—about half the students made a major mistake by using noncount data, for which the Poisson distribution was inappropriate. For example, one student used suicide rates in the 50 states and tried to fit the Poisson distribution to rates (for example, treating the rate of 5 per 100 000 as an observation of 5) even though it is intended to model counts. The students who did pick count data found some interesting examples, however, including the number of major earthquakes in each of 25 years, the number of law enforcement officers killed in each of 120 months, and the number of people killed in witch trials in Geneva in each of 155 years.

Such an assignment could of course be adapted to other discrete or continuous probability distributions or even for stochastic processes such as random walks or Markov chains. The emphasis here is not on statistical modeling so much as on the understanding of the properties of probability distributions. For example, Poisson data must be discrete counts, binomial data must have an upper bound n, gamma data must be continuous and positive, and so forth.

15.5 Organizing student projects

Midway through the course, students form groups to work on projects. The aim is to introduce the students to research and to enable them to discover a subfield of probability on their own. Group work is structured toward genuine collaborations, with each student contributing his or her strengths, supporting and being supported by the others. They choose the direction in which they want to head, starting with introductory material covered in the seminar.

We have used three different formats for these projects. In one format, projects consist of a series of well defined problems that build on each other. We call them structured projects (see Section 15.6). The aim is to have the students discover the proof of a nontrivial and interesting result that requires a more sustained and deep effort than standard undergraduate work. With a few modifications, the challenging problems in Section 15.3 can double as structured projects.

An alternative to the structured project has students work on a set of general questions from a subfield of probability that they have not yet studied. Again we offer students a set of questions to guide their studies, but these questions are less well defined (see Section 15.7 for examples of these unstructured projects). Finally, we also have students read original research papers for their projects. They choose the paper from an annotated bibliography that we hand out (see Section 15.8). It works best when we give the students a set of questions, as with the structured projects, to guide them in reading the paper. This project helps students learn how to read mathematics papers.

In the last weeks of class, we spend time both in and out of class meeting with the students and monitoring progress. Students write up their projects in papers (one per group) of 8–10 pages. We sometimes ask them to present their findings to the class.

These projects are very demanding, but the students rise to the challenge and ultimately find a great sense of accomplishment in their work.

15.6 Examples of structured projects

This section contains the material we hand out to students for two of the structured projects. The first is on the arcsine approximation to various distributions connected with the symmetric random walk. The development exploits symmetry and sample path properties. Students enjoy figuring out how to cut and paste bits of sample paths to arrive at the results. They learn to write out mathematical proofs showing that their pictorial arguments worked; and lastly, they use some real analysis to do the asymptotics.

The second structured project is on the recurrence and transience of Markov chains, with special attention to random walks. The project starts with the definition of a Markov chain with a countable state space, and then leads students to a proof of the theorem that connects transience and recurrence with the convergence or divergence of a series of n-step transition probabilities. This theorem is applied to study the recurrence and transience of random walks, both symmetric and asymmetric, in one or more dimensions. Like the project on arcsine laws, this one relies heavily on clever uses of elementary probability and real analysis to prove theorems.

15.6.1 Fluctuations in coin tossing—arcsine laws

Let $X_1, X_2, \cdots$ be independent random variables, and for all i, let X_i take the values 1 and -1 with probability 1/2 each. Let $S_0 = 0$, and for $n \geq 1$, let

$$S_n = S_{n-1} + X_n = X_1 + X_2 + \cdots + X_n.$$

The sequence S_n is called a simple symmetric *random walk* starting at 0.

We will think of $(S_0, S_1, S_2, \cdots, S_n)$ as a polygonal path, whose segments are

$$(k-1, S_{k-1}) \rightarrow (k, S_k).$$

Preliminaries.

1. How many possible paths are there from $(0, 0)$ to (n, x)?
2. Find $\Pr(S_{2n} = 0)$.
3. *Stirling's formula—almost!* Let n be a positive integer. Show that there is a constant c such that

$$n! \sim cn^n e^{-n}\sqrt{n}.$$

(Hint: compare $\log(n!)$ with $\int_1^n \log x dx$. It helps to draw a picture.) In fact $c = \sqrt{2\pi}$. It takes a little work to prove this, but you can assume it.

4. Let $u_{2n} = \Pr(S_{2n} = 0)$. Show that

$$u_{2n} \sim \frac{1}{\sqrt{\pi n}}.$$

What does this say about the statement, "In the long run you expect about half heads and half tails"?

Arguments by translation and reflection.

5. *Warm-up: the reflection principle.* Let z and w be positive integers. Show that the number of paths from $(0, z)$ to (n, w) that touch or cross the x-axis is equal to the number of paths from $(0, -z)$ to (n, w).

6. Show that
$$u_{2n} = \Pr(S_1 \geq 0, S_2 \geq 0, \cdots, S_{2n} \geq 0).$$
To do this, consider all paths of length $2n$. Let the paths that end at level 0 be called "Type A" paths, and the ones that never get below the x-axis be "Type B." The idea is to establish a one-to-one correspondence between paths of these two types. Here is how to start. Take a Type A path. If it never gets below the x-axis, leave it alone. If it does, let $(k, -m)$ be the coordinates of the leftmost lowest point of the path. Make this point your new origin. Take the initial portion from $(0,0)$ to $(k, -m)$, reflect it about the vertical line $x = k$, chop it off, and add it on to the end of the path. Draw lots of pictures!
7. Show that
$$2 \cdot \Pr(S_1 > 0, S_2 > 0, \cdots, S_{2n} > 0) = \Pr(S_1 \geq 0, S_2 \geq 0, \cdots, S_{2n} \geq 0).$$
8. Show that
$$\Pr(S_1 \neq 0, S_2 \neq 0, \cdots, S_{2n} \neq 0) = u_{2n}.$$

Arcsine laws.

9. Let $k < n$. In terms of the u's, find $\alpha_{2k,2n}$, defined by
$$\alpha_{2k,2n} = \Pr(S_{2k} = 0, S_{2k+1} \neq 0, S_{2k+2} \neq 0, \cdots, S_{2n} \neq 0).$$
10. Get a good approximation to $\alpha_{2k,2n}$ and sketch a graph of this approximation as a function of k for fixed n. What does the graph say about the time of the last zero in a random walk of $2n$ steps?
11. *An arcsine law.* Let x be between 0 and 1, and let the random walk go for a very long time. Show that the probability that the last zero occurs before a fraction x of the total time is approximately
$$\frac{2}{\pi} \arcsin \sqrt{x}.$$
12. *Arcsine law for the time of the first maximum.* Consider a random walk of n steps. Say that the first maximum occurs at time k if
$$S_0 < S_k,\ S_1 < S_k, \ldots, S_{k-1} < S_k,$$
and
$$S_{k+1} \leq S_k,\ S_{k+2} \leq S_k, \ldots, S_n \leq S_k.$$
Show that the time of the first maximum also follows an arcsine law. (Hint: Consider the "dual" random walk; the first step is X_n, the second step is X_{n-1}, and so forth.)

15.6.2 Recurrence and transience in Markov chains

A stochastic process $\{X_0, X_1, X_2, X_3, \ldots\}$ is called a *Markov chain* if, for each n, the conditional distribution of X_{n+1} given $(X_0, X_1, \ldots, X_n)$ is the same as the conditional distribution of X_{n+1} given just X_n.

Assume each X_i takes values in a countable set S, called the state space of the process. For ease of notation, we will assume that S is the set of integers. Then $\{X_0, X_1, X_2, \ldots\}$ is a Markov chain if for every n and all integers $i_0, \ldots, i_{n+1}$,

$$\Pr(X_{n+1} = i_{n+1} | X_0 = i_0, \ldots, X_n = i_n) = \Pr(X_{n+1} = i_{n+1} | X_n = i_n).$$

The probability on the right side of the equation is called a *transition probability*, because it is the probability that the process makes a transition to state i_{n+1} at time $n+1$ given that it was in state i_n at time n. In general, the transition probability,

$$\Pr(X_{n+1} = j \mid X_n = i),$$

depends on i, j, and n. But in many interesting examples, it depends only on i and j. A processes with such transition probabilities is called *stationary*, and you can write the probability as a function of just i and j:

$$\Pr(X_{n+1} = j \mid X_n = i) = p_{ij}.$$

As the notation suggests, these transition probabilities can be arranged in a *transition matrix* P whose (i, j)th element is p_{ij}.

In what follows, $\{X_0, X_1, X_2, \ldots\}$ is an integer-valued Markov chain with stationary transition matrix P. The process is said to start at time 0 with value X_0.

1. *The n-step transition probabilities.* Let $p_{ij}^{(n)}$ be the probability that the chain is at state j at time n, given that it started at state i. Show that $p_{ij}^{(n)}$ is the (i, j)th element of the matrix P^n. (To avoid irritation later, define P^0 to be the identity matrix, and check that this definition makes intuitive sense.)

2. *First-hitting probabilities.* Let $f_{ij}^{(n)}$ be the probability that the chain visits state j *for the first time* at time n, given that it started at state i. That is,

$$f_{ij}^{(n)} = \Pr(X_1 \neq j, X_2 \neq j, \cdots, X_{n-1} \neq j, X_n = j \mid X_0 = i).$$

Which is bigger: $f_{ij}^{(n)}$ or $p_{ij}^{(n)}$?

Let f_{ij} be the probability that the chain ever visits state j, given that it started at i. Find a formula for f_{ij} in terms of the $f_{ij}^{(n)}$'s.

Recurrence and transience. A state i is called *recurrent* if $f_{ii} = 1$, and *transient* if $f_{ii} < 1$.

So i is recurrent if, given that the chain started at i, it is certain to return to i. If there is some probability that it does not return to i, then i is transient.

Consider the simple random walk determined by independent steps, each with probability p of increasing by 1 and probability of $1-p$ of decreasing by 1.

Suppose the walk starts at 0. For which values of p do you think the state 0 will be transient? For which values will it be recurrent? (For now just say what your intuition tells you—by the end of this project you will have a rigorous answer!)

Back to the Markov chain. For brevity, write $P_i(A)$ for the probability of A given that the chain started in state i. That is,

$$\text{define } P_i(A) = \Pr(A|X_0 = i) \text{ for any event } A.$$

Theorem. The main theorem of this project set says that transience of the state i is equivalent to

$$P_i(\text{infinitely many visits to } i) = 0,$$

which is also equivalent to

$$\sum_n p_{ii}^{(n)} < \infty.$$

Conversely, recurrence of the state i is equivalent to

$$P_i(\text{infinitely many visits to } i) = 1,$$

which is also equivalent to

$$\sum_n p_{ii}^{(n)} = \infty.$$

Before you prove this theorem, use it to check your intuitive answer to the question about random walks above. Exercises 3–6 constitute a proof of the theorem.

3. In terms of the f's defined in Exercise 2, find a formula for the probability that the chain makes at least m visits to j, given that it started at i. Hence find a very simple formula for P_i (infinitely many visits to j), in terms of the f's. Specialize this to the case $j = i$.

4. Show that $\sum_n p_{ii}^{(n)}$ is the expectation of a certain random variable. Deduce that

$$\sum_n p_{ii}^{(n)} < \infty$$

implies

$$P_i(\text{infinitely many visits to } i) = 0.$$

5. Show that $f_{ii} < 1$ implies $\sum_n p_{ii}^{(n)} < \infty$, in the following steps.

a) Show

$$p_{ij}^{(n)} = \sum_{s=0}^{n-1} f_{ij}^{(n-s)} p_{jj}^{(s)}.$$

b) Specialize to the case $j = i$, and sum both sides above, to show

$$\sum_{t=1}^{n} p_{ii}^{(t)} \le \sum_{s=0}^{n} p_{ii}^{(s)} f_{ii}.$$

c) Now complete the exercise.

6. Show that the theorem has been proved!
7. More than one dimension. Consider a simple symmetric random walk in d dimensions, defined as follows. Start at the origin. At each stage, toss d coins, and change the ith coordinate by $+1$ or -1 according to whether the ith coin lands heads or tails. Show that the origin is recurrent in one and two dimensions, but transient in three or more dimensions.
8. Recurrent (or transient) chains. Call a Markov chain *irreducible* if every state can be reached with positive probability from every other state—that is, if for all i and j there is some n for which $p_{ij}^{(n)} > 0$.

a) Give an example of a Markov chain that is *not* irreducible.

b) Show that in an irreducible chain, either all states are recurrent, or none is. Thus, for example, we can say, "the random walk is recurrent" once we have checked that the origin is a recurrent state.

15.7 Examples of unstructured projects

Included below are the introductions to four projects. They are very informal, and provide ideas for interesting avenues to explore, but students are urged to investigate other areas and not to limit themselves to those listed here. The project descriptions below are written to give directly to students.

In these projects, students are more self-directed than in the structured projects. They decide which project they want to work on after we have given introductions to each topic to the entire class. They spend a couple of days looking over the materials and making their choice. As the groups develop their material, they meet regularly with the instructor, where they receive direction, supplemental reading, and fact sheets on related topics. They make periodic progress reports in class, and each group keeps a notebook of its findings.

15.7.1 Martingales

After one introductory lecture on martingales, students try their hands at proving an optional stopping theorem. They begin by considering specific problems such as the expected time it will take a monkey sitting at a keyboard to type ABRACADABRA. Here are the instructions we give to the students.

Loosely speaking, a martingale is a process in which the expected value of the process tomorrow, given all information up through today, equals the value of the process today. In that sense, it can be thought of as your sequence of fortunes in a fair game, though as you have seen, this doesn't necessarily imply that you'll break even in the long run!

Martingales have beautiful properties: under mild regularity conditions they converge in various ways, and there are bounds on how wild they can get. Many interesting problems, apparently unrelated to martingales, can be solved using *optional stopping theorems*, which say the following: it's easy to check that the expected value of a martingale at every fixed time is the same, but the same is also true at certain well-behaved random times, known as stopping times. This allows you to compute all kinds of probabilities and expectations, as you will see.

It is often fun to "hunt the martingale" in problems so that the powerful martingale methods can be used in solutions. See for example the connection with the branching processes project.

The study of martingales is central to modern probability theory. To do it right, you have to be careful about exactly what is meant by "expected values given all the information up through today," etc. This involves analysis and measure theory, which are good areas for you to get acquainted with anyway.

Suggested starting place. Read the lecture note sketches, and for the moment don't worry that there are gaping holes in them (you'll plug those holes later). Trust that the optional stopping theorem is true and use it to get the expected duration of the game in the gambler's ruin problem, and the expected amount of time a monkey will take to type ABRACADABRA. This last is not as silly a problem as it might seem; similar issues become very important in data compression. Once you have done this, read Williams (1991) for a proof of the optional stopping theorem.

15.7.2 Generating functions and branching processes

After one lecture on branching processes, we set our students loose on the problem of trying to determine the long-run behavior of the population; for example, will it die out?

This concerns a particular population growth model: start with one individual, who then has a random number of offspring according to a certain distribution; each child in turn has a random number of offspring, with the same distribution as the offspring distribution of the original parent. All individuals reproduce independently of each other. It is natural to draw the "family tree," hence the name "branching process."

Questions of interest concern the long-run behavior of the population: does it die out? does it explode? And so on. Answers are very elegant.

But it helps to have some mathematical technique, namely the theory of generating functions. This is a general technique involving power series, which often makes short work of otherwise intractable problems. For example, you can tackle the asymmetric random walk, which doesn't have the pretty symmetries of the symmetric walk. The methods are analytical, rather than probabilistic, but well worth learning for the mathematics student as they pop up in all kinds of places.

Suggested starting place. Glance through the branching process lecture, first ignoring the math, and try to sift out the results. You might try taking a very simple offspring distribution and see if you can figure out yourself whether your population will die out. Then read Chapter 0 in Williams (1991). Then start learning about generating functions.

Connections with other projects. There's a useful martingale hidden in the branching process.

15.7.3 Limit distributions of Markov chains

In this project on Markov chains, we encourage students to find more than one way to establish the stationary distribution. Our introductory lecture introduces them to the probabilistic method of coupling, and we give them the following instructions to start them on their project.

Suppose you run a Markov chain for a long time. How might it behave? You can guess at some of the behavior: because of the Markov property which says the most recent state is the only one that matters, the chain should somehow forget where it started. In fact it does, under certain conditions, and it has a limiting distribution that is easy to compute and has useful properties.

The big theorem is that under certain conditions on the one-step transition probabilities p_{ij}, the n-step transition probabilities converge: that is, for any i, $\lim_{n\to\infty} p_{ij}^{(n)} = \pi_j$, and this limit does not depend on the starting state i. It is then easy to check that the vector π satisfies the equations $\pi P = \pi$, and hence you have a way of computing it. Explicitly computing π in special cases is a lot of fun and yields interesting results.

The "invariant distribution" π has many other interpretations and uses. It allows you to compute things like the expected number of visits to i before the next visit to j, the long-run fraction of time the chain spends at any given state, etc.

There is more than one way of proving the big theorem; after all, in some sense it's just a theorem about matrices and shouldn't involve any probability. But the proof outlined in the project is very "probabilistic" and uses a method called "coupling." This is a general method which works as follows. Suppose you are trying to prove that your Markov chain has a certain asymptotic behavior. And suppose you know of some other "nice" Markov chain that clearly has that same asymptotic behavior. Set them both running, and try to show that the two paths are bound to meet somewhere. If you can do this, you are home free, because after the meeting-point your Markov chain is probabilistically equivalent to the other one, and therefore must have the same asymptotic behavior!

Suggested starting place. The statement of the big theorem is at the bottom of page 10 of Williams (1991). You'll quickly see that you need to learn some terminology, etc. At this point you can simply start reading from page 1, and mark the places where things aren't proved. Most are proved in the appendix, but it's best to start filling in the gaps on your own. It will be good preparation for proving the big theorem.

15.7.4 Permutations

The project on random permutations revisits a problem given to the students in the first week of class (see Section 15.2.2). There is not a large literature on this subject that is accessible to undergrads, and so we provide more details in our handout than for the other projects of this type.

Consider the $n!$ permutations of the integers 1 through n. A *random permutation* of 1 through n is obtained by choosing one of these permutations at random so that all $n!$ permutations are equally likely. Random permutations appear in the letter-matching problem and the locked-box problem in the last example in Section 15.2, though they are couched in traditionally colorful language. Suppose there are seven locked boxes and keys arranged as follows:

$$\begin{pmatrix} \text{Box} & 1 & 2 & 3 & 4 & 5 & 6 & 7 \\ \text{Key} & 5 & 1 & 7 & 4 & 6 & 2 & 3 \end{pmatrix}.$$

A cycle representation for this permutation is (1562)(37)(4). There are three cycles, one of length 4, one of length 2, and one of length 1. The cycle of length 1 represents a match of box and key. Another way to express the cycle representation for the permutation that does not require the use of parentheses is to use the convention that each cycle is written so as to end in its smallest number, and cycles are written in the order of increasing smallest last numbers. This representation is called the Hungarian map, and for the above permutation the Hungarian map is 5621734.

It may be interesting to investigate the properties of the number of cycles of length 1, the length of the cycle that contains 1, or the total number of cycles in a random permutation.

What is the probability the permutation is an involution; that is, that it contains only cycles of length 1 and 2? What is the probability that all cycle lengths are divisible by d? To answer these questions, it may be helpful to consider the cycle vector for a permutation. It is the vector $(l_1, l_2, \ldots, l_n)$, where l_i is the number of cycles of length i in the permutation. For our example, the cycle vector is (1,1,0,1,0,0,0). Notice that $\sum_{i=1}^{n} il_i = n$. Show that the probability mass function for the cycle vector is,

$$p(l_1, \ldots, l_n) = \prod_{i=1}^{n} \frac{1}{(l_i)!\, i^{l_i}}.$$

Can you find the probability generating function for the cycle vector? To do this you will need to learn about multivariate probability mass functions and generating functions.

Another interesting avenue to explore is the connection between cycles and records. Consider the independent records, $R_1, R_2, \ldots, R_n$, where R_i has a discrete uniform distribution on $\{1, 2, \ldots, i\}$. Try generating the Hungarian map in reverse order by first choosing the R_nth element from the ordered set $\{1, 2, \ldots, n\}$; then choose the R_{n-1}th element from $\{1, 2, \ldots, n\} - R_n$, and so on. What do you find?

There is also a connection between Polya's urn and cycles. Consider the urn with 1 red and 1 black ball. Mark the red ball with the number 1. Then draw a ball at random from the urn, and return it along with an additional ball of the same color to the urn. But, before putting the new ball into the urn, mark it with the number 2. Continue drawing balls in this fashion. That is, on the ith draw, a ball is chosen at random from the urn, then the ball is returned to the urn along with another ball that is the same color and that is marked $i+1$. What is the connection between the numbers on the red balls in the urn and cycles? What happens to the proportion of red balls in the urn as the number of draws goes to infinity?

Suggested starting place. The problems in the first week's handout (see Section 15.2) are a good place to start. Can you rephrase these questions in terms of cycles and Hungarian maps?

Connections with other projects. Polya's urn has a hidden martingale in it; and there is an interesting connection to Markov chains.

15.8 Research papers as projects

Many students enjoy reading an original account of a research problem, rather than prepared textbook material. These papers can cover a wide range of topics.

With assistance from the instructor, each group of students chooses a paper that suits their interests and mathematical preparation. Some groups are also given supplemental reading to prepare them for reading their research papers.

We have experimented with alternatives to writing papers on the research papers. We sometimes ask the groups to give an oral presentation of their research paper, to write a detailed abstract for the paper which we collect in a scrapbook and distribute to all of the students, or to make up a worksheet of problems (and their solutions) that are related to the paper.

We provide here an annotated list of some research papers that are suitable for undergraduate mathematics students to read. This list can be distributed to the students.

Dorfman, R., "The detection of defective members of large populations," *Annals of Mathematical Statistics 14*, 436–440 (1943). This self-contained paper is an application of two-stage testing for a disease by pooling samples of blood. It requires work with conditional expectations and discrete probability. For further reading, a three-stage procedure is discussed in Finucan, "The blood testing problem," *Applied Statistics 13*, 43–50 (1964).

Ferguson, T. S., "A characterization of the geometric distribution," *American Mathematical Monthly 72*, 256–260 (1965). This technical article develops a characterization of two independent discrete random variables that can only hold if they have geometric distributions. It works with the independence properties of the minimum and the difference of two random variables.

Halley, E., *Philosophical transactions—giving some account of the present undertakings, studies and labors of the ingenious, in many considerable parts of the world.* (1693) Halley's work on the uses of life tables is reprinted in Newman, J. R., *The World of Mathematics* (1956). An excellent supplement to this reading is pages 131–143 in Hald, A., *A History of Probability and Statistics and Their Applications Before 1750* (1989). There are many interesting problems to work on from Halley's methods, such as the relation between tail sums and expectation.

Hendricks, W., "A single-shelf library of N books," *Journal of Applied Probability 9*, 231–233 (1972). Some additional background reading in Markov chains is required. A simple move-to-the-front scheme for reshelving books is put in a Markov chain framework. The stationary distribution is determined by induction.

Kingman, J., "The thrown string (with discussion)," *Journal of the Royal Statistical Society B 44*, 109–138 (1982). A rigorous treatment of the extension of Buffon's needle discussed in Section 15.3.

Knuth, D., "The toilet paper problem," *American Mathematical Monthly 91*, 465–470 (1984). This problem is related to the classical ballot problem (Feller, 1968). Recurrence relations are used to find a generating function, which leads to large sample behavior for the problem.

Mendel, G., *An Experiment in Plant Hybridization* (1865). Mendel's work is also described in Newman, J. R., *The World of Mathematics* (1956). Mendel's paper is quite elementary, but it offers many avenues for exploration. Feller (1968, Section V.8) contains a variety of problems including some related to the Hardy–Weinberg equilibrium and sex-linked traits.

Moran, P. A. P., "A mathematical theory of animal trapping," *Biometrika 38*, 307–311 (1951). This paper develops a probability model for estimating the size of an animal population using maximum likelihood estimation and Stirling's approximation.

Newell, G. F., "A theory of platoon formation in tunnel traffic." *Operations Research 7*, 589–598 (1959). This paper constructs a probability model for the flow of traffic through a tunnel; it includes finding the distribution of a maximum of independent random variables.

Ramakrishnan, S., and Sudderth, W. D., "A sequence of coin tosses for which the strong law fails," *American Mathematical Monthly 95*, 939–941 (1988). Some measure theory is required for this paper.

Student, "On the error of counting with a haemacytometer," *Biometrika 5*, 351–360 (1906). This paper derives the Poisson process in the plane from first principles. It offers good practice with infinite series and moment generating functions. Any supplemental reading from an undergraduate text on the Poisson process is helpful for synthesizing the results presented here.

Tversky, A., and Gilovich, T. "The cold facts about the hot hand in basketball," *Chance 2* (1) 16–21 (1989). Plus additional articles in *Chance* in 1989 by Larkey, Smith, and Kadane, issue 4, 22–30, and Tversky and Gilovich, issue 4, 31–34. This series of articles provides a very accessible analysis of streak shooting from actual basketball games and designed experiments. Additional reading on the distribution of run tests helps focus the project for a mathematics class; see, for example, Lehmann, E., *Nonparametrics: Statistical Methods Based on Ranks* (1975).

16

Directed projects in a mathematical statistics course

In this chapter we present a model for using case studies in undergraduate mathematical statistics courses. Statistics teachers at many institutions have created innovative data-based homework assignments; we focus on these case studies because we hope the suggestions in this chapter will be helpful for you in implementing your own data-analysis assignments.

These case studies have more depth than most examples found in typical mathematical statistical texts. In each case study, we provide students with a real-world question to be addressed. The question is raised in the context of an application, so the students have a sense that there is a good reason for trying to answer it. We also provide students with background information, data to address the problem, and a list of suggestions for investigating the problem. An important goal of this approach is to encourage and develop statistical thinking.

We provide here examples of a case study (in Section 16.3) and a directed project (in Section 16.4) along with a discussion of how we have changed the class to fit these activities in the traditional material covered in a mathematical statistics course. In essence, we make the case studies the centerpiece of the course; that is, finding the answer to the question raised in the case study motivates the develoment of related statistical theory. As a result, the curriculum, lectures, and assignments are significantly different from a traditional mathematical statistics course.

Our reason for teaching mathematical statistics this way is that we have found that students have difficulty bringing the mathematical statistics learned in the classroom to independent projects in school or on the job. (To be honest, this can be a problem for professional statisticians too.) It can also be difficult for students to make the transition from reading and understanding a critical review of a statistical analysis to successfully working on their own problem. With case studies, students develop their quantitative reasoning and problem-solving skills in a broad multidisciplinary setting. They communicate their ideas orally and in writing, and they learn to use statistical software.

Table 16.1 A sample set of eight lab assignments for a mathematical statistics course.

Lab	Option A	Option B	Topics
Smoking and infant health I		×	summary statistics
Student use of video games	×		simple random sampling, ratio estimator
AIDS in hemophiliacs	×		2×2 tables, Mantel–Haenszel test
Patterns in DNA		×	estimation and testing, Poisson process
Crab molting	×		regression, prediction
Snow gauge calibration		×	simple linear model, inverse regression
Smoking and infant health II	×		multiple regression, indicator variables
Monitoring quality control		×	analysis of variance, random effects

16.1 Organization of a case study

After trying several approaches, we settled on a model for the organization of a case study, where each case has five parts: introduction, data, background, investigations, and theory. Sometimes we include an extension section for a more advanced analysis of the data and related theoretical material.

1. *Introduction.* First we state a clear scientific question and give motivation for answering it. The question is presented in the context of the scientific problem, not as a request to perform a particular statistical analysis. We avoid questions suggested by the data, and attempt to orient the case study around the original question raised by researchers who collected the data.
2. *Data.* Documentation for the data collected to address the question is provided to the students, including a detailed description of the study protocol.
3. *Background.* We provide students with background material to put the problem in context. This information is gathered from a variety of sources and presented in nontechnical language. The idea is to present a picture of the field of interest that is accessible to a broad college audience.
4. *Investigations.* Suggestions for answering the question posed in the introduction are also given. These suggestions are written in the context of the problem, using very little statistical terminology. The ideas behind the suggestions vary in difficulty and we sometimes group them into subsets to pace the assignment. Also included are suggestions on how to write up the results, for example, as an article for a widely read magazine, a memo to the head of a research group, or a pamphlet for consumers.
5. *Theory.* After the students have had a chance to look over the data, and think about how to answer the motivating question, we discuss methodology and theory that may be useful in attacking the problem. We present this material in a general framework, and use the case as a specific example.

Examples of case studies with these five parts are sketched in Sections 16.3 and 16.4.

16.2 Fitting the cases into a course

We use case studies extensively in two mathematical statistics courses, one for statistics and mathematics majors and one for engineering students. The enrollments for these courses varies from 20 to 60 students. (We have also used versions of these cases with simpler investigations and less advanced theoretical material in a course for social and life science majors.)

We cover about eight cases in a semester. They are chosen by the instructor according to the theoretical topic, area of application, and the background of the students. We divide the eight cases into two types: those discussed primarily in lecture and those that require students to do extensive analyses outside of class and write short papers on their findings.

Typically we ask students to write reports for four cases. Table 16.1 provides two example sets of assignments. Option A could be used in a class for students in the social and life sciences, and option B could be used for engineering students. In addition to covering the core material, many cases also include special topics in statistics, as listed in Table 16.1.

16.2.1 Covering the cases in lectures

In class, students work in groups on the case studies. Whether the analysis of a case is an out-of-class assignment, or solely an in-class topic, we find it useful for lectures to include discussion of the background to the particular problem, where students who have taken courses in related fields can bring their expertise to the discussion. In addition, for cases where students write reports, we hold regular question-and-answer periods in class to give students an opportunity to raise questions about their work. If a few students are having problems with the assignment, then everyone receives the benefit of hearing the instructor respond to these problems. We sometimes post our responses to these questions on the Web for students to look over outside the class when they are analyzing the data.

We spend about one class period in three on these kinds of activities, with the remainder being more traditional presentations of theoretical results (with reference to the current case). Motivation of the statistical theory comes out of these discussions. We develop the mathematical statistics for the methods used to analyze the data after we discuss the case in detail, and begin to address the problem.

16.2.2 Group work in class

To facilitate group work in class, we use handouts. One sort of handout is an abbreviated list of investigations, where students work in groups to come up with a plan of attack for addressing the questions. That is, they think of ways to summarize or operate on the data, but they are not expected to do any actual computations. With this exercise, we bring to class results from several possible analyses of the data that we have prepared in advance in anticipation of their comments. When we discuss with the class the ideas that the groups came up with, we sketch our previously prepared results on the board or put them on an overhead, and we talk about the pros and cons of the approaches and what we found.

In another kind of handout, we provide students with a set of tables and graphs. Students break into groups and we ask them to further summarize and interpret the output with the goal of addressing the questions from the investigation. We have groups write their solutions on the board, and the instructor leads the class in a discussion of the analysis. We have found that our students really enjoy working on problems in class because they receive immediate feedback from the instructor and also because they get to see the great variety of ideas their classmates come up with.

Another format for group work with which we have had some success has groups of students work on different cases, and lead a class discussion on their findings. Each group prepares a presentation of their results, including handouts for the class. These presentations are synchronized with the curriculum and scheduled throughout the semester. Students are given leeway in designing their presentation; some lead discussions modeled after our in-class activities, and others use role-playing as a research team to present their new research, or they hold a debate where they present conflicting statistical evidence to make their arguments. To help students prepare their presentations, we schedule two appointments with each group, one to discuss their investigations and one to review their handouts and plan the class presentation. In this sort of group work, we expect a more thorough level of analysis, and a more in-depth report than for the other written assignments (see Section 16.2.3).

16.2.3 Cases as reports

For the out-of-class assignments, students work on a case for about two weeks. Their time is spent analyzing data and preparing reports on their findings. The datasets are complex, the analysis is open-ended, and students must synthesize their findings coherently on paper. The students find this work very challenging, and we typically allow them to work in groups of two or three.

It can be difficult to grade the reports, because the investigations allow students to be creative in their solutions to the problem. We usually break down the score into four equal parts: (1) composition and presentation, including statistically sound statements; (2) a basic minimum set of analyses; (3) relevant and readable graphs and tables; and (4) advanced analyses. For the advanced analyses, we make a list of several possible avenues of pursuit, and look for a subset of these in the write-up. Sometimes we also request an appendix to the report for technical material.

The reports typically constitute up to half of the course grade, the remaining being grades for homework and exams. We also think it is important to include questions related to the cases on exams, because it maintains consistency with the approach we have taken in teaching theoretical statistics through applications.

16.2.4 Independent projects in a seminar course

If the class is small, say fewer than 20 students, then we run the course in a seminar style. Students are responsible for reading the background materials on their own, and class time includes discussion and brainstorming on how to solve the problems in the cases. We set assignments on a daily or weekly basis to respond to questions and problems raised by the students. As a result, students are very engaged in the course and feel class time is essential.

An excellent accompaniment to this style of teaching is the independent project. After practicing statistics on our case studies, the students are ready and enthusiastic to try their hands on a project of their own invention. Students present their projects in class and we use their projects as a source for exam questions. Details on how to organize this kind of project can be found in Section 11.4.

16.3 A case study: quality control

In this section, we give an outline of how we organize a specific case study in our courses. This case examines the quality control of a manufacturing process. In our introduction of the problem, we explain that a computer manufacturer has designed a new piece of equipment—a head on a tape drive used in computer backups. This head is to be manufactured by a supplier, and plans are to install a machine, which we call a tester, at the supplier's production facility to test the heads before shipment. The main question is whether the tester is good enough for monitoring the supplier's manufacturing process. There is variation in the manufacturing of the tape-drive head, and there is variation in the measurement made by the tester for the performance of the head. The measurement error needs to be small relative to the process variation in order for the tester to be able to detect variation in the process.

The data used to address the question were collected as follows. Nine heads were selected at random from the manufacturing process. Each head was tested on six magnetic tapes, which were also selected at random. Two repeated measures were made on each head/tape combination for a total of 108 measurements. We provide a table of data so students can look over the numbers and see the design. The measurement taken is a ratio of two voltages measured at two different frequencies as the drive writes information to the tape in the reverse direction. This ratio ideally should be 1.

For background, we provide information on the equipment being manufactured and on the test process. For example, we describe why tapes are required for backing up the disk drives on computer, how information is transferred from the computer to the tape, the function of the head in this transfer, what can go wrong if the head does not function properly, and how the tester is able to detect a malfunction of the head.

Next, we make suggestions for investigating the data. These suggestions are purposely vague in that they do not refer to specific statistical methods to be applied to the data. For example, we begin by asking the student to consider three sources of variation and how to estimate each of them: measurement error (the variation in repeated measurements taken on the same head/tape combination), the variation that might be attributed to differences in tapes, and the process variation in the making of the heads. We point out that the average of the two measurements taken for each of the tapes on one of the drives includes two sources of variation, from measurements and from different tapes. We ask the student to compare each part to the combined variation due to measurement error and tape reproducibility, and we point out that when making this comparison it is important to consider the accuracy of their estimates of these different sources of variation. One final suggestion is to consider a simulation study to estimate the uncertainty in the estimates of variation. We instruct the students to write up their findings, taking on the role of a statistician working in the quality control department of the computer manufacturer who has been asked to advise a production manager whether the tester should be installed in the supplier's production facility.

The accompanying theoretical material we present depends on what the students have already studied. For example, they may already be familiar with one-way and two-way analysis of variance and manipulating sums of squares. Assuming this to be the case, we would begin by introducing the random effects model,

$$y_{ijk} = \mu + \alpha_i + \beta_j + \gamma_{ij} + \epsilon_{ijk},$$

where $i = 1, \ldots, I$, $j = 1, \ldots, J$, $k = 1, \ldots, K$, μ is an unknown constant, and the α_i, β_j, γ_{ij}, and ϵ_{ijk} are independent and normally distributed with mean 0 and standard deviations σ_α, σ_β, σ_γ, σ_ϵ, respectively. The various σ's in the model are related to repeatability (σ_ϵ) and reproducibility ($\sqrt{\sigma_\beta^2 + \sigma_\gamma^2}$). Finally, a useful topic to cover is how to estimate these quantities using mean square errors. For example,

$$\frac{1}{I(J-1)} \sum_{i=1}^{I} \sum_{j=1}^{J} (\bar{y}_{ij} - \bar{y}_i)^2$$

is an unbiased estimate of $\sigma_\beta^2 + \sigma_\gamma^2 + \sigma_\epsilon^2/K$. The students would be expected to apply these ideas to the quality control problem to derive and compare estimates of standard deviations such as σ_α and $\sqrt{\sigma_\beta^2 + \sigma_\gamma^2 + \sigma_\epsilon^2}$.

16.4 A directed project: helicopter design

16.4.1 General instructions

We present here an experiment we assign as a group project. We give students copies of Fig. 16.1 (a diagram for making a helicopter from a half-sheet of paper) and Fig. 16.2 (the flight times for 10 flights from two helicopters with specified wing widths and lengths). The average of all 20 flight times is 1.66 seconds, and the standard deviation is 0.04 seconds. The helicopters were dropped from a height of approximately eight feet. We ask our students to build a better one one that takes *longer* to reach the ground.

When we hand out these materials, we also pass out some comments made by previous students on helicopter building:

Rich creased the wings too much and the helicopters dropped like a rock, turned upside down, turned sideways, etc.

Helis seem to react very positively to added length. Too much width seems to make the helis unstable. They flip-flop during flight.

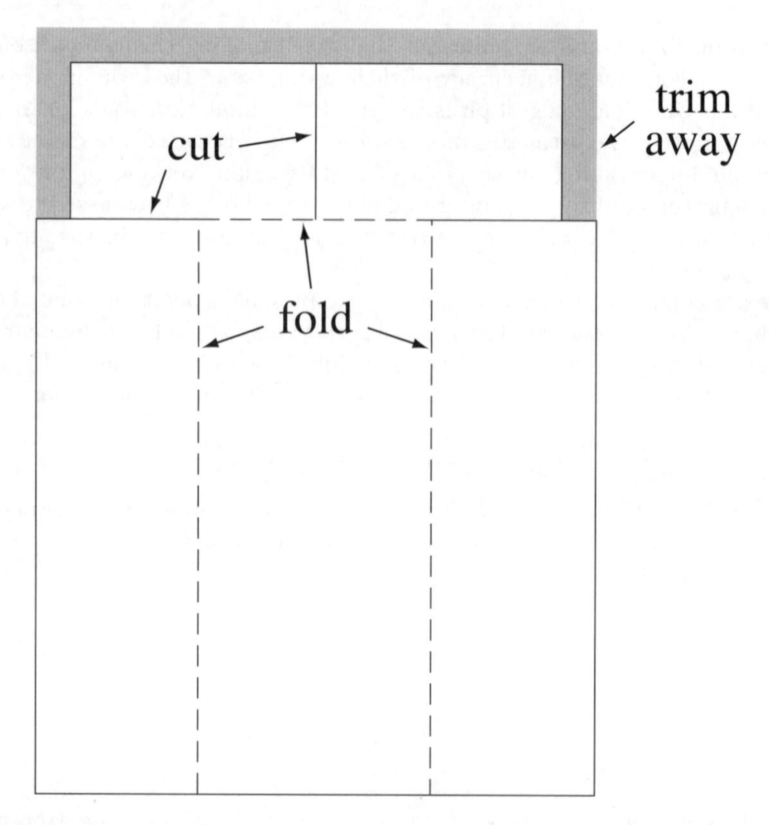

Fig. 16.1 Diagram for making a "helicopter" from half a sheet of paper and a paper clip. The long segments on the left and right are folded toward the middle, and the resulting long 3-ply strip is held together by a paper clip. One of the two segments at the top is folded forward and the other backward. The helicopter spins in the air when dropped.

Andy proposes to use an index card to make a template for folding the base into thirds. After practicing, we decided to switch jobs. It worked better with Yee timing and John dropping. 3 – 2 – 1 – GO.

The helicopters are restricted to have the general form shown in Fig. 16.1. No additional folds, creases, or perforations are allowed. The wing length and the wing width of the helicopter are the only two design variables, that is, the only two measurements on the helicopter that can be changed. The body width and length must remain the same for all helicopters. A metal paper clip is attached to the bottom of the helicopter. We hand out 25 half-sheets of paper and 10 paper clips to each group of students. They may use only these materials in the construction of their helicopters. We also lend each group of students a stopwatch that measures time in hundredths of a second.

Students work in groups on the experiment. We give each group a notebook to record their data. This notebook serves as the group report. It must contain all

Helicopter #1:
1.64 1.74 1.68 1.62 1.68 1.70 1.62 1.66 1.69 1.62
Helicopter #2:
1.62 1.65 1.66 1.63 1.66 1.71 1.64 1.69 1.59 1.61

Fig. 16.2 Flight times, in seconds, for 10 flights each of two identical helicopters (wing width 4.6 centimeters and wing length 8.2 centimeters), for the project described in Section 16.4.

data, statistical calculations, diagrams, plots, and a narration of the experimental process, including problems encountered, reasons for any decisions made, and final conclusions. We warn them to be careful, and to avoid making conclusions of the following sort:

Our data are very suspicious.

We made an extremely vital mistake.

Since we are out of paper, we will just ...

Great things were expected from our contour analysis, but unfortunately, fell far short of our goals.

16.4.2 Designing the study and fitting a response surface

Before we set our students loose on this project, we launch a few helicopters in class and discuss the sources of variability in flight time. We use the following list of questions as a guide in this process.

1. *How reproducible are the measurements of flight time?* We put a string across the room to mark the height from which to drop the helicopters, and then stand on a desk to fly a helicopter a few times. We have one of the students time the flights, and another records them on the board. We discuss how it would be best to fly helicopters multiple times and compare mean flight times. The question of how many times to fly a helicopter we leave to them to decide. These helicopters wear out quickly: after a few flights the wings get floppy and the flights become more erratic.
2. *Is there any variability between "identical" helicopters?* That is, how similar are the flight times for two helicopters built according to the same specifications? Here we repeat the above procedure with a second helicopter made to the same specifications as the first.
3. *Are there noticeable person-to-person differences or other environmental effects?* Here we discuss the experimentation process, and how to set up a procedure that reduces other sources of variability.
4. *Which wing dimensions lead to viable helicopters?* We launch a few funny helicopters with exceptionally wide wings or tiny wings and watch them flutter or plummet to the ground.
5. *How big do the changes in the dimensions of the helicopters need to be to create noticeable differences in flight time?* Here we emphasize the importance of making a good estimate for the standard error for the mean

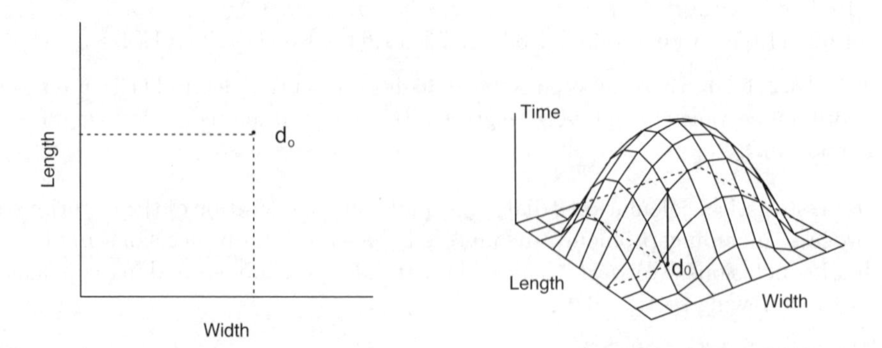

Fig. 16.3 Examples of (a) a design space and (b) a response surface for the helicopter project described in Section 16.4.

response. We fly two helicopters that have very similar wing specifications and two that are very different.

We hand out the following set of instructions to guide the students in their experimentation.

1. Choose a set of wing dimensions to begin your search for the best helicopter. Mark this helicopter on the design space (Fig. 16.3a) and call it d_0. Each helicopter has an expected flight time. If this were plotted above the width and length of the helicopter, then the flight times might look like a surface map above the design space (Fig. 16.3b). The flight time is called the helicopter's response, and the map of flight times is called the *response surface*.
2. Form a rectangle around the design point d_0, with d_0 at the center. Label the corners d_1, d_2, d_3, and d_4, as in the left plot in Fig. 16.4. Build four helicopters according to the dimensions of the four corners of the rectangle. Fly each many times, and record the flight times in your notebook.
3. Make a linear fit of flight to the dimensions of the helicopter wings,

$$\text{time} \;=\; b_0 \,+\, b_1 \cdot \text{width} \,+\, b_2 \cdot \text{length} \,+\, \text{error}.$$

4. This fit can be used to determine the direction of future searches. Draw contour lines on the design region that correspond to the contours of the fitted plane. Next draw a line from the center of d_0 perpendicular to these contour lines. This perpendicular indicates the path of steepest ascent.
5. Choose a design point along this line and repeat the previous steps.

If the linear model is inadequate, or in the final phase of the search, a quadratic model may be fit. To do this, we add four more design points to the

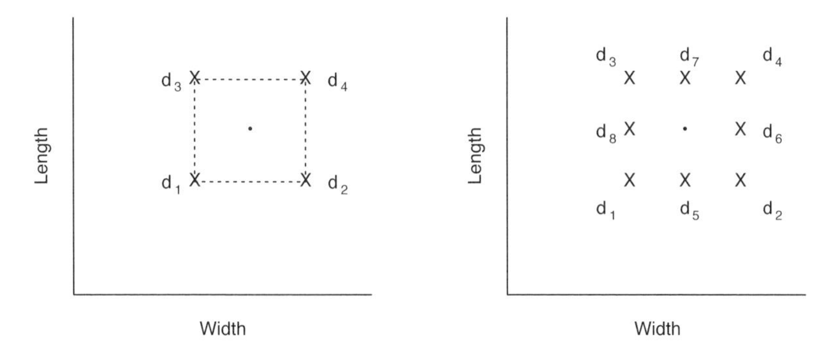

Fig. 16.4 Examples of (a) a rectangular design and (b) a rectangular plus star design.

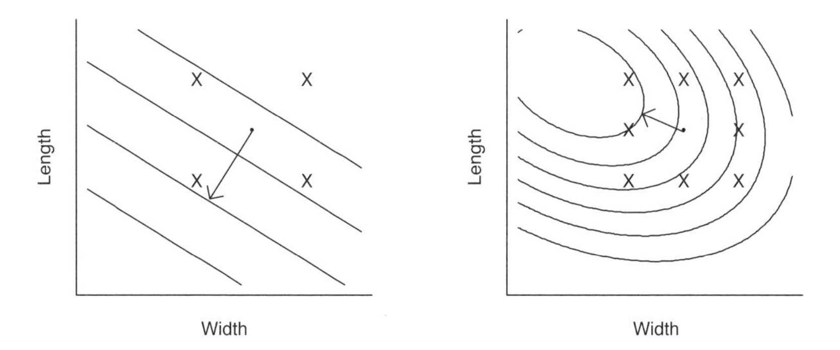

Fig. 16.5 Examples of (a) a linear contour plot and (b) a quadratic contour plot.

rectangular design (Fig. 16.4). Once the additional data are collected, multiple regression can be used to fit the quadratic model:

$$\begin{aligned}\text{time} = b_0 \;+\; b_1 \cdot \text{width} \;&+\; b_2 \cdot \text{length} \\ &+\; b_3 \cdot \text{width}^2 \;+\; b_4 \cdot \text{length}^2 \;+\; b_5 \cdot \text{width} \cdot \text{length} \;+\; \text{error}.\end{aligned}$$

We provide the students with a plotting routine so they can graph the contours of the quadratic response surface (Fig. 16.5) and determine the peak region.

Before the students start on their experiment, we give them some last words of advice taken from previous student reports.

We realized that we did this wrong, because we used [whole] centimeter increments instead of [whole] millimeters, so our design rectangle was too large. We started over.

Instead of testing regions in a contiguous fashion, perhaps we should have moved further along the steepest ascent vector at each round. We had wanted to have complete coverage of the test path, but our test method was too slow and used up too many helicopters along the way.

We figured out posthumously why our analyses were so bad is that we were using the wrong proportions for our path of steepest ascent. (Actually no one died.)

Notes and further reading

Much of the material in this book appeared, in somewhat different form, in Gelman and Nolan (1997, 2002a, b, c, d, e), Gelman (1997, 1998), Adhikari and Nolan (1997), Nolan and Speed (1999, 2000), and Gelman and Glickman (2000).

Preface. For information on the statistical packages R and S-Plus, see Becker, Chambers, and Wilks (1988), MathSoft (2000), and R Project (2000). For information on Stata, see StataCorp (2000).

Section 1.1. See Cobb (1992) and Gnanadesikan et al. (1997) for some discussion of the benefits of class participation in statistics classes, and Bligh (2000a, b) for discussion about class participation in general. Hawkins, Jolliffe, and Glickman (1992) discuss methods of teaching statistical concepts.

Section 1.2. Chapter 11 of this book has further discussion and references of ways of keeping students involved in statistics classes.

Section 1.4. The Texas school reforms and the Toxics Release Inventory are described by Liebman and Sabel (2002) and Sabel, Fung, and Karkkainen (1999), respectively.

Section 2.1. A version of this age-guessing demonstration appears in Gelman and Glickman (2000); it was adapted from Charlton and Williamson (1996, p. 76). See George and Hole (1995) for discussion and further references on biases in estimating ages.

Section 2.2. A version of this demonstration appears in Gelman (1998). The cancer map in Fig. 2.3 follows Manton et al. (1989). Gelman and Price (1999) discuss the general problem of artifacts in interpreting maps.

Section 2.3. Estimating orders of magnitude has been discussed by many authors, notably by Paulos (1988) in his discussion of innumeracy. This particular activity appears in Gelman and Glickman (2000). There are about 360 000 school buses in the United States and about 1620 Smiths in Oakland, California.

Section 2.4. The clippings in Fig. 2.5 are reprinted with permission of the *San Francisco Chronicle* and the *San Francisco Examiner*.

Section 2.5. The questionnaire in Fig. 2.6 is adapted from the Edinburgh Handedness Inventory (Oldfield, 1971). Various ways of measuring handedness are discussed in Hardyck and Petrinovich (1977). For more on the topic, see the book by Coren (1992). Handedness also comes up in Sections 3.3.2 and 10.2.3 of this book.

Section 3.2.1. The mile run example is adapted from Gelman and Nolan

(2002e). It is also discussed in Anderson and Loynes (1987, pp. 128–130). If your students might be interested in other work on world record times of men and women in running and swimming, see Wainer, Njue, and Palmer (2000). Carlin and Gelfand (1993) discuss the statistical analysis of data on record breaking.

Section 3.3.1. The table of variables labeled as categorical comes from Utts and Heckard (2001, p. 15), a textbook with a large collection of interesting examples.

Section 3.3.2. Data on the distribution of handedness in the general population occasionally appear in the psychological and medical literature; for example, see Oldfield (1971), Tan (1983), Lindesay (1987), Schachter, Ransil, and Geschwind (1987), and Ellis, Ellis, and Marshall (1988).

Sections 3.3.3 and 3.4.1. The data on soda drinking in Fig. 3.3 come from a survey by the U.S. Department of Agriculture (1995). The other information on beverage consumption is in the *Statistical Abstract of the United States.*

Section 3.4.2. Our information about Quetelet comes from Stigler (1986).

Section 3.5. For more on displaying data multidimensionally, see Cleveland (1985), Tufte (1983), and Wainer (1997).

Section 3.5.1. The example of guessed exam scores appears in Gelman (1997). Lightman and Sadler (1993) report a study in which physics teachers were far too optimistic in predicting their students scores on a short test.

Section 3.5.2. The actual results from the 1971 Gallup poll were,

	Adults with: Grade school education	High school education	College education	Total adults
% for withdrawal of U.S. troops (doves)	80%	75%	60%	73%
% against withdrawal of U.S. troops (hawks)	20%	25%	40%	27%
Total	100%	100%	100%	100%

Students are generally surprised that less-educated people were the ones most likely to be doves. This example comes from Loewen (1995, pp. 297–303), along with a very interesting discussion of college students' reactions. For an example of the ongoing confusion about this topic, see the quotation from the *Economist* on page 149.

Section 3.6.1. The discussion of the heights of men and women is adapted from Gelman and Nolan (2002e). The height distributions appear in Brainard and Burmaster (1992). Schilling, Watkins, and Watkins (2001) provide a detailed discussion of the common misconception that the distribution of heights is bimodal.

Section 3.6.2. The data on heights of conscripts, along with the explanation of the apparent bimodality, appear in Stigler (1986). To see how the rounding could affect the histogram, the following table shows the height ranges in centimeters that convert to 60–64 inches.

Height range (centimeters)	Estimated height (centimeters)	Estimated height (inches)	Estimated height range (inches)
152–153	152.5	60.03	60–61
153–154	153.5	60.43	60–61
154–155	154.5	60.82	60–61
155–156	155.5	61.22	61–62
156–157	156.5	61.61	61–62
157–158	157.5	62.01	62–63
158–159	158.5	62.40	62–63
159–160	159.5	62.79	62–63
160–161	160.5	63.18	63–64
161–162	161.5	63.58	63–64
162–163	162.5	63.97	63–64

The interval 60–61 inches contains only two of these bins, while the other intervals each contain three bins.

Section 3.6.3. See Smith (1997) for more on correlation and the regression effect with exam scores.

Section 3.7.2. Cobb, Halstead, and Rowe (1995) discuss problems with GDP and alternative measures of the size of the economy. Madrick (1998) discusses different ways of adjusting for inflation.

Section 3.8.2. The estimates of world population over time in Table 3.12 come from McEvedy and Jones (1978) and U.S. Census Bureau (2000).

Section 3.8.3. Figure 3.14 of metabolic rates and body masses of animals is reproduced from Schmidt-Nielsen (1984, p. 57). The two books by Schmidt-Nielsen (1978, 1984) have dozens of plots of this sort. Schmidt-Nielsen (1984, pp. 58–61) discusses why it is reasonable for the slope of the line in the graph to be 3/4 rather than 2/3. Dodds, Rothman, and Weitz (2001) discuss the history of the problem and suggest that the 3/4 power is not so appropriate.

Section 4.1.2. The height–earnings regression exampleas presented here is adapted from Gelman and Nolan (2002e). The data come from the Work, Family, and Well-Being Survey (Ross, 1990). An interesting analysis of the relation between height and wages appears in Persico, Postlewaite, and Silverman (2001). More generally, economists have modeled income data in many ways; see, for example, Geweke and Keane (2000). McDonald and Moffitt (1980) discuss the general use of regression models for nonnegative data such as income.

Section 4.2.1. Pearson and Lee (1903) give data on body measurements of university students and their families. Scheaffer et al. (1996, pp. 323–329) provide another example of this activity.

Section 4.3.1. Anderson and Loynes (1987, p. 140) and Lawrance (1996) discuss directed projects in which students design experiments on short-term memory.

Section 4.3.2. The last section of Kahneman and Tversky (1973) discusses, from a psychological perspective, the persistent confusion about regression to the mean.

Section 5.1. Many of the examples of survey sampling in this section appear in Gelman and Nolan (2002c). Paranjpe and Shah (2000) discuss sampling to estimate frequencies of letters.

Section 5.1.1. The idea of sampling from the telephone book has appeared in various statistics texts. The telephone book clipping in Fig. 5.2 appears courtesy of Verizon.

Section 5.1.2. Benford's law of the distribution of first digits was first discovered by Newcomb (1881), who noticed the pattern in the use of tables of logarithms. Benford (1938) rediscovered it and fit it to many datasets. A recent mathematical discussion appears in Hill (1998), and Matthews (1999) summarizes the history of Benford's law and many of its applications.

Section 5.1.3. The survey sampling template in Fig. 5.6 comes from Lohr (1999).

Hemenway (1997) uses the sort of conditional probability reasoning discussed in Section 7.5.2 to explain why surveys tend to overestimate the frequencies of rare events. Misclassifications that induce seemingly minor biases in estimates of certain small probabilities can lead to large errors in estimated frequencies. Hemenway discusses this effect in the context of traditional medical risk problems and then argues that this bias has caused researchers to drastically overestimate the number of times that guns have been used for self defense. Direct extrapolations from surveys suggest 2.5 million self-defense gun uses per year in the United States, but Hemenway shows how response errors could be causing this estimate to be too high by a factor of 10.

Section 5.1.5. Groves (1989) gives a thorough and practical presentation of sources of error in surveys.

Section 5.1.6. The family size example also appears in Madsen (1981), Gelman and Glickman (2000), and Gelman and Nolan (2002e). On related topics, Gelman and Little (1998) show how to correct for the fact that persons in larger households are less likely to be picked in a telephone survey, and Cohen and Cohen (1984) discuss related biases of size-based sampling in psychiatry research. See also the selection bias examples in Section 10.2. Size-based sampling in waiting times is a standard example in probability theory; a classroom demonstration based on this idea appears in Scheaffer et al. (1996, pp. 157–160). Also see Anderson and Loynes (1987, pp. 123–124).

Section 5.2. The class project on sample surveys is adapted from Gelman and Nolan (2002b). Handouts and additional questionnaires can be found at `www.stat.berkeley/users/nolan/surveys/`.

Section 5.2.1. For more guidance on questionnaire design, see Fowler (1995). For more on sampling, see Lohr (1999).

Section 5.2.2. The Rand Corporation survey, "Combining Service and Learning in Higher Education," is at `www.rand.org/publications/MR/MR998/`.

Section 5.3.1. The United Nations experiment appears in Gelman and Glickman (2000) and is adapted from a description in Kahneman and Tversky (1974). The actual proportion of countries in the United Nations that are in Africa is $53/189 = 28\%$. For many other examples of cognitive illusions (such as the anchoring heuristic illustrated here), see, for example, Kahneman, Slovic, and Tversky (1982) and Goldstein and Hogarth (1997). For another classroom demonstration in which an experiment is embedded in a survey, see Scheaffer et al. (1996, pp. 284–288).

Section 5.3.2. The experiment of altering the order of the questions on the midterm exam is described in Gelman (1997).

Section 5.3.3. Details of the taste-testing experiment appear in Chapter 5 of Nolan and Speed (2000). The original experiment conducted by Fisher is described in Fisher (1966) and Box (1978).

Section 5.4.1. See U.S. Surgeon General's Office (1964) for the full report.

Section 5.4.3. Briggs (2001) provides a thorough discussion of the effects of test preparation on SAT scores. See Gladwell (2001) for an interesting journalistic overview of SAT coaching.

Chapter 6. This chapter on statistical literacy assignments is adapted from Gelman and Nolan (1997). Some statistics teaching ideas based on newspaper clippings appear in Chatterjee, Handcock, and Simonoff (1995), Finn and Snell (1992), and Pearl and Stasny (1992). For general discussion of statistical literacy and education, see Bessant (1992) and Wallman (1993). Course packets for the 11 articles listed in Table 6.1 (which include the four articles in Section 6.6) appear at `www.stat.columbia.edu/~gelman/literacy/`.

Section 6.6.1. The article in Fig. 6.3 is reprinted with permission of the *New York Times*. The excerpted journal article appears with permission of the *New England Journal of Medicine*.

Section 6.6.2. The excerpted article appears with permission of *Pediatrics*.

Section 6.6.3. The articles in Fig. 6.4 are reprinted with permission of the *San Francisco Examiner* and *Science News*. The excerpted journal article appears with permission of *Pediatrics*.

Section 6.6.4. The article in Fig. 6.5 is reprinted with permission of the *San Francisco Examiner*.

Section 7.3.1. Estimating the probability of sequences of boys and girls is discussed in Kahneman and Tversky (1972). The Elvis Presley problem is Exercise 1.6 in Gelman et al. (1995).

Section 7.3.2. This form of the coin-flipping demonstration is adapted from Gelman and Glickman (2000). Other versions appear in Maxwell (1994), Revesz (1978), Schilling (1990), and Gnanadesikan et al. (1997). Kahneman and Tversky (1974) discuss people's general belief that any finite sequence of coin flips should have a haphazard pattern. Gilovich, Vallone, and Tversky (1985) show how apparent hot and cold streaks of basketball players are actually consistent with random fluctuations. In Fig. 7.1, the sequence on the left is real and the one on the right is fake. The left sequence has a few long sequences of 0's and 1's, which are typical in real coin flips but are almost never produced by students trying to produce realistic fake sequences.

Section 7.3.3. The lottery examples, and the method of discussing them in a statistics class, come from Hanley (1984, 1992), who considers the case of the woman who won the lottery twice and several other examples of extreme probability estimates that are "too good to be true." Articles of this sort appear occasionally in the statistics teaching literature as well as in newspapers; for example, Wetzel (2001) considers the probability of three sisters giving birth on the same day.

Section 7.4.1. The probability distribution of the length of the World Series is a well-known example, dating back at least to Mosteller (1952). A thorough discussion and references appear in Stern (1998a).

Section 7.4.2. Voting power and the probability of a decisive vote have been much studied in political science; Penrose (1946), Shapley and Shubik (1954) and Banzhaf (1965) are important early references. Gelman and Katz (2001) study the probability of decisiveness under various probability models (and argue against the models of Penrose, Shapley and Shubik, and Banzhaf); see also Mulligan and Hunter (2001). Gelman, King, and Boscardin (1998) consider rare events in the context of estimating the probability that your vote will be decisive in a Presidential election.

Section 7.4.3. Martz and Zimmer (1992) describe various estimates of the probability of space shuttle failure. Fischhoff et al. (1978) discuss other examples of this sort. The World Cup example comes from Doward (1998), a newspaper article about how insurance companies establish coverage for promotional events.

Section 7.5.1. The problem of the three cards is a standard elementary probability example. The Monty Hall problem is described by Selvin (1975) and Morgan et al. (1991), and the related problem of the three prisoners appears in Gardner (1961) and Mosteller (1965).

Section 7.5.2. The lie-detection demonstration appears in Gelman and Glickman (2000). The conditional probability of a false positive test is a standard example in probability teaching (see, for example, Moore, 1990, pp. 123–124), but we have not seen it elsewhere in the form of a demonstration. Tversky and Kahneman (1982) discuss the cognitive error of neglecting the base rate. Manly and Thomson (1998) describe a more realistic example of using statistics to discover theft.

Hemenway (1997) has used conditional probability to assess the bias in survey estimates of the frequency that guns are used in self-defense; see the Note to Section 5.1.3 on page 268.

Section 7.6. The biased coin demonstration appears in Gelman and Nolan (2002d). The story of King Olaf is told in Ekeland (1993), and Jay (2000) has many other interesting stories of the history of biased dice. Kerrich (1946) describes his coin-flipping experiences.

Section 7.6.2. The three news clips on the Euro coin come from Denny and Dennis (2002), Henderson (2002), and MacKenzie (2002).

Section 7.6.3. The physical model for coin flipping leading to the equal-probability rule is discussed by Keller (1986), Peterson (1990, 1997), and Jaynes (1995).

Section 8.1. Variations of the subjective sampling demonstration appear in the statistical literature; for example, Scheaffer et al. (1996, pp. 149–156) present a version based on rectangles drawn on paper. The example of weighing a random sample of candy is adapted from Gelman and Nolan (2002e).

Section 8.2.1. Dugger (2001) reports on varying sex ratios of births as an indication of selective abortion or infanticide.

Section 8.3.3. The idea of estimating the proportion of land area from a globe is described in Burrill and Cobb (1994) and Johnson (1997). This version of the demonstration appears in Gelman and Glickman (2000).

Section 8.3.4. Ansolabehere and Belin (1993) discuss the relative importance of sampling and nonsampling errors in poll differentials for Presidential elections.

Section 8.3.5. The data from golf putts appear in Berry (1995, pp. 80–82). The golf putting activity is described in Gelman and Nolan (2002a). Further information on golf putting appears in Pelz (1989).

Section 8.4.1. Moore (1990, pp. 130–131) illustrates the use of stacked intervals to picture the sampling distribution of a confidence interval.

Section 8.5.3. The p-values for the taste-testing experiment are derived by Fisher (1966).

Section 8.6.1. Miller (1956) is a fascinating article summarizing and interpreting studies on the number of items that can be memorized from a short exposure. It has been suggested that the estimated memory capacity of seven items is the reason why telephone numbers in the United States are seven digits long, but this has been disputed; Bailey (2000) provides a brief critical overview summarizing more recent developments.

Section 8.6.2. Anderson and Loynes (1987, pp. 163–164) discuss how to estimate the number of people in the country with your surname. McCarty et al. (2001) use the frequency of certain first names to estimate the size of personal networks.

Section 8.7.1. The basketball shooting example appears in Gelman and Glickman (2000). For more on the probability of making a basketball shot, see Gilovich, Vallone, and Tversky (1985) and some related articles mentioned on page 253 at the very end of Chapter 15 of this book.

Section 8.7.2. This data-dredging activity appears in Gelman and Glickman (2000). The assumed normal distribution for IQ scores is discussed by Reese (1997).

Section 8.7.3. The articles about prayer are Bowen (1999) and Harris et al. (1999).

Section 9.1. See the notes for Section 4.1.2 on page 267.

Section 9.2.2. Gulliksen (1950) and Lord and Novick (1968) develop various methods for checking exam questions by comparing to scores on other parts of the test. Wainer (1983) has an example of a graphical check.

Section 9.3. The data from golf putts appear in Berry (1995, pp. 80–82). The probability modeling appears in Gelman and Nolan (2002a).

Section 9.4. The linear regression of y on x_1 and x_2, where $y^2 = x_1^2 + x_2^2$, is a well-known example; we do not remember where we first saw it.

Chapter 10. Huff (1954) is the classic book on lying with statistics. It is still interesting and informative despite being a bit dated.

Section 10.1. Friendly (2000) and Wainer (1984, 1997) provide many interesting examples of statistical graphics in the news media and elsewhere.

Section 10.1.1. Senator McCarthy's lists of Communists are described by Rovere (1959). The articles in Fig. 10.1 are reproduced with permission of the *San Francisco Examiner.*

Section 10.1.2. The quotation is from the *Economist*, October 6, 2001, p. 36.

Section 10.1.3. The map in Fig. 10.2 is reproduced with permission of the *Daily Californian.*

Section 10.1.5. The article in Fig. 10.3 is reprinted with permission of the *Boston Herald.* The *Boston Globe* article is Wen (1989). The articles in Fig. 10.4 are reproduced with permission of the *San Francisco Weekly* and the *San Francisco Examiner.*

Section 10.2.2. Here are the explanations of the selection bias puzzles:

1. Deaths of students are rare, but when a student dies, he or she is usually young. Most students survive school and move to other professions.
2. As people age, their palms get more wrinkly. Hence, people who happen to live a long time have wrinkly palms when they die.
3. Patients with chronic cases will see psychiatrists more often, compared to patients with short-term problems. Thus, any given psychologists will tend to spend more time with chronic patients. This is similar to the family-size survey described in Section 5.1.6.
4. If a person has more friends, he or she is more likely to be someone's friend also. Hence, a random person's friend is likely to be a person with many friends.
5. It is logical for the survivors of a fight to blame it on the dead person, who is in no position to deny it.

Wainer, Palmer, and Bradlow (1998) present several thought-provoking and amusing examples of selection bias, including examples 1, 2, and 5 above. Example 3, and its practical importance, are discussed by Cohen and Cohen (1984), who label it the "clinician's illusion." Example 4 is discussed by Newman (2002) in the context of studying connections in personal networks.

Section 10.2.3. For more about the studies of the early deaths of left-handers, see Porac, Coren and Duncan (1980), Coren and Halpern (1991), and Halpern and Coren (1991). Salive, Guralnik, and Glynn (1993) argue that these findings are statistical artifacts. Further debate appears in Ellis (1990) and Coren (1990).

Section 10.3.2. The newspaper article in Fig. 10.5 is reprinted with permission of the *Chicago Sun-Times.* Its two errors are:

1. The graph is misleading. The first thing you see when you look at the graph is a big drop from 1993 to 1994, but this is just because the first three

months of 1994 are compared to all of 1993.

2. The claim in the figure caption that "Chicago is on pace for 150 fire-related deaths in 1994," is wrong. Presumably they got this by multiplying 38 deaths by 4 (since Jan 1 to Mar 29 is 1/4 of the year). But fire deaths are more common in the winter, so taking these winter deaths and multiplying by 4 gives an overestimate. If you read the article, it says that in 1993 there were 37 deaths in the first 3 months of the year and 104 total. This suggests that Chicago was on pace for $38 \cdot (104/37) = 107$ fire-related deaths in 1994.

Section 10.4. Statistics on the number of marriages and divorces in the United States can be found in the *Statistical Abstract of the United States*. Whitehead and Popenoe (2001) contains a discussion on the probability that a marriage ends in divorce.

Section 10.5.1. More information on the National Surgical Adjuvant Breast Project and Dr. Poisson's data can be found on the Web at
`www.lij.edu/education_and_research/research2000/Sessions/sess_04.ppt`

Section 10.5.5. Quantitative studies of the rates of police stops by ethnic group include New York State Attorney General's Office (1999), Harris (1999), and Persico, Knowles, and Todd (2001).

Herman Chernoff told us about the idea of fractional grades for multiple-choice questions. More formally, grades can be estimated using the likelihood function with stochastic models that implicitly are allocating points based on the estimated probability of guessing; see Thissen and Wainer (2001).

Chapter 11. Some ideas about encouraging class participation, including the use of student activities in lectures and discussions, appear in the books by Davis (1993), Lowman (1995), McKeachie (1999), Bligh (2000a, b), Lovett and Greenhouse (2000).

Section 11.3. Some of the methods of using exams to teach statistical concepts appear in Gelman (1997).

Section 11.4. Short and Pigeon (1998) discuss how to introduce to students the idea of a formal data collection protocol. Students might also be interested in the Roberts (2001) article on self-experimentation. Weinberg and Abramowitz (2000) discuss how to implement examples as case studies for students.

Section 11.5.5. The end-of-class feedback sheet is described by Mosteller (1988, pp. 97–98). The effectiveness of student evaluations of teachers, and the information used in these evaluations, is discussed by Ambady and Rosenthal (1997).

Chapter 12. This course syllabus mostly follows Moore and McCabe (1998), which is the textbook we use when teaching introductory statistics. We augment Moore and McCabe by including more material on probability, adding some more difficult homework assignments throughout the course, including a project, and covering "lying with statistics" (see Chapter 10) at the end of the semester.

Chapter 13. Much of the material in this chapter appeared in Gelman (1998). Textbooks in applied decision analysis and Bayesian statistics, from various perspectives, include Watson and Buede (1987), Clemen (1996), Berry (1995), Gelman et al. (1995), and Carlin and Louis (2000).

Section 13.1.2. The concept of behavior that is not consistent with any utility function is discussed and modeled by Kahneman and Tversky (1979), Kahneman, Slovic, and Tversky (1982), and Thaler (1992).

Section 13.1.4. There is a large social-science literature on the equating of dollars and lives; two interesting books on the topic are Rhoads (1980) and Dorman (1996). The example and true values for the risk estimates in Fig. 13.3 come from Fischhoff et al. (1981), Slovic, Fischhoff, and Lichtenstein (1982), and Rodricks (1992) and are, approximately, 2, 200, 800 000, 11 000, 40 000, 450, 90 000, 90.

Section 13.1.5. The Brier score was introduced by Brier (1950); it and related probability scoring rules are discussed by Dawid (1986). When applied to exam scoring, many methods have been proposed to take into account students' uncertainties in their answers; see, for example, Coombs, Milholland, and Womer (1956) and DeFinetti (1965).

Section 13.1.6. Stern (1997, 1998b) analyzes betting odds and point spreads for several professional sports. The books by Rombola (1984) and Jimmy "the Greek" Snyder (1975) are full of fascinating anecdotes of sports gambling.

Romer and Romer (2000) use regression models to evaluate different organizations' predictions of inflation rates.

Section 13.2.1. The map in Fig. 13.4 follows Manton et al. (1989). Gelman and Price (1999) discuss this problem in general and explains why the map of posterior means also has problems, in that it overemphasizes the high-population counties. Clayton and Bernardinelli (1992) discuss this sort of Bayesian inference for epidemiology.

Section 13.2.2. Figure 13.5 with the too-narrow subjective intervals is adapted from Hynes and Vanmarcke (1977) and is a well-known example in risk analysis. An interesting discussion of overconfidence, with experimental results, appears in Alpert and Raiffa (1984). The question about the total number of eggs comes from that paper. The true values of the uncertain quantities in Fig. 13.6 are 12.1, 64.6 billion, 814, 48.7, 2, 6.25 million, 39.4 million, 4.1 million, 1.5 million, 35 500.

Section 13.2.4. The fitted gamma distribution is estimated based on information in Manton et al. (1989). Gelman and Price (1999) also discuss this model.

Chapter 14. We use Lohr (1999) as a textbook when we teach survey sampling.

Section 14.4. The idea of Clusterville is adapted from Schwarz (1997).

Section 14.5. The article in Fig. 14.3 appears courtesy of the *New York Times*.

Section 14.6.1. Descriptions of the complex surveys can be found on the Web at the following locations:

National Survey of Family Growth, National Center for Health Statistics,

www.cdc.gov/nchs/nsfg.htm
Youth Risk Behavior Surveillance, Centers for Disease Control,
www.cdc.gov/nccdphp/dash/yrbs
National Health Information Survey, National Institutes of Health,
www.cdc.gov/nchs/nhis.htm
Consumer Expenditure Survey, Bureau of Labor Statistics,
www.bls.gov/cex/home.htm
Current Population Survey, Bureau of Labor Statistics,
www.bls.census.gov/cps/cpsmain.htm
American Housing Survey, Department of Housing and Urban Development,
www.census.gov/hhes/www/ahs.html
Commodity Flow Survey, Census Bureau,
www.census.gov/econ/www/se0700.html
National Survey of Speeding and Other Unsafe Driving Actions, Bureau of Transportation, www.nhtsa.dot.gov/people/injury/aggressive/unsafe/methods
Survey of Income and Program Participation, Census Bureau,
www.sipp.census.gov/sipp
National Crime Victimization Survey, Department of Justice,
www.ojp.usdoj.gov/bjs

Section 14.6.2. The StatCity population and R code are available at www.stat.berkeley.edu/users/nolan/.

Section 14.6.3. The article in Fig. 14.6 appears courtesy of the *San Francisco Chronicle.*

Chapter 15. Much of the material in this chapter appeared in Adhikari and Nolan (1997).

Section 15.3. Some good sources of probability problems, at different levels of difficulty, are Mosteller (1965), Feller (1968), and Williams (1991). The development of the number theory example in this section is modeled after that in Williams's text (1991). For more on the stereology example at the end of Fig. 15.2, see, for example, Baddeley (1982).

Section 15.6.1. These instructions follow the treatment of the arcsine rule in Feller (1968).

Chapter 16. The material in this chapter is adapted from Nolan and Speed (1999, 2000). The book by Nolan and Speed (2000) gives several individual case studies in depth. Data and instructions for a set of labs, including those in Table 16.1, are available at www.stat.berkeley.edu/users/nolan/statlabs/. Another source of case studies on a wide range of topics in statistics is Chatterjee, Handcock, and Simonoff (1995).

Recently Cobb and Moore (1997) called for the design of a better one-semester statistics course for mathematics majors that both strengthens their mathematical skills and integrates data analysis into the curriculum. Others have called for similar courses (Foster and Smith, 1969; Hogg, 1985, 1992; Kempthorne, 1980; Moore and Roberts, 1989; Mosteller, 1988; Petruccelli, Nandram, and Chen,

1995; Whitney and Urquhart, 1990), and many consider training in statistical thinking important (see, for example, Daisley, 1979; Joiner, 1989; Schuyten, 1991; Riffenburgh, 1995; and Nash and Quon, 1996).

Section 16.3. The data for the quality control project are available at `www.stat.berkeley.edu/users/statlabs/`. This case study is described in Dolezal, Burdick, and Birch (1998). More details on the theoretical material can be found in Vardeman and VanValkenburg (1999).

Section 16.4. The helicopter design project is described by Bisgaard (1991). Details appear in Chapter 12 of Nolan and Speed (2000).

References

Adams, G. (2001). Voting irregularities in Palm Beach, Florida. *Chance* **14** (1), 22–24.

Adhikari, A., and Nolan, D. (1997). Probability and stochastic processes. In *Women in Mathematics: Scaling the Heights*, ed. Nolan, D., MAA Notes **46**, 27–39. Washington, D.C.: Mathematical Association of America.

Alpert, M., and Raiffa, H. (1984). A progress report on the training of probability assessors. In *Judgment Under Uncertainty: Heuristics and Biases*, ed. Kahneman, D., Slovic, P., and Tversky, A., 294–305. Cambridge University Press.

Ambady, N., and Rosenthal, R. (1997). Judging social behavior using "thin slices." *Chance* **10** (4), 12–18, 51.

Anderson, C. W., and Loynes, R. M. (1987). *The Teaching of Practical Statistics.* New York: Wiley.

Babbie, E. (1999). *The Basics of Social Research.* Belmont, Calif.: Wadsworth.

Baddeley, A. (1982). Stochastic geometry: An introduction and reading-list. *International Statistics Review* **50**, 179–193.

Bailey, B. (2000). Reducing reliance on superstition. *UI Design Update Newsletter*, September. `www.humanfactors.com/library/sep002.htm`

Banzhaf, J. (1965). Weighted voting doesn't work: a mathematical analysis. *Rutgers Law Review* **19**, 317–343.

Becker, R. A., Chambers, J. M., and Wilks, A. R. (1988). *The New S Language: A Programming Environment for Data Analysis and Graphics.* Pacific Grove, Calif.: Wadsworth.

Benford, F. (1938). The law of anomalous numbers. *Proceedings of the American Philosophical Society* **78**, 551–572.

Berry, D. (1995). *Statistics: A Bayesian Perspective.* Belmont, Calif.: Duxbury.

Bessant, K. C. (1992). Instructional design and the development of statistical literacy. *Teaching Sociology* **20**, 143–9.

Bisgaard, S. (1991). Teaching statistics to engineers. *American Statistician* **45**, 274–283.

Bligh, D. A. (2000a). *What's the Point in Discussion?* Exeter, England: Intellect.

Bligh, D. A. (2000b). *What's the Use of Lectures?* San Francisco: Jossey-Bass.

Bowen, J. (1999). Faith healing: can prayer do anything more than make you feel better? *Salon*, November 3.
`www.salon.com/health/feature/1999/11/03/prayer`

Box, J. F. (1978). *R. A. Fisher: The Life of a Scientist.* New York: Wiley.

Brainard, J., and Burmaster, D. E. (1992). Bivariate distributions for height and weight of men and women in the United States. *Risk Analysis* **12**, 267–275.

Brier, G. W. (1950). Verification of forecasts expressed in terms of probability. *Monthly Weather Review* **78**, 1–3.

Briggs, D. C. (2001). The effect of admissions test preparation: evidence from NELS:88 (with discussion). *Chance* **14** (1), 10–21.

Burrill, G., and Cobb, G. (1994). Everyone's favorite subject: what's new. *Stats*, **12**, 8–21.

Carlin, B. P. and Gelfand, A. E. (1993). Parametric likelihood inference for record breaking problems. *Biometrika* **80**, 507–515.

Carlin, B. P., and Louis, T. A. (1996). *Bayes and Empirical Bayes Methods for Data Analysis.* London: Chapman and Hall.

Case, B. A., ed. (1989). *Responses to the Challenge: Keys to Improved Instruction by Teaching Assistants and Part-Time Instructors.* Washington, D.C.: Mathematical Association of America.

Charlton, J., and Williamson, R. (1996). *Practical Exercises in Applied Statistics.* Oxford University Press.

Chatterjee, S., Handcock, M. S., and Simonoff, J. S. (1995). *A Casebook for a First Course in Statistics and Data Analysis.* New York: Wiley.

Clayton, D., and Bernardinelli, L. (1992). Bayesian methods for mapping disease risk. In *Geographical and Environmental Epidemiology: Methods for Small-Area Studies*, ed. P. Elliott, J. Cusick, D. English, and R. Stern, 205–220. Oxford University Press.

Clemen, R. T. (1996). *Making Hard Decisions*, second edition. Belmont, Calif.: Duxbury.

Cleveland, W. S. (1985). *The Elements of Graphing Data.* Monterey, Calif.: Wadsworth.

Cobb, C., Halstead, T., and Rowe, J. (1995). If the GDP is up, why is America down? *Atlantic* **278** (4), 59–78.

Cobb, G. (1992). Teaching statistics. In *Heeding the Call for Change: Suggestions for Curricular Action*, ed. L. A. Steen, 3–34. Mathematical Association of America.

Cobb, G., and Moore, D. (1997). Mathematics, statistics, and teaching. *American Mathematical Monthly* **104**, 801–823.

Cohen, P., and Cohen, J. (1984). The clinician's illusion. *Archives of General Psychiatry* **41**, 1178–1182.

Coombs, C. H., Milholland, J. E., and Womer, J. F. B. (1956). The assessment of partial knowledge. *Educational and Psychological Measurement* **15**, 337–352.

Daisley, P. (1979). Statistical thinking rather than statistical methods. *Statistician* **28**, 231–239.

Dawid, A. P. (1986). Probability forecasting. In *Encyclopedia of Statistical Sciences*, Vol. 7, ed. S. Kotz, N. L. Johnson, and C. B. Read, 210–218. New York: Wiley.

Davis, B. G. (1993). *Tools for Teaching*. San Francisco: Jossey-Bass.

Denny, C., and Dennis, S. (2002). Heads, Belgium wins—and wins. [London] *Guardian*, January 4.

DeFinetti, B. (1965). Methods for discriminating level of partial knowledge concerning a test item. *British Journal of Mathematical and Statistical Psychology* **18**, 87–123.

Dodds, P. S., Rothman, D. H., and Weitz, J. S. (2001). Re-examination of the '3/4-law' of metabolism. *Journal of Theoretical Biology* **209**, 9–27.

Dolezal, K. K., Burdick, R. K., and Birch, N. J. (1998) Analysis of a two-factor R & R study with fixed operators. *Journal of Quality Technology* **30**, 163–170.

Dorman, P. (1996). *Markets and Mortality: Economics, Dangerous Work, and the Value of Human Life*. Cambridge University Press.

Doward, J. (1998). Truth behind that 'win a million' offer. [London] *Observer*, May 17, Business section, p. 8.

Dugger, C. W. (2001). Modern Asia's anomaly: the girls who don't get born. *New York Times*, May 6, Section 4, p. 4.

Ekeland, I. (1993). *The Broken Dice*. University of Chicago Press.

Feller, W. (1968). *An Introduction to Probability Theory and Its Applications*, Volume 1, third edition. New York: Wiley.

Finkelstein, M. O., and Levin, B. (2001). *Statistics for Lawyers*, second edition. New York: Springer-Verlag.

Finn, J., and Snell, J. L. (1992). A course called Chance. *Chance* **5**, 12–16.

Fischhoff, B., Lichtenstein, S., Slovic, P., Derby, S. L., and Keeney, R. L. (1981). *Acceptable Risk*. Cambridge University Press.

Fischhoff, B., Slovic, P., and Lichtenstein, S. (1978). Fault trees: sensitivity of estimated failure probabilities to problem representation. *Journal of Experimental Psychology: Human Perception and Performance* **4**, 330–344.

Fisher, R. A. (1966). *The Design of Experiments*, eighth edition. New York: Hafner.

Foster, F. G., and Smith, T. M. F. (1969). The computer as an aid in teaching. *Applied Statistics* **18**, 264–269.

Friendly, M. (2000). Gallery of data visualization: the best and worst of statistical graphics. Statistical Consulting Service, York University. `www.math.yorku.ca/SCS/Gallery/`

Gardner, M. (1961). *The Second Scientific American Book of Mathematical Puzzles and Diversions*. New York: Simon and Schuster.

Gastwirth, J. L., ed. (2000). *Statistical Science in the Courtroom.* New York: Springer-Verlag.

Gelman, A. (1997). Using exams for teaching concepts in probability and statistics. *Journal of Educational and Behavioral Statistics* **22**, 237–243.

Gelman, A. (1998). Some class-participation demonstrations for decision theory and Bayesian statistics. *American Statistician* **52**, 167–174.

Gelman, A., Carlin, J. B., Stern, H. S., and Rubin, D. B. (1995). *Bayesian Data Analysis.* London: Chapman and Hall.

Gelman, A., and Glickman, M. E. (2000). Some class-participation demonstrations for introductory probability and statistics. *Journal of Educational and Behavioral Statistics* **25**, 84–100.

Gelman, A., and Katz, J. (2001). How much does a vote count? Voting power, coalitions, and the Electoral College. Technical report, Department of Statistics, Columbia University.

Gelman, A., King, G., and Boscardin, W. J. (1998). Estimating the probability of events that have never occurred: When is your vote decisive? *Journal of the American Statistical Association* **93**, 1–9.

Gelman, A., and Little, T. C. (1998). Improving upon probability weighting for household size. *Public Opinion Quarterly* **62**, 398–404.

Gelman, A., and Nolan, D., with Men, A., Warmerdam, S., and Bautista, M. (1997). Student projects on statistical literacy and the media. *American Statistician* **52**, 160–166.

Gelman, A., and Nolan, D. (2002a). A probability model for golf putting. *Teaching Statistics*, to appear.

Gelman, A., and Nolan, D. (2002b). A class project in survey sampling. *College Teaching*, to appear.

Gelman, A., and Nolan, D. (2002c). Some statistical sampling and data collection activities. *Mathematics Teacher*, to appear.

Gelman, A., and Nolan, D. (2002d). You can load a die but you can't bias a coin. *American Statistician*, to appear.

Gelman, A., and Nolan, D. (2002e). Double takes: some statistical examples with surprise twists. *Journal of Statistics Education*, submitted.

Gelman, A., and Price, P. N. (1999). All maps of parameter estimates are misleading. *Statistics in Medicine* **18**, 3221–3234.

George, P. A., and Hole, G. J. (1995). Factors influencing the accuracy of age estimates of unfamiliar faces. *Perception* **24**, 1059–1073.

Gilovich, T., Vallone, R., and Tversky, A. (1985). The hot hand in basketball: on the misperception of random sequences. *Cognitive Psychology* **17**, 295–314.

Gladwell, M. (2001). What Stanley H. Kaplan taught us about the S.A.T. *New Yorker*, December 17, p. 86.

Gnanadesikan, M., Scheaffer, R. L., Watkins, A. E., and Witmer, J. A. (1997). An activity-based statistics course. *Journal of Statistics Education* **5** (2).

Goldstein, W. M., and Hogarth, R. M. (1997). *Research on Judgment and Decision Making*. Cambridge University Press.

Gopen, G. D., and Swan, J. A. (1990). The science of scientific writing. *American Scientist* **78**, 550–558.

Gulliksen, H. O. (1950). *Theory of Mental Tests*. New York: Wiley.

Hanley, J. A. (1984). Lotteries and probabilities: three case reports. *Teaching Statistics* **6**, 88–92.

Hanley, J. A. (1992). Jumping to coincidences: defying odds in the realm of the preposterous. *American Statistician* **46**, 197–202.

Harris, D. A. (1999). The stories, the statistics, and the law: why "driving while black" matters. *Minnesota Law Review* **84**, 265–326.

Harris, W. S., Gowda, M., Kolb, J. W., Strychacz, C. P., Vacek, J. L., Jones, P. G., Forker, A., O'Keefe, J. H., McCallister, B. D. (1999). A randomized, controlled trial of the effects of remote, intercessory prayer on outcomes in patients admitted to the coronary care unit. *Archives of Internal Medicine* **159**, 2273–2278.

Hawkins, A., Jolliffe, F., and Glickman, L. (1992). *Teaching Statistical Concepts*. London: Longman.

Hemenway, D. (1997). The myth of millions of annual self-defense gun uses: a case study of survey overestimates of rare events. *Chance* **10** (3), 6–10.

Henderson, C. (2002). Hussain's flipping fillip. *BBC Sport*, January 4.

Hill, T. P. (1998). The first digit phenomenon. *American Scientist* **86**, 358–363.

Hogg, R. V. (1985). Statistical education for engineers: an initial task force report. *American Statistician* **39**, 21–24.

Hogg, R. V. (1992). Report of workshop on statistical education. In *Heeding the Call for Change: Suggestions for Curricular Action*, ed. L. A. Steen, 34–43. Washington, D.C.: Mathematical Association of America.

Hollander, M., and Proschan, F. (1984). *The Statistical Exorcist: Dispelling Statistics Anxiety*. New York: Dekker.

Huff, D. (1954). *How to Lie with Statistics*. New York: Norton.

Hynes, M. E., and Vanmarcke, E. (1977). Reliability of embankment performance predictions. In *Mechanics in Engineering*, 367–384. University of Waterloo Press.

Jay, R. (2000). The story of dice: gambling and death from ancient Egypt to Los Angeles. *New Yorker*, December 11, 91–95.

Jaynes, E. T. (1995). *Probability Theory: The Logic of Science*, pp. 1003–1007. `www-laplace.imag.fr/ENGLISH/PRESENTATION/index.html`

Joiner, B. L. (1989). Statistical thinking: What to teach and what not to teach managers. *American Statistical Association Proceedings* **150**, 448–461.

Johnson, R. (1997). Earth's surface water percentage? *Teaching Statistics* **19**, 66–68.

Kahneman, D., Slovic, P., and Tversky, A. (1982). *Judgment Under Uncertainty: Heuristics and Biases.* Cambridge University Press.

Kahneman, D., and Tversky, A. (1972). Subjective probability: a judgment of representativeness. *Cognitive Psychology* **3**, 430–454. In *Judgment Under Uncertainty: Heuristics and Biases*, ed. D. Kahneman, P. Slovic, and A. Tversky, 32–47. Cambridge University Press.

Kahneman, D., and Tversky, A. (1973). On the psychology of prediction. *Psychological Review* **80**, 237–251. In *Judgment Under Uncertainty: Heuristics and Biases*, ed. D. Kahneman, P. Slovic, and A. Tversky, 48–68. Cambridge University Press.

Kahneman, D., and Tversky, A. (1974). Judgment under uncertainty: heuristics and biases. *Science* **185**, 1124–1131. In *Judgment Under Uncertainty: Heuristics and Biases*, ed. D. Kahneman, P. Slovic, and A. Tversky, 3–20. Cambridge University Press.

Kahneman, D., and Tversky, A. (1979). Prospect theory: an analysis of decision under risk. *Econometrica* **47**, 263–291.

Keller, J. B. (1986). The probability of heads. *American Mathematical Monthly* **93**, 191–197.

Kempthorne, O. (1980). The teaching of statistics: content versus form. *American Statistician* **34**, 17–21.

Kerrich, J. E. (1946). *An Experimental Introduction to the Theory of Probability.* Copenhagen: J. Jorgensen.

Lawrance, A. J. (1996). A design of experiments workshop as an introduction to statistics. *American Statistician* **50**, 156–158.

Liebman, J. S., and Sabel, C. F. (2002). A public laboratory Dewey barely imagined: the emerging model of school governance and legal reform. *NYU Journal of Law and Social Change*, to appear.

Lightman, A., and Sadler, P. (1993). Teacher predictions versus actual student gains. *Physics Teacher* **31**, 162–167.

Loewen, J. W. (1995). *Lies My Teacher Told Me: Everything Your American History Textbook Got Wrong.* New York: Norton.

Lohr, S. (1999). *Sampling: Design and Analsis.* Pacific Grove, Calif.: Duxbury.

Lord, F. M., and Novick, M. R. (1968). *Statistical Theories of Mental Test Scores.* Reading, Mass.: Addison-Wesley.

Lovett, M. C., and Greenhouse, J. B. (2000). Applying cognitive theory to statistics instruction. *American Statistician* **54**, 196–206.

Lowman, J. (1995). *Mastering the Techniques of Teaching*, second edition. San Francisco: Jossey-Bass.

MacKenzie, D. (2002). Euro coin accused of unfair flipping. *New Scientist*, January 4.

Madsen, R. W. (1981). Making students aware of bias. *Teaching Statistics* **3**, 2–5.

Magel, R. C. (1996). Increasing student participation in a large introductory statistics class. *American Statistician* **50**, 51–56.

Magel, R. C. (1998). Using cooperative learning in a large introductory statistics class. *Journal of Statistics Education* **6** (3).

Manly, B. F., and Thomson, R. (1998). How to catch a thief. *Chance* **11** (4), 22–25.

Manton, K. G., Woodbury, M. A., Stallard, E., Riggan, W. B., Creason, J. P., and Pellom, A. C. (1989). Empirical Bayes procedures for stabilizing maps of U.S. cancer mortality rates. *Journal of the American Statistical Association* **84**, 637–650.

Martz, H. F., and Zimmer, W. J. (1992). The risk of catastrophic failure of the solid rocket boosters on the space shuttle. *American Statistician* **46**, 42–47.

Matthews, R. (1999). The power of one. *New Scientist*, July 10.

MathSoft (2000). S-Plus. `www.splus.mathsoft.com/`

Maxwell, N. P. (1994). A coin-flipping exercise to introduce the P-value. *Journal of Statistics Education* **2** (1).

McCarty, C., Killworth, P. D., Bernard, H. R., Johnsen, E., and Shelley, G. A. (2001). Comparing two methods for estimating network size. *Human Organization* **60**, 28–39.

McDonald, J. F., and Moffitt, R. A. (1980). The uses of tobit analysis. *Review of Economics and Statistics* **62**, 318–321.

McEvedy, C., and Jones, R. (1978). *Atlas of World Population History*. London: Penguin.

McKeachie, W. J. (1999). *Teaching Tips*, tenth edition. Boston: Houghton Mifflin.

Miller, G. A. (1956). The magical number seven, plus or minus two: some limits on our capacity for processing information. *Psychological Review* **63**, 81–97.

Moore, D. S. (1990). Uncertainty. In *On the Shoulders of Giants: New Approaches to Numeracy*, ed. L. A. Steen, 95–137. Washington, D.C.: National Academy Press.

Moore, D. S., and McCabe, G. P. (1998). *Introduction to the Practice of Statistics*, third edition. New York: Freeman.

Moore, T., ed. (2000). *Teaching Resources for Undergraduate Statistics*. Washington, D.C.: Mathematical Association of America.

Moore, T. L., and Roberts, R. A. (1989). Statistics at liberal arts colleges. *American Statistician* **43**, 80–85.

Morgan, J. P., Chaganty, N. R, Dahiya, R. C., and Doviak, M. J. (1991). Let's make a deal: the player's dilemma. *American Statistician* **45**, 284–289.

Mosteller, F. (1952). The World Series competition. *Journal of the American Statistical Association* **47**, 355–380.

Mosteller, F. (1965). *Fifty Challenging Problems with Solutions*. New York: Dover.

Mosteller, F. (1988). Broadening the scope of statistics and statistical education. *American Statistician* **42**, 93–99.

Mulligan, C. B., and Hunter, C. G. (2001). The empirical frequency of a pivotal vote. National Bureau of Economic Research Working Paper 8590.

Nash, J. C., and Quon, T. K. (1996). Issues in teaching statistical thinking with spreadsheets. *Journal of Statistics Education* **4** (1).

Newcomb, S. (1881). Note on the frequency of the use of digits in natural numbers. *American Journal of Mathematics* **4**, 39–40.

Newman, M. E. J. (2002). Ego-centered networks and the ripple effect, or why all your friends are weird. *Social Networks*, to appear.

New York State Attorney General's Office (1999). *Stop and Frisk Report.* www.oag.state.ny.us/press/1999/dec/dec01a_99.htm

Nolan, D., and Speed, T. (1999). Teaching statistics: theory after application? *American Statistician* **53**, 370–376.

Nolan, D., and Speed, T. (2000). *Stat Labs: Mathematical Statistics Through Applications.* New York: Springer-Verlag.

Oldfield, R. C. (1971). The assessment and analysis of handedness: the handedness inventory. *Neuropsychologia* **9**, 97–114.

Ortiz, D. (1984). *Gambling Scams.* New York: Lyle Stuart.

Paranjpe, S. A., and Shah, A. (2000). How may words in a dictionary? Innovative laboratory teaching of sampling techniques. *Journal of Statistical Education* **8** (2).

Paulos, J. A. (1988). *Innumeracy: Mathematical Illiteracy and Its Consequences.* New York: Hill and Wang.

Pearl, D. K., and Stasny, E. A. (1992). *Experiments in Statistical Concepts.* Dubuque: Kendall/Hunt.

Pearson, K. and Lee, A. (1903). On the laws of inheritance in man. *Biometrika* **2** (4), 357–462.

Pelz, D. (1989). *Putt Like the Pros.* New York: Harper Collins.

Penrose, L. S. (1946). The elementary statistics of majority voting. *Journal of the Royal Statistical Society* **109**, 53–57.

Persico, N., Knowles, J., and Todd, P. (2001). Racial bias in motor-vehicle searches: theory and evidence. *Journal of Political Economy* **109**, 203–229.

Persico, N., Postlewaite, A., and Silverman, D. (2001). The effect of adolescent experience on labor market outcomes: the case of height. Technical report, Department of Economics, University of Pennsylvania.

Peterson, I. (1990). *Islands of Truth.* New York: Freeman.

Peterson, I. (1997). A penny surprise. *MathTrek*, MAA Online, December 15. www.maa.org/mathland/mathtrek_12_15.html

Petruccelli, J. D., Nandram, B., and Chen, M.-H. (1995). Implementation of a modular laboratory and project-based statistics curriculum. *American Statistical Association, Proceedings of the Section on Statistical Education*, 165–170.

Porac, C., Coren, S., and Duncan, P. (1980). Life-span age trends in laterality. *Journal of Gerontology* **35**, 715–721

R Project (2000). The R project for statistical computing. `www.r-project.org`

Reese, R. A. (1997). IQ—abnormal thoughts. *Chance* **10** (4), 49–51.

Revesz, P. (1978). Strong theorems on coin tossing. *Proceedings of the International Congress of Mathematicians* (Helsinki, 1978), ed. O. Lehto, 749–754. Helsinki, Finland.

Rhoads, S. E. (1980). *Valuing Life: Public Policy Dilemmas*, Boulder, Colo.: Westview.

Riffenburgh, R. H. (1995). Infusing statistical thinking into clinical practice. *American Statistical Association, Proceedings of the Section on Statistical Education*, 5–8.

Roberts, S. (2001). Surprises from self-experimentation: sleep, mood, and weight (with discussion). *Chance* **14** (2), 7–18.

Rodricks, J. V. (1992). Calculated risks: the toxicity and human health risks of chemicals in our environment. Cambridge University Press.

Rombola, F. (1984). *The Book on Bookmaking*. Pasadena, Calif.: Pacific Book and Printing.

Romer D. H., and Romer, C. D. (2000). Federal Reserve information and the behavior of interest rates. *American Economic Review* **90**, 429–457.

Ross, C. E. (1990). Work, family, and well-being in the United States. Survey data available from Inter-university Consortium for Political and Social Research, Ann Arbor, Mich.

Rossman, A., and Von Oehsen, J. B. (1997). *Workshop Statistics: Discovery with Data and the Graphing Calculator*. New York: Springer-Verlag.

Rovere, R. H. (1959). *Senator Joe McCarthy*. New York: Harper and Row.

Sabel, C. F., Fung, A., and Karkkainen, B. (1999). Beyond backyard environmentalism (with discussion). *Boston Review*, **24**, October/November, 1–12.

Scheaffer, R. L., Gnanadesikan, M., Watkins, A., and Witmer, J. (1996). *Activity-Based Statistics: Instructor Resources*. New York: Springer-Verlag.

Schilling, M. F. (1990). The longest run of heads. *College Mathematics Journal* **21**, 196–207.

Schilling, M. F., Watkins, A. E., and Watkins, W. (2001). Is human height bimodal? Submitted to *American Statistician*.

Schmidt-Nielsen, K. (1978). *How Animals Work*. Cambridge University Press.

Schmidt-Nielsen, K. (1984). *Scaling: Why is Animal Size So Important?* Cambridge University Press.

Schuyten, G. (1991). Statistical thinking in psychology and education. *Proceedings of International Conference on Teaching Statistics*, 486–489.

Schwarz, C. J. (1997). StatVillage: An on-line www-accessible, hypothetical city based on real data for use in an interactive class on surve sampling. *Journal of Statistics Education* **5** (2).

Selvin, S. (1975). Letter. *American Statistician* **29**, 67.

Shapley, L. S., and Shubik, M. (1954). A method for evaluating the distribution of power in a committee system. *American Political Science Review* **48**, 787–792.

Short, T. H., and Pigeon, J. G. (1998). Protocols and pilot studies: taking data collection projects seriously. *Journal of Statistics Education* **6** (1).

Slovic, P., Fischhoff, P., and Lichtenstein, S. (1982). Facts versus fears: understanding perceived risk. In *Judgment Under Uncertainty: Heuristics and Biases*, ed. D. Kahneman, P. Slovic, and A. Tversky, 463–489. Cambridge University Press.

Smith, G. (1997). Do statistics test scores regress to the mean? *Chance* **10** (4), 42–45.

Snyder, J., with Herskowitz, M., and Perkins, S. (1975). *Jimmy the Greek, by Himself.* Chicago: Playboy.

Sprent, P. (1988). *Taking Risks: The Science of Uncertainty*. London: Penguin.

StataCorp (2000). *Stata Statistical Software: Release 7.0.* College Station, Tex.: Stata Corporation.

Stern, H. S. (1997). How accurately can sports outcomes be predicted? *Chance* **10** (4), 19–23.

Stern, H. S. (1998a). Best-of-seven playoff series. *Chance* **11** (2), 46–49.

Stern, H. S. (1998b). How accurate are the posted odds? *Chance* **11** (4), 17–21.

Stigler, S. (1986). *The History of Statistics: The Measurement of Uncertainty Before 1900.* Harvard University Press.

Stilgoe, J. R. (1998). *Outside Lies Magic: Regaining History and Awareness in Everyday Places.* New York: Walker.

Tanur, J. M., Mosteller, F., Kruskal, W. H., Link, R. F., Pieters, R. S., and Rising, G. R. (1972). *Statistics: A Guide to the Unknown.* New York: Holden-Day. Third edition (1989). Belmont, Calif.: Wadsworth.

Thaler, R. H. (1992). *The Winner's Curse: Paradoxes and Anomalies of Economic Life.* New York: Free Press.

Thissen, D., and Wainer, H. (2001). *Test Scoring*. Hillsdale, N.J.: Lawrence Erlbaum Associates.

Tollefson, S. (1988). Encouraging student writing. Office of Educational Development, University of California, Berkeley.

Tufte, E. R. (1983). *The Visual Display of Quantitative Information.* Cheshire, Conn.: Graphics Press.

Tversky, A., and Kahneman, D. (1982). Evidential impact of base rates. In *Judgment Under Uncertainty: Heuristics and Biases*, ed. D. Kahneman, P. Slovic, and A. Tversky, 153–160. Cambridge University Press.

U.S. Census Bureau (2000). World population information. `www.census.gov/ipc/www`

U.S. Department of Agriculture (1995). Continuing survey of food intakes by individuals and diet and health knowledge survey.

U.S. Department of Commerce (annual). *Statistical Abstract of the United States.* Washington, D.C.: Government Printing Office.

U.S. Surgeon General's Office (1964). *Reducing the health consequences of smoking.* `www.cdc.gov/tobacco/sgrpage.htm`

Utts, J. M., and Heckard, R. F. (2001). *Mind on Statistics.* Pacific Grove, Calif.: Duxbury.

Vardeman, S. B., and VanValkenburg, E. S. (1999). Two-way random-effects analyses and gauge R&R studies. *Technometrics* **41** (3), 202–211.

Wainer, H. (1983). Pyramid power: Searching for an error in test scoring with 830,000 helpers. *American Statistician* **37**, 87–91.

Wainer, H. (1984). How to display data badly. *American Statistician* **38**, 137–147.

Wainer, H. (1997). *Visual Revelations.* New York: Springer-Verlag.

Wainer, H., Njue, C., and Palmer, S. (2000). Assessing time trends in sex differences in swimming and running (with discussion). *Chance* **13** (1), 10–21.

Wainer, H., Palmer, S., and Bradlow, E. T. (1998). A selection of selection anomalies. *Chance* **11** (2), 3–7.

Wallman, K. K. (1993). Enhancing statistical literacy: enriching our society. *Journal of the American Statistical Association* **88**, 1–8.

Watson, S. R., and Buede, D. M. (1987). *Decision Synthesis: The Principles and Practice of Decision Analysis.* Cambridge University Press.

Weinberg, S. L., and Abramowitz, S. K. (2000). Making general principles come alive in the classroom using an active case studies approach. *Journal of Statistical Education* **8** (2).

Wen, P. (1989). Hub students improve in national test; nine grades at or above average. *Boston Globe*, 9 June, Metro/Region, 19.

Wetzel, N. (2001). Three sisters give birth on the same day. *Chance* **14** (2), 23–25.

Whitehead, B. D., and Popenoe, D. (2001). *The State of Our Unions.* The National Marriage Project. `marriage.rutgers.edu/publicat.htm`

Whitney, R. E., and Urquhart, N. S. (1990). Microcomputers in the mathematical sciences: effects on courses, students, and instructors. *Academic Computing* **4**, 14.

Williams, D. (1991). *Probability with Martingales.* Cambridge University Press.

Author Index

Note: with references having more than three authors, the names of all authors appear in this index, but only the first authors appear in the in-text citations.

Subject Index